DATE DUE

DEC 0 9 97		
MAY 02 '97		
DEC 0 3 1999		
OCT 0 8 2000		
DEC 0 8 2000		
JUL 0 8 2001		
APR 2 8 2006		
APR 3 0 2007		
MAY 0 2 2013		
MAY 0 2 2013		
OCT 2 7 2014		

Applied Numerical Methods
for Engineers

Applied Numerical Methods for Engineers

Terrence J. Akai
University of Notre Dame

WITHDRAWN

John Wiley & Sons, Inc.
New York • Chichester • Brisbane • Toronto • Singapore

Acquisitions Editor	Charity Robey
Marketing Manager	Susan Elbe
Senior Production Editor	Savoula Amanatidis
Text Designer	Lynn Rogan
Cover Designer	Joanne Birdsell
Illustration Coordinator	Jaime Perea
Manufacturing Manager	Inez Pettis

This book was typeset in Times Roman by Publication Services and printed and bound by Malloy Lithographing Company. The cover was printed by Phoenix Color Corporation.

Library of Congress Cataloging-in-Publication Data

Akai, Terrence J., 1949–
   Applied numerical methods for engineers / Terrence J. Akai.
      p.   cm.
   Includes index.
   ISBN 0-471-57523-2
   1. Engineering mathematics.   2. Numerical analysis.   I. Title.
TA335.A38   1993
620$'$.001$'$5194—dc20   93-6374
                                                                          CIP

Printed and bound by Malloy Lithographing, Inc.

Printed in the United States of America
10 9 8 7 6 5 4 3 2 1

To my parents, Rena and John Akai, who gave their children the opportunities to develop a love of learning

About the Author

Terrence J. Akai is Assistant Dean of Engineering at the University of Notre Dame and is Concurrent Associate Professor of Computer Science and Engineering and of Aerospace and Mechanical Engineering, and also serves as Associate Chairman of Computer Science and Engineering. He obtained bachelor's degrees in Aeronautics and Astronautics and in Mathematics from the University of Washington in 1971, and M.S. and Ph.D. degrees in Aeronautical and Astronautical Engineering in 1973 and 1976, respectively, from the University of Illinois. His past research has been in computational fluid dynamics in the areas of nonlinear free-surface hydrodynamics, unsteady aerodynamics of turbine and compressor cascades, rarefied gas dynamics, and Navier–Stokes solvers. Since coming to Notre Dame in 1976, he has won numerous teaching awards that include the Dow Outstanding Young Faculty Award for the Illinois–Indiana section of ASEE, the University's Madden Award for Outstanding Teaching of Freshmen, and awards from the College of Engineering and the Departments of Aerospace and Mechanical Engineering and of Computer Science and Engineering. He has also participated in many programs for minorities and was the winner of the first two Outstanding Faculty Awards from Notre Dame's Minority Engineering Program.

Preface

The major objective of this text is to present a wide range of numerical methods for solving the types of problems that occur in engineering. These problems are generally those that do not have closed-form solutions; however, they may also include those problems whose closed-form solutions are beyond the mathematical expertise of the engineer. To illustrate the relevance of numerical methods to engineering, more than 75% of the case studies, examples, and exercise problems is drawn from engineering disciplines, and several design problems are posed in the last chapter.

Solutions of applied problems involve several steps — interpreting the problem, modeling it, deciding on a method of solution, and implementation of the method. The focus of the text is on the last two steps. The choice of a method depends on what methods are available, their salient features and suitability for a given problem, and how they are actually used on a computer. The interpretation and modeling steps cannot be divorced from the problem-solving process; however, these aspects are given less attention because they are taught more properly in courses that deal with a particular discipline.

The text is aimed mainly at students in junior/senior courses in the undergraduate engineering curriculum. However, some of the more elementary material is appropriate for freshman/sophomore courses, and some of the more advanced material is suitable for an overview or introductory graduate course. It is generally assumed that students have had a year or more of the standard mathematics and science courses, and that they have at least a passing acquaintance with the material that is typically taught in elementary engineering courses.

The student in a course that uses this text should acquire the kind of understanding of the course topics that comes with hands-on programming experience. This type of comprehension is useful in choosing methods from the variety of existing software libraries and applications that deal with numerical methods. It is unlikely that an undergraduate student will cover the entire text or even all of the methods in a particular chapter; therefore, the text also serves as a reference for other methods that might be required later for a particular problem.

LAYOUT

The text is laid out in four major parts. **Part I: FOUNDATIONS** introduces the computer and programming languages as tools, and covers the basic methods for solving linear and nonlinear algebraic equations. Sections 1.3 and 1.4 of Chapter 1 are particularly useful for inexperienced programmers and those who have only a cursory knowledge of computer arithmetic with finite precision. Chapter 2 provides a comprehensive treatment of solving systems of linear algebraic equations. It contains a section on fundamental linear algebra because the mathematical content of most engineering curricula provide only a limited treatment of that topic. Chapter 3 treats the solutions of nonlinear algebraic equations, including root-solving for systems and polynomials. Although the methods of Chapters 2 and 3 go beyond elementary topics, they are placed in Part I because many other methods rely on the availability of solvers for algebraic equations.

     Part II: DATA ANALYSIS deals with statistical and least-squares methods in Chapter 4, and it deals with more general curve fitting in Chapter 5. These methods are pertinent to the analysis and manipulation of raw data that might arise from experimental or other sources. Elementary statistics is clearly relevant to engineering and technology; yet, many texts on numerical methods omit statistical methods.

     Part III: NUMERICAL CALCULUS deals with numerical differentiation and integration in Chapter 6, and it continues with numerical solutions of ordinary differential equations in Chapter 7. Chapter 6 includes treatments of improper and multidimensional integrals. Chapter 7 includes adaptive methods, methods for stiff equations, and methods for boundary-value problems.

     Part IV: ADVANCED TOPICS deals with matrix eigenproblems in Chapter 8, and it provides a basic introduction to numerical methods for solving partial differential equations in Chapter 9. Eigenproblems are placed in Part IV because they generally arise from the more advanced types of engineering problems and because the theory of eigenproblems is not an elementary topic (even though the algorithms in the text are not particularly difficult). The material in Chapter 8 is allied closely to the material that is covered in Chapter 2; instructors might therefore choose to move directly from Chapter 2 to Chapter 8, depending on the methods they'd like to cover. The exception is that a solver for polynomial equations (covered in Chapter 3) is required for the Faddeev–Leverrier method. Chapter 9 on partial differential equations has the iterative methods of Chapter 2 and the concepts of Chapter 7 as prerequisites.

     Part IV also includes Chapter 10, which deals with constrained optimization within the context of design. The problems for this chapter are really design projects, which combine the optimization methods with one or more of the methods from the preceding chapters. These projects are more substantial than the usual exercises at the end of each preceding chapter, and they give the student the opportunity to make choices in the methods they wish to use.

OTHER FEATURES

The text presents algorithms in pseudocode. These pseudocodes employ the same structures that exist in the commonly used programming languages, and they are detailed enough to reveal the traps that are often hidden when algorithms are presented as descriptive procedures. The pseudocodes are written in a style that promotes structured coding. Modularity in programming is emphasized, and adherence to this principle will be especially useful for those who tackle the more advanced material. Experienced programmers can, of course, use the features of a particular programming language to implement an algorithm more elegantly and efficiently than is shown in the general-purpose pseudocodes.

The procedures shown for the implementation of the methods are for standard computer architectures, such as those found on the desktop computers that are generally available to students. The method of choice for such architectures may be different from the method of choice for other architectures such as those for vector and parallel computing, or it may require modification to take advantage of the features of advanced architectures.

Most of the problems in the text are such that inefficient applications will not cause severe difficulties with modern computer systems; nevertheless, the student should recognize that inefficiency can be costly for more complicated applications. Efficiency (in terms of both computer and human resources) is therefore encouraged, but not to the extent that it overrrides correct application of a method.

Some aspects of style warrant some comment. Worked problems are presented in two ways — as case studies if they are longer problems that are used in the discussion of a method, or as examples if they are shorter problems that are intended to highlight a particular feature. An in-line rather than a built-up style is used for equations, because the in-line style is compatible with the way in which statements are coded in a program.

Results for case studies and examples are reported to several significant digits. Although it is not sensible to use so many digits in engineering practice, the precision level allows students who work through these examples to discern when the coding of an algorithm is not completely correct.

Finally, the background information for the design projects is presented in a descriptive style, which is not neatly blocked. This approach is deliberate — it is in keeping with the way in which a problem might be introduced to an engineer. It is then up to the engineer to sort out the material and to devise a logical plan of attack.

TO THE INSTRUCTOR

There is more material in the text than can be covered in a one-semester course; nevertheless, the instructor of the course has a great amount of flexibility in designing the course content. A truly elementary course can use material from Chapter 1 and from the early sections of Chapters 2 though 6 in almost any order. Intermediate courses can use more of the material from Chapters 2 through 7, provided that the topics requiring matrix solvers and solvers for nonlinear algebraic equations follow the relevant prerequisites in Chapters 2 and 3. An advanced course can use in-depth coverage of the material in Chapters 2 and 3, and in Chapters 6 through 9. Chapter 10 contains projects that are suitable for all but the most elementary courses.

The course instructor has a significant influence on what the student gains from the course. The instructor chooses the topics and decides on the emphasis of the material. The instructor is invited to modify the exercise problems at the end of each chapter and the design problems in Chapter 10, or to use these problems as springboards to additional problems. The use of this text does not preclude the use of software libraries and applications. The instructor is also invited to have students compare solutions from their own programs with those from the software that is available on a local computer system.

A solutions manual is available to course instructors. It includes ASCII source codes and data files on diskette.

TO THE STUDENT

Final comments are directed to the student. The use of computational methods is partly an art form, which cannot be truly learned by a "recipe" approach. There should be a willingness to experiment with codes, with the parameters of a problem, and with different methods of solving the same problem. The experience can provide considerable insight into both the problem and the method. An effort should also be made to understand how a problem is to be approached before an attempt at writing a code is made. In other words, forget about the programming language until a plan of attack has been selected — it is impossible to instruct a machine to perform tasks if the tasks are not well defined.

Verification of solutions should always be performed. The student should learn to recognize when a solution cannot possibly be associated with a physically real problem. Beyond this stage, solutions may be completely verified for simple problems (for example, when finding a root of an equation), or they may be partially verified by carrying out sample calculations for the more difficult problems. Verifications can also be partially performed by solving a similar problem with a known solution.

Acknowledgments

The genesis of this text lies in the explorations of its concepts with Gene Davenport (formerly with Wiley). However, the real development of those concepts was initiated by Executive Editor Charity Robey, who is the sponsoring editor of the text. I am grateful to Charity for nurturing the project throughout its duration, for the trust that she placed in my abilities, and also for many interesting conversations.

The reviewers of samples of preliminary work on the text were Prof. Ismail Celik, West Virginia University; Dr. Oren Masory, Florida Atlantic University; Prof. Don Riley, University of Minnesota; Dr. John White, University of Lowell; and Prof. Amde Wolde–Tinsae, University of Maryland. Their comments and suggestions were encouraging and extremely helpful in setting the level of the text, in choosing its contents, and in choosing the sequence of material.

Dr. Masory was also a reviewer of the final manuscript along with Prof. Michael Papadakis, Wichita State University; Prof. Tom Shih, Carnegie-Mellon University; Prof. Virgil Snyder, Michigan Technological University; and one other who asked not to be listed in the acknowledgments. Their reviews included many subtle points that I had overlooked and led to various enhancements throughout the text.

The publisher assigned a wide variety of resources to producing and marketing the text. I have relied heavily on the professional expertise of the people who were involved in these phases of the project. Among them are Savoula Amanatidis, Senior Production Editor, who was in constant communication with me and was my source of explanations during production, and Suzanne Ingrao, who took on the painstaking task of copyediting the manuscript.

Among those in the College of Engineering at Notre Dame, I would like to thank Shirley Wills, my secretary, for managing those things to which I could not devote much time, and Joanne Birdsell, Supervisor of Graphics and Publications, for lending her time and artistic talent to the design of the book cover. My thanks also go to several colleagues who constantly demonstrated their support and enthusiasm for the project.

My family, of course, provided the kind of support that allowed me to focus on the writing of the manuscript. My daughters, Carol Akai and Jenny Carroll, helped with some of the proofreading. My wife, Becky, assumed many extra responsibilities, and as a professional in education, gave several helpful suggestions on pedagogical aspects.

South Bend, Indiana   Terrence J. Akai

Contents

PART I FOUNDATIONS   1

1 **Introduction   3**

1.1 Numerical Methods in
     Engineering   3

1.2 Computing Tools   4

1.3 Fundamental Programming
     Concepts   9

   1.3.1 Data Types and
            Structures   10

   1.3.2 Data Control   11

   1.3.3 Modularity in
            Programming   18

1.4 Data Representation   23

   1.4.1 Fixed-Point Numbers   23

   1.4.2 Floating-Point
            Numbers   27

   1.4.3 Some Consequences of
            Finite Data
            Representation   31

1.5 Closure   36

1.6 Exercises   37

2 **Systems of Linear Algebraic
     Equations   43**

2.1 Fundamentals of Linear
     Algebra   44

   2.1.1 Notation and
            Definitions   45

   2.1.2 Operations   45

   2.1.3 Square Matrices   47

   2.1.4 The Determinant of a
            Square Matrix   49

   2.1.5 The Matrix Equation for
            Linear Algebraic
            Systems   51

   2.1.6 Computational Techniques
            for Basic Operations   53

2.2 Direct Methods for Linear
     Systems   54

   2.2.1 Methods for Triangular
            Matrices   54

   2.2.2 Cramer's Rule   56

   2.2.3 Gauss Elimination with
            Row Pivoting   57

XV

2.2.4   The Gauss–Jordan
        Method   65

2.2.5   LU Decomposition   69

2.2.6   Applicability of General
        Direct Methods   76

2.2.7   Cholesky Decomposition
        for Symmetric
        Matrices   81

2.2.8   The Thomas Algorithm for
        Tridiagonal Matrices   86

2.2.9   Block Matrix
        Methods   88

2.3   Iterative Methods for Linear
      Systems   90

2.3.1   The Method of Residual
        Correction   91

2.3.2   Jacobi and Gauss–Seidel
        Iterations   92

2.3.3   Successive
        Overrelaxation   98

2.3.4   The Conjugate Gradient
        Method   105

2.4   Closure   107

2.5   Exercises   108

**3   Nonlinear Algebraic
      Equations   113**

3.1   Methods for Equations in a
      Single Variable   114

3.1.1   The Incremental-Search
        Method   114

3.1.2   Fixed-Point Iteration   116

3.1.3   The Bisection
        Method   118

3.1.4   The False-Position
        Method   122

3.1.5   The Newton–Raphson
        Method   124

3.1.6   The Secant Method   129

3.1.7   Convergence
        Criteria   130

3.2   Systems of Nonlinear
      Equations   132

3.3   Roots of Polynomials   137

3.4   Closure   143

3.5   Exercises   144

PART II   DATA ANALYSIS   149

**4   Statistics and Least-Squares
      Approximation   151**

4.1   Elementary Statistics   151

4.1.1   Statistical Quantities from
        Individual
        Measurements   152

4.1.2   Statistical Quantities from
        Grouped Data   156

4.1.3   Prediction of Behavior from
        Statistical Quantities   161

4.1.4   The Chi-Squared
        Distribution   167

4.1.5   Testing for Goodness of
        Fit   170

4.2   The Least-Squares
      Approximation   171

4.2.1   Linear Regression   172

4.2.2   Linear Combinations of
        Functions   175

4.2.3   Nonlinear Models   176

4.3   Closure   177

4.4   Exercises   178

5   Curve Fitting   181

5.1   Polynomial Interpolation   181

5.1.1   Lagrange
        Interpolation   182

5.1.2   Newton's General
        Interpolating
        Formula   183

5.1.3   Neville's Algorithm   186

5.2   Cubic Splines   187

5.3   The Discrete Fourier
      Transform   190

5.4   Closure   193

5.5   Exercises   194

**PART III   NUMERICAL
             CALCULUS   197**

6 **Differentiation and
     Integration   199**

6.1   Numerical
      Differentiation   199

  6.1.1   Difference Formulas   200
  6.1.2   Truncation and Round-Off
          Errors   203

6.2   Numerical Integration   205

  6.2.1   The Trapezoidal
          Rule   206
  6.2.2   Simpson's Rule   209
  6.2.3   Romberg Integration   210
  6.2.4   Gaussian Quadrature   213
  6.2.5   Improper Integrals   216
  6.2.6   Multidimensional
          Integrals   220

6.3   Closure   222

6.4   Exercises   223

7 **Ordinary Differential
     Equations   229**

7.1   Single First-Order Equations
      with Initial Values   230

  7.1.1   Basic Concepts and the
          Euler Method   230
  7.1.2   Some Simple Second-Order
          Methods   234
  7.1.3   Runge–Kutta
          Methods   238
  7.1.4   The Adams
          Methods   242

  7.1.5   The Milne–Simpson and
          Hamming Methods   246
  7.1.6   The Gragg Method   248
  7.1.7   Adaptive Methods   249

7.2   Systems of First-Order
      Equations   251

7.3   Stiff Equations   254

7.4   Boundary-Value
      Problems   257

  7.4.1   The Shooting
          Method   257
  7.4.2   The Finite-Difference
          Method   259

7.5   Closure   261

7.6   Exercises   262

**PART IV   ADVANCED
            TOPICS   269**

8 **Matrix Eigenproblems   271**

8.1   The Faddeev–Leverrier
      Method   272

8.2   The Power Method   278

8.3   The Jacobi Method for
      Symmetric Matrices   282

8.4   Closure   285

8.5   Exercises   286

9 **Introduction to Partial
     Differential Equations   289**

9.1   Preliminary Concepts   289

9.2   Methods for Parabolic
      Equations   292

  9.2.1   Analysis of an Explicit
          Method   294
  9.2.2   The Crank–Nicolson
          Method   297
  9.2.3   The DuFort–Frankel
          Method   299

9.2.4   Predictor–Corrector
        Methods   300
9.2.5   Other Considerations for
        Parabolic Equations   302
9.3   Elliptic Equations   303
9.4   Methods for Hyperbolic
      Equations   305
9.4.1   The Lax–Wendroff
        Method   305
9.4.2   The Wendroff Implicit
        Method   307
9.4.3   First-Order Systems   308
9.4.4   Direct Treatment of the
        Wave Equation   309
9.5   Hyperbolic Equations in Two
      Space Dimensions   310
9.6   Orthogonal Coordinate
      Transformations   310
9.7   Closure   312
9.8   Exercises   313

10   **Design and
     Optimization   317**
10.1   The Design Process   317
10.1.1   A Problem
         Description   318
10.1.2   Modeling the
         Problem   320
10.2   Optimization Methods   324
10.2.1   Exhaustive
         Searching   324
10.2.2   Searching on Constraint
         Boundaries   327
10.2.3   Lagrange
         Multipliers   329
10.2.4   Other Strategies   332
10.3   General Discussion   333
10.4   Design Projects   334
10.4.1   Problem 1: Design of an
         Airship Gas
         Envelope   335

10.4.2   Problem 2: Design of a
         Plane Truss   337
10.4.3   Problem 3: Design of a
         Four-Bar Linkage   339
10.4.4   Problem 4: Design of a
         Rack and Pinion Steering
         Linkage   342
10.4.5   Problem 5: Design of a
         Water Park Slide   345
10.4.6   Problem 6: Design of a
         Ventilation System   348
10.4.7   Problem 7: Design of a
         Software
         Application   349
10.4.8   Problem 8: Design of a
         Rocket Launch
         Configuration   351

Appendix A   **Introduction to the
               Great Mathemati-
               cians   355**

Appendix B   **Summary of
               Pseudocode
               Structures   367**
B.1   Elementary
      Constructions   367
B.2   Selection   368
B.3   Loops   369
B.4   Modules   369

Appendix C   **Useful
               Mathematical
               Relations   377**
C.1   Matrix Norms and
      Spectra   377
C.2   Newton–Cotes
      Integration
      Formulas   379
C.3   Quadratic Equations   380
C.4   Series Expansions for
      Common Functions   380
C.5   Special Functions   381

C.6   Sums of Powers of
        Integers   382
C.7   Taylor Series   383

Appendix D   **Physical
                Models   385**
D.1   Arcs and Bodies of
        Revolution   385
D.2   Particle Dynamics   386
D.3   Resistor Networks   387
D.4   Rocket Thrust   388

D.5   Stream Functions and
        Streamlines   390
D.6   Truss Equilibrium   391

References   395

Bibliography   397

**Answers to Selected
Problems   399**

Index   405

PART I

FOUNDATIONS

1

Introduction

Numerical methods are used to produce quantitative approximations to solutions of mathematical problems; they have become increasingly important in engineering as access to computing tools has become easier. Our major goals are to present a large selection of numerical methods and to demonstrate their relevance and application to problems that arise in engineering. We introduce the subject in this chapter by discussing what engineers do, the roles of computers and programming languages as tools for their activities, and the fundamentals of programming and data representation. The theory and application of numerical methods are described in Chapters 2 through 9 with the aid of worked examples and exercises. Longer examples and explanatory problems serve as case studies which are integrated with the text. Chapter 10 is devoted to design problems in which techniques of constrained optimization are coupled with appropriate methods from preceding chapters.

1.1 NUMERICAL METHODS IN ENGINEERING

Engineers use the laws, materials, and energy sources of nature to produce goods and services for the benefit of people. These products form a societal infrastructure on which we rely in our daily lives. Some broad examples of the components of that infrastructure are systems for power transmission, communications, data management, transportation, and manufacturing.

The problem-solving traits and modeling background of engineers enable them to be successful in both technical and nontechnical endeavors. The traditional technical activities of engineers include research, design, testing, manufacturing, and

3

education. Design (used in a wide sense so that it encompasses the tasks related to producing this book) is the essential activity of engineering. It is through design that concepts are transformed into useful, safe, and reliable products.

Engineers depend heavily on mathematical modeling in performing their tasks. These models must, of course, yield physically meaningful solutions if they are to be useful. Solutions may be obtained by standard mathematical techniques if the models are simple enough. Even in such cases, however, the values of the solutions may require a large amount of calculation. Computers are ideal tools for such tasks because they can perform calculations very quickly and because, unlike people, they are neither subject to tedium nor prone to clerical mistakes. The direct methods of Chapter 2 for solving systems of simultaneous linear equations are examples of numerical methods for calculating solutions from mathematically well-established forms.

Other types of numerical methods are those aimed at solving problems for which solutions can not be obtained by ordinary mathematical manipulation. In these cases, we replace the mathematical equations to be solved by discrete processes and try to obtain sufficiently accurate numerical approximations of the solutions. Many of these methods require iteration of a process until convergence to a solution is achieved. An example of such a method is the Newton–Raphson method of Chapter 3 for solving nonlinear equations. Others, like the methods of Chapter 7 for solving ordinary differential equations, require us to proceed in a step-by-step fashion from an initial condition to a specified destination.

Implementation of numerical methods on computers requires a programming language through which we can provide instructions to the computer. We also need to recognize the limitations of computers so that we do not inadvertently demand or expect more than they can produce. A discussion of computing tools is given in Section 1.2 and is followed by a more detailed discussion of fundamental concepts in Sections 1.3 and 1.4.

1.2 COMPUTING TOOLS

A history of computing devices (Ref. 1, 2, 3) usually begins with the abacus. In reality, the abacus is a data storage system based on bead positions; the instructions for operating the abacus are stored in the user's brain, and the user's fingers carry out the instructions by moving the beads in prescribed ways. True computing machines were not invented until the seventeenth century. Among the notable inventors (see Appendix A for biographical sketches) were Blaise Pascal (1623–1662) and Gottfried von Leibniz (1646–1716). Pascal's machine performed addition and subtraction through mechanical gears. Leibniz improved on Pascal's machine by developing one that could also perform multiplications.

The English mathematician Charles Babbage (1792–1871) and his associate Augusta Ada Byron (1815–1852), the Countess of Lovelace and herself a mathematician, are considered to be the true inventors of the digital computer. Byron also devised a program for computing Bernoulli numbers and is thus credited as the first programmer. Babbage abandoned ten years of work on an earlier design to focus on a better idea that he called the Analytical Engine. Although lack of funding and inadequate mechanical technology prevented Babbage from completing the Analytical Engine, Ada Byron's description of it preserves for us the concepts it embodied. Features of Babbage's design with counterparts in modern computers were an input device using punched cards, a "store" for data storage, a "mill" for arithmetic operations, and a printer to produce hard copy so that clerical transcription errors could be avoided.

Babbage obtained the idea of using punched cards for input by observing the invention of Joseph Marie Jacquard (1752–1834). Jacquard designed an automatic loom in which the needles used to weave complicated patterns were controlled by perforations in thin boards. A similar idea was used by the statistician Herman Hollerith (1860–1929) to process data for the U.S. census of 1890. The punched card concept survived as a mainstay of the modern computing era (through card and paper tape readers) until the early 1980s.

Machines with electronically controlled mechanical relays appeared in the 1940s. One of the first was the Mark I (or Automatic Sequence-Controlled Calculator) whose construction was led by Howard Aiken (1900–1973) of Harvard University. With help from IBM (International Business Machines), Aiken's team completed the Mark I in 1944.

Electromechanical machines were immediately superseded by purely electronic digital computers, which began appearing at the same time. Among the latter was the ABC or Atanasoff–Berry Computer, which was proposed from 1937 to 1938 by John Atanasoff of Iowa State University (then known as Iowa State College) and built with his graduate assistant Clifford Berry from 1939 to 1941.

Another electronic computer project was the ENIAC (Electronic Numerical Integrator and Computer) project at the University of Pennsylvania's Moore School of Electrical Engineering. This project was proposed by John Mauchly when he joined the University of Pennsylvania faculty in 1942. Mauchly and his colleague John Presper Eckert built ENIAC from 1943 to 1945. A court case involving ENIAC arose later when Atanasoff claimed that basic techniques he had developed were used by Mauchly (who had met Atanasoff in 1940); the result was that ENIAC's patent was invalidated.

ENIAC was programmed by an externally wired plugboard, and it did not use binary arithmetic. During its construction, John von Neumann of Princeton became involved with the design of the successor to ENIAC. A major development that arose from this collaboration was the ***stored-program computer***, in which both a program's data and instructions are stored in memory. The architecture proposed by von Neumann for a stored-program computer employing ***binary arithmetic*** became the structural framework for most later computer designs.

The evolution in computing since the 1940s is marked by tremendous and rapid advances in technology and by diversification of computer types (e.g., personal computers, desktop workstations, minicomputers, mainframes, and supercomputers). Modern computers typically consist of a central processing unit (cpu), which contains a control unit and an arithmetic/logic unit, primary storage (memory) and input/output units connected by an internal bus to the cpu, and external or peripheral devices such as disk drives, tape drives, displays, and printers. Except for advances in technology and placement of the primary memory and input/output units outside the cpu, modern computers have the same structural framework as proposed by von Neumann. A schematic of a computer system is shown in Fig. 1–1.

Computers operate with binary digits or *bits* whose patterns are used for instructions and data representation. Communication with the hardware is accomplished with various programs, which are generally termed *software*. One major suite of programs constitutes the *operating system* for the computer. Operating systems perform several functions such as controlling operations and communications within the system, managing of resources, task scheduling, and interpreting the user's commands. The operating system that is used on a computer depends on the manufacturer. Some, like DOS (Disk Operating System) and Unix, are used by several manufacturers; others are proprietary operating systems of the manufacturer.

Other commercial software products that are used by engineers include word processors, CAD (Computer Aided Design) and graphics tools, simulators, spreadsheets, equation solvers, symbolic mathematics tools, and programming languages. Numerical methods are embedded in some of these products for special purposes; for more general coding of numerical methods, we rely on programming languages to communicate instructions to the computer. Libraries of numerical routines are commercially available for specific languages and computer platforms.

One might wonder at this point why we bother to study numerical methods if codes for the more common methods already exist in commercial packages. Some reasons for doing so are listed below.

- Routines may not be available for the language or the type of computer we are using.
- An affordable commercial package may not contain all of the methods we need.

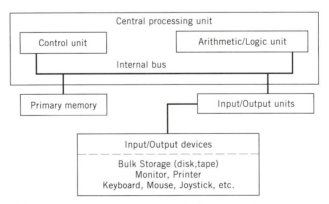

Figure 1–1   Schematic diagram of a basic computer system.

- Adaptation of methods to fit special situations may be impossible to accomplish with prewritten tools.

- Programming our own routines helps us to understand the steps of a process more fully; it also allows us to explore a problem beyond the mere determination of a solution. Such exploration may provide insights that can expand our view of a particular problem or help us in a different problem.

- Even if prewritten codes are available, an understanding of the numerical processes allows us to choose the best method among several for performing a task.

- We become better educated in the tools of our profession.

Communication with a computer is actually done through a *machine language* whose instructions are encoded in bit patterns. The next step up from machine language is a *low-level* language such as assembly language, in which the binary instructions of the machine language are replaced by symbolic instructions. Languages that are more oriented to humans are called *high-level* languages; their structure and syntax more closely resemble the language people use to communicate with each other. Common high-level languages for engineering are FORTRAN, BASIC, C, Pascal, and Ada. Programs written in these languages are often called *source codes*.

Source codes are not executed directly; they must undergo a translation to machine language. Two methods of translation are compilation and interpretation. A *compiled* language is completely translated to a low-level target code, which might be the computer's machine or assembly language, or even a generic intermediate language. A successful compilation means that the translator understood the structure and syntax used in the source code and was able to produce a low-level version of the instructions; it does not necessarily mean that the instructions were correct or that they could be executed.

Translation of an *interpreted* language occurs during execution of the program as each individual source statement is encountered. This process usually slows down execution; however, an error is immediately detected when execution of an offending instruction is attempted, and it can be corrected interactively. The ease of detecting and correcting errors is a feature of interpreted languages that is helpful to novice programmers who are likely to make many programming mistakes.

Discussions among the *cognoscenti* on the merits and failings of different programming languages sometimes degenerate into heated debates that are conducted with near religious fervor. Such arguments are inevitable because the commonly used programming languages were developed at different stages of computing history and for different purposes. We do not intend to join the arguments, especially since this book is intended to be independent of language; rather, we provide some of the salient features of the languages that we mentioned earlier.

FORTRAN (from FORmula TRANslator) is the oldest high-level language. It was developed to meet scientific needs by a committee sponsored by IBM (International Business Machines) and headed by John Backus. It is a compiled language whose executable code is among the most efficient. FORTRAN was first released in 1956, and its availability when engineering schools introduced programming in the 1960s made it the "engineering" language. As users and teachers of FORTRAN became more knowledgeable about programming and about what they wanted to accomplish through programming, early versions of FORTRAN began to receive much criticism about their deficiencies.

The responses to the criticism were standardization and evolution. The latter included high-level programming structures for conditions, improved handling of textual material, and routines that could accept generic data types for its arguments. The FORTRAN 90 version that was most recently standardized by ANSI (American National Standards Institute) includes other Pascal-like and C-like conveniences such as a while loop, a case statement, recursion, multistatement lines, in-line comments, provisions for defining new data types, pointers, and low-level bit manipulation. It also features new intrinsic operations for vector and matrix arithmetic.

FORTRAN is still one of the most widely used languages in engineering. Some of the factors that keep it alive are its efficiency, its evolution, intrinsic capabilities for dealing with complex numbers, the wealth of scientific subroutine libraries that have been developed during its long life, availability of compilers for every class of computers, and its status as the first language with universal implementation on supercomputers in vectorized and sometimes parallelized forms.

BASIC (Beginner's All-purpose Symbolic Instruction Code) was developed by John Kemeny and Thomas Kurtz of Dartmouth College and was first used in 1964. The intent was to produce a language that retained the formula translation characteristics of FORTRAN, but which was easier to learn and use. BASIC is generally an interpreted language that is suitable for novice programmers (although compiled versions now exist).

Ease of use made BASIC a popular choice for personal computers, and it is through those machines that BASIC acquired its popularity. Indeed, many younger programmers learned BASIC as their first serious programming language on a personal computer. The growth of the personal computer market also spawned a variety of BASIC dialects. These dialects produced many powerful enhancements to the original language; unfortunately, they appeared without the benefits of standardization. The variety of dialects and the relative inefficiency of the interpreted versions seem to have offset BASIC's ease of use and to have prevented its adoption as a major programming language in engineering schools.

C was developed in 1974 by Brian Kernighan and Dennis Ritchie of Bell Labs as a language for the Unix operating system; it allows more low-level control of the computer than most other languages. Because of this background, C has a somewhat terse style which can make codes written by one programmer nearly incomprehensible to another in the absence of good documentation. The terseness and low-level features of C have led detractors to rename it "Cryptic" or to describe it as a hacker's language.

Although C was not intended to be a teaching or general purpose language, it has emerged as an important language for a wide range of applications. For engineering purposes, the attractive features are its powerful capabilities, the efficiency of its executable code, and its "host language" status for machines operating under Unix (which has *de facto* become a standard for scientific computers and high-powered workstations). Evidence of C's importance is its availability across all computer platforms from personal computers to supercomputers.

Pascal (named in honor of Blaise Pascal) was introduced in 1971 by Niklaus Wirth, a Swiss computer scientist. Its parent ALGOL (ALGOrithmic Language) was developed by the ACM (Association for Computing Machinery) in 1958 to correct FORTRAN's perceived flaws and was used widely in Europe during the 1960s and 1970s for scientific applications. Most versions of Pascal are compiled, but interpreted versions also exist. Some enhanced versions supply extensions to the standard language and combine compiled features with an interactive environment.

Pascal was developed as a teaching language that could enforce good programming habits. It has a block-structure style and requires all variables to be declared. The latter feature makes Pascal a ***strongly typed*** language as opposed to FORTRAN and BASIC, which are weakly typed. Pascal also contains high-level control structures that are intended to eliminate explicit jumps (such as the GO TO statement in FORTRAN). Although jumps are necessary in programming, abuse of jump instructions can make codes difficult to read and can make it difficult to find and correct errors.

Pascal's form and its ability to deal with a variety of data structures helped it to crack FORTRAN's dominance in engineering colleges and BASIC's dominance in high schools. Its block-structure style is inherited by the Ada programming language, and many of its features have beneficially enhanced FORTRAN's evolution. Despite its obvious influence on other languages, Pascal is somewhat limited and is not powerful enough to be considered a production language. Its commercial replacement is Wirth's Modula-2, which was introduced in 1980.

FORTRAN, BASIC, C, and Pascal are all ***procedural*** languages in which the procedures for manipulating data are separate from the data. Enhanced versions of C and Pascal are ***object-oriented*** languages. In an object-oriented language, the programmer encapsulates or groups the data together with the procedures for manipulating the data into objects. The programmer can also define new objects that inherit all of the data and procedures of an existing object.

Ada (named for Augusta Ada Byron) is an object-oriented language that was developed in 1982 by the U.S. Department of Defense in an attempt to standardize all of its military software. Ada has the block-structured features of its parent Pascal, but it is considerably more powerful. Ada's importance among programming languages is increasing because of the large amount of software that is developed for the military, and Ada compilers are now available for supercomputer architectures.

1.3 FUNDAMENTAL PROGRAMMING CONCEPTS

Good programming is based on fundamental concepts that are independent of the programming language that is used. We must be aware of the kind of data that we are processing, the structures that are available for storing the data, and the structures that allow us to control operations on the data. Although it is presumed that users of this book either already have received or are concurrently receiving instruction in a programming language, we review some of the fundamental programming concepts in the following subsections.

1.3.1 Data Types and Structures

The dominant data used in numerical methods are, of course, numbers. Numbers are generally classified as *fixed-point* or *floating-point*. Fixed-point numbers are whole numbers (which have no fractional part); they are often called *integers*. Floating-point numbers may contain a fractional part and are often called *real numbers*. Fixed-point and floating-point data are stored differently and therefore require different processing for arithmetical operations. We shall discuss storage concepts in Section 1.4; for now, we merely acknowledge that the two classes of numbers are different.

A number may be stored in memory under a *variable name*, which identifies the memory cell containing the bit pattern that corresponds to the value of the number. Strongly typed programming languages, like Pascal, require us to declare the type of number stored under a given name; others, like FORTRAN, have default typing which may be overridden. Even if a language is weakly typed, declaration of all variables is considered to be good programming practice. Complete variable declaration helps us to acquire the discipline of completely specifying the task to be programmed before we attempt to put it in coded form; it also simplifies debugging (or correcting) a program and program maintenance.

The simple variable we have just described is associated with a single quantity. For example, we may use the name V to store the value of a given velocity. In other instances, we may want to identify several related quantities of a given data type by the same name. A data structure that can be used in such cases is an *array.* For example, we may store ten measurements of a height under the common name H and identify individual measurements or array *elements* by a subscript j ranging from 1 to 10; thus H_j refers to the j-th element of the array H. Declaration of an array is used to tell the computer what kind of data is to be stored and how many memory cells are to be reserved for the array. The pertinent data for the latter are the number of subscripts and the range of values for each subscript.

When we wish to store more than one piece of information for each of several entities, we may use *parallel* arrays. Therefore, we may use three different arrays to store three kinds of data for each of several cars. Let us consider the array NAME for storing the make and model of a car as a character string, the array HP for storing the maximum engine horsepower as an integer, and the array RPM for storing the engine speed at which maximum power occurs as an integer. We would subscript the arrays identically so that $NAME_j$, HP_j, and RPM_j all contain information about the j-th car.

An alternative to parallel arrays is a single array of records. A *record* can store several pieces of information for a given entity in partitions, which are known as *fields*. In our previous example, a single record would have three fields called NAME,

HP, and RPM. An array of these records might be named CAR, and the fields of the array element CAR_j would contain the information for the j-th car in a single record structure. Some languages or versions of languages do not contain provisions for defining records.

The array structure is ***statically*** allocated in memory; in other words, storage space is fixed by the information given in the declaration of the array. This storage must be allocated even if we do not use all of the memory cells that are reserved. Advanced programmers who use languages that contain pointers (data types that store the addresses of target memory cells) are aware of structures that can be ***dynamically*** allocated. Dynamic allocation allows us to acquire memory as needed during execution of a code, and it also allows us to release memory cells that are no longer required. The subjects of pointers and associated dynamic structures are outside the scope of our introduction; they are treated extensively in many text books and manuals for programming languages.

1.3.2 Data Control

Procedural programming languages control data through statements and control structures. We shall discuss data control in terms of ***pseudocode*** constructions, which are somewhat arbitrary and do not use the syntax of a particular language. The pseudocode elements that we shall develop here are summarized in Appendix B.

The most elementary pseudocode constructions are the four types described below.

Construction	Description
\ Text	\ indicates a comment used for documentation.
Initialize: List	Set values of input quantities appearing in List. This may be done by reading values from an external data source or by assigning the values within the code.
LeftSide ← RightSide	Assign the value of RightSide to the variable LeftSide.
Write: List	Write the values of the quantities in List on paper, on a display screen, or to an output file.

These elementary constructions are used in simple programs, which receive input data, use the data to produce required results through a fixed sequence of computations and assignments, and display the results on an output device.

Consider the problem of computing the maximum height h_{max} attained by a projectile that is launched with a given initial speed v at a given angle α above the horizontal. If we neglect air resistance, h_{max} is given in terms of v and α by

$$h_{max} = (v \sin \alpha)^2/(2g); \qquad g = \text{gravitational acceleration}$$

We recognize that the units in the expression for h_{max} must be consistent; that is, the same length and time units must be used throughout the expression. The pseudocode

Pseudocode 1–1   Example of a simple code

\ Provide input data
Initialize: v, α, g

\ Assign a value to h_{max}
$h_{max} \leftarrow (v \sin \alpha)^2 / (2g)$

\ Provide output of the result
Write: h_{max}

---

for a program to provide the value of h_{max} may then be written as in Pseudocode 1–1.

The most general form of the projectile code is obtained by interpreting the initialization statement as an instruction to read the values from an external source. In this way, we can execute the code for various input data sets, with arbitrarily chosen consistent units for each set, without modifying the code.

Selection of a section of code to be executed is done with an **If** structure whose general pseudocode form is given below.

If Structure	Description

If [Condition$_0$] **then**
    Group$_0$
Else If [Condition$_1$] **then**
    Group$_1$
    $\vdots$
	Execute *only* Group$_j$ among Group$_0$ through Group$_k$ if Condition$_j$ is the *first* condition that is found to be true; otherwise, if none of Condition$_0$ through Condition$_k$ is true, execute Alt-Group.

Else If [Condition$_k$] **then**
    Group$_k$
Else
    Alt-Group
End If
	The minimum structure consists of the first **If**, Group$_0$, and the **End If** terminator. The **Else If** and **Else** parts of the structure are used as dictated by logic.

Each of the conditions in the **If** structure is a ***Boolean*** quantity whose value may be either **true** or **false**. Simple Boolean expressions are typically constructed from the six ***relational*** operators in the set $\{<, \leq, =, \neq, \geq, >\}$; these are used to compare values of two operands. More complex expressions may be formed by combining Boolean quantities through ***logical*** operators from the set {**not, and, or, xor**}. The last operator in the set is the "exclusive or" operator; it returns a **true** result if one and ***only*** one of its two Boolean operands is **true**. The actions of the logical operators are described in the truth table of Table 1–1.

To illustrate the use of the **If** structure, we consider the evaluation of a function $f(t)$ for a given value of t in the range $(0 \leq t \leq 6)$. The function is as follows and may be evaluated by Pseudocode 1–2.

$$f(t) = \begin{cases} 0.9t; & 0 \leq t < 3 \\ 2.7 + 1.6(t-3); & 3 \leq t < 5 \\ 5.9 + 2.2(t-5); & 5 \leq t \leq 6 \end{cases}$$

Table 1–1 Truth Table for Logical Operators

Boolean Quantity		Logical Operation			
A	B	**not** A	A **and** B	A **or** B	A **xor** B
true	true	false	true	true	false
true	false	false	false	true	true
false	true	true	false	true	true
false	false	true	false	false	false

Pseudocode 1–2 contains an **If** structure *nested* within the **Else** clause of an outer **If** structure, which is used to establish the validity of the input value for t. If the t value is in the required range, the inner **If** structure is used to select the appropriate expression for computing the function value $f(t)$.

Iterations or repetitions of a process are accomplished with *loop* structures. Two of the three major loop structures are the **While** and **Repeat–Until** loops described below.

Loop Structure	Description
\ While loop: **While** [Condition] Process **End While**	Process is executed repeatedly as long as Condition is **true**. Variable quantities that are used in Condition must be initialized prior to the loop, and they must also undergo alteration as a part of Process. The latter requirement makes it possible for Condition to become **false** and thus prevents infinite loops (which contain no mechanisms for ending the iterations).
\ Repeat–Until loop: **Repeat** Process **Until** [Condition]	Process is executed repeatedly until Condition is found to be **true**. Process is executed at least once, and it must include a mechanism to make Condition **true**.

Pseudocode 1–2 Illustration of the If structure

```
\ Get input value for t
Initialize: t

If [(t < 0) or (t > 6)] then          \ t is out of range

    Write: 'Value of t is out of range'

Else                                   \ for a valid t value

    If [t < 3] then
        f ← 0.9t                       \ for 0 ≤ t < 3
    Else If [t < 5] then
        f ← 2.7 + 1.6(t − 3)           \ for 3 ≤ t < 5
    Else
        f ← 5.9 + 2.2(t − 5)           \ for 5 ≤ t ≤ 6
    End If

    Write: f                           \ output function value

End If
```

While and **Repeat–Until** structures are typically used when the number of iterations is arbitrary or cannot be determined *a priori*. Consider the summation of an arbitrary number of nonzero values. Procedures using both the **While** and the **Repeat–Until** loops are given in Pseudocode 1–3.

In both pseudocodes, the variable Answer, which is used to store the result of the summation, is initially set to zero. The result is accumulated by adding new x values as they are received. A zero input value, which would not normally occur in the data to be summed, is used to indicate the end of the data; it is called a ***sentinel*** value.

The **While** loop in the first pseudocode requires the condition variable x to be initialized prior to the loop. Other values of x, if any, are obtained within the loop. When a sentinel value of zero is obtained for x at the end of the loop process, the loop condition ($x \neq 0$) is found to be **false** before the next iteration and the loop is discontinued at that point. Note that the **While** structure allows us to consider the degenerate case in which there are no values to be added. The first input value of x is zero for this case, and the loop is never entered.

The **Repeat–Until** loop in the second pseudocode obtains all x values within the loop. When a sentinel value of zero is encountered, it is added to Answer before the loop is ended by the condition ($x = 0$). Because we are adding and because the sentinel value is zero, inclusion of the sentinel value in Answer does not affect the result. The **Repeat–Until** loop is executed at least once; therefore, it is entered even in the degenerate case in which there are no values to be added.

Pseudocode 1–3   Illustrations of loop structures

```
\ Summation with a While loop:
  Write: 'Enter 1st value or enter 0 to quit:'
  Initialize: x                      \ 1st value; required for loop condition
  Answer ← 0                         \ Initial value (before any addition)
  While [x ≠ 0]
      Answer ← Answer + x        \ Accumulate total
      Write: 'Enter next value or enter 0 to quit:'
      Initialize: x              \ New value
  End While
  Write: Answer          \ Result of summation

\ Summation with a Repeat–Until loop:
  Answer ← 0                         \ Initial value (before any addition)
  Repeat
      Write: 'Enter new value or enter 0 to quit:'
      Initialize: x              \ New value
      Answer ← Answer + x        \ Accumulate total
  Until [x = 0]
  Write: Answer          \ Result of summation
```

It is important to recognize the difference between the **While** and **Repeat–Until** loops. In the **While** loop, the loop condition appears at the *top* of the structure and a **true** value causes continuation of the loop. In the **Repeat–Until** loop, the loop condition is at the *bottom* of the structure and a **true** value causes the loop to end.

Although either type of loop seems suitable for the summation problem we considered, let us see what happens when we try to compute the product of an arbitrary number of nonzero x values. Suppose we use the summation pseudocodes with minimal modifications; namely, we set the initial value of Answer to 1 instead of 0, and multiply Answer by x instead of adding x to Answer within the loops. The modified code with the **While** loop would be correct, but the modified code with the **Repeat–Until** loop would give an incorrect product equal to zero. The zero result with the **Repeat–Until** loop occurs because the sentinel value of zero is included in the product *before* the loop condition is tested.

The major errors that occur in the coding of **While** and **Repeat–Until** loops are usually associated with the logic of ending the loop at the correct stage and with the sequence of steps within the loop process. Another error that frequently occurs is the use of a **While** loop when an **If** structure is intended. For example, the **While** loop

While $[|x| \le 1]$
    $y \leftarrow$ arcsin x
End While

in place of the intended **If** structure

If $[|x| \le 1]$ **then**
    $y \leftarrow$ arcsin x
End If

would cause an infinite loop when the absolute value of x is at most equal to 1.

The third major type of loop is the **For** loop (or DO loop in FORTRAN) whose pseudocode structure is now described.

For Loop Structure	Description
For $[k = k_1, k_2, \ldots, k_n]$ Process **End For**	Execute Process for k taking on the values k_1 through k_n in succession. The k_j values are such that the increment $(k_j - k_{j-1})$ is constant for all j values from 2 through n.

The **For** loop is typically used when the loop variable k has known limits and a known increment. The implementation of the **For** loop varies with language. The most general implementations allow the loop variable to belong to arbitrary data types and the increment to be of arbitrary size; other implementations restrict the loop variable to *ordinal* data types (such as integers or characters whose set of values is arranged in a unique sequence) and permit incrementing only to the next value in

Pseudocode 1–4 While equivalent of a For loop

\ Set initial, limiting, and increment values for the loop variable k
Initialize: $k_{initial}, k_{limit}, k_{inc}$
$k \leftarrow k_{initial}$ \ Set starting value of k
If $[(k_{limit} - k_{initial})(\text{sign}[k_{inc}]) < 0]$ **then**
    Write: 'Error: Increment direction incompatible with limit value'
Else
        While $[(k - k_{initial})/(k_{limit} - k_{initial}) \leq 1]$
            Process
            $k \leftarrow k + k_{inc}$ \ Reset k for the next iteration
        End While
    End If

---

the sequence. We recommend the use of ordinal data types for the loop variable even if the language permits nonordinal data types. The reason for this recommendation is that machine precision limitations can cause errors that are easy to overlook.

To illustrate the difficulty that can occur with nonordinal loop variables, we first show the equivalent of a general **For** loop with a numerical loop variable in Pseudocode 1–4. The starting value of k is $k_{initial}$, and the loop is executed as long as incremented values of k do not go beyond the limiting value k_{limit}. A **direction** is implied in the code; that is, k_{limit} must neither exceed $k_{initial}$ when k_{inc} is negative nor be less than $k_{initial}$ when k_{inc} is positive.

Now consider the summation of the quantities $(1.4, 1.8, \ldots, 9.4)$ with the code

```
Sum ← 0
For [k = 1.4, 1.8, ..., 9.4]
    Sum ← Sum + k
End For
```

in which the **For** loop is implemented by the equivalent structure described previously. The values of $k_{initial}$, k_{limit}, and k_{inc} in the equivalent code are 1.4, 9.4, and 0.4, respectively. With exact arithmetic, the final k value for which the **While** loop is executed would be 9.4. With arithmetic involving *finite* precision, the values stored in memory for k and k_{limit} may both contain errors in representation. If these errors cause the condition for continuing the loop to be **false** when k should be 9.4, the last intended execution of the loop process would not be performed. The result for Sum would then contain an error that is easy to overlook unless we are aware of the potential for it to occur.

We may perform the summation that we have just discussed with exact integer representation of the loop variable by either of the following pseudocodes. The first version employs a nonunit increment and loop variable values that are easily converted to the values to be summed through multiplication by a power of 10.

```
\ Sum of 1.4 + 1.8 + ⋯ + 9.4
\ with an integer loop variable and a nonunit increment
    Sum ← 0.0
    For [k = 14, 18, ..., 94]          \ with increments of 4
        Sum ← Sum + 0.1(k)
    End For
```

The second version uses unit incrementing which requires us to do some preliminary calculations. We wish the loop variable k to range from 0 to n in increments of 1, and we wish to obtain the numbers to be summed from the linear transformation

$$\text{number} = \text{first number} + k \text{ (number increment)}$$

We know that the first number is 1.4 and that the number increment is 0.4. To obtain n, we apply the transformation for the last number, which is equal to 9.4, to get

$$9.4 = 1.4 + 0.4n \qquad \text{or} \qquad n = 20$$

The second pseudocode is then

```
\ Sum of 1.4 + 1.8 + ⋯ + 9.4
\ with an integer loop variable and a unit increment
    Sum ← 0.0
    For [k = 0, 1, ..., 20]          \ to sum 21 values
        Sum ← Sum + (1.4 + 0.4k)
    End For
```

Control elements other than loops and **If** structures are calls to **modules** to perform particular tasks, exits and returns from a module, halting execution of the code, and jumps. We close this subsection with a discussion of jumps and treat modularity in the next subsection.

Jumps allow us to move directly from one code address to another location that is marked by a label. The pseudocode jump instruction is

Go To Label

Jumps are essential in low-level programming and in some high-level languages. For example, a language that does not incorporate a formal high-level **While** loop structure must use **If** structures and jumps to achieve the same effect. The **If** and jump form of the **While** loop is

```
1: If [condition] then
        Process
        Go To 1
    End If
```

The use of the jump to establish the loop above is legitimate; however, the jump can be abused to produce ***spaghetti*** code, which is difficult to read and thus to debug. The code below to find the square root of the absolute value of $(x - y)$ is an example of horrible coding. A trace of the path through the code when x is less than y should make the origin of the term "spaghetti code" obvious.

```
\ Example of a poor code to find the square root of abs(x − y)
Initialize: x, y
z ← x − y
If [z < 0] then
    Go To 5
End If
1: root ← square root of z
    Go To 9
5: z ← (−z)
    Go To 1
9: Write: root
```

Because of potential abuses, jumps are sometimes prohibited by instructors of certain languages that contain high-level alternatives to explicit jumps. The stain of programming heresy that jumps have acquired is unfortunate because there are situations in which the jump is elegant and the alternative to it is clumsy.

1.3.3   Modularity in Programming

Modules are code segments for performing specific tasks; they are variously known as procedures, functions, subroutines, etc., depending on the language. Some advantages of modularity in programming are as follows.

- We can deal with a complicated problem more easily if we can break it down into relatively small, manageable pieces.
- A code segment for a task that must be performed at different stages of the problem has to be written only once.
- Modules may be tested independently of other parts of the code; in this way, debugging (or error detection and correction) is simplified.
- Well-chosen modules for production codes simplify program maintenance and upgrading because modification of one module either leaves other modules unaffected or minimizes any necessary changes in other modules.

Our pseudocode equivalent of a call to a module is the name of the module with a list of names of quantities that are exchanged between the calling and called modules.

We illustrate the use of modules by considering the problem of determining the vertices of a triangle (assumed to exist) formed by three arbitrarily specified straight lines. Be aware that some of the discussion may be irrelevant to some languages, and that the syntax and structure for implementing modules in actual languages may vary considerably.

Let the straight lines be specified in Cartesian (x, y) coordinates by

$$\text{Line } j: y = a_j x + b_j; \qquad j = 1, 2, 3$$

The problem is to find the intersections of lines 1 and 2, lines 2 and 3, and lines 3 and 1. The same task is to be performed three times with different slopes and intercepts in each case; we may therefore delegate this task to a module, which is called as needed. The task to be performed by the module is to find the intersection of *any* two lines

$$y = c_1 x + d_1; \qquad y = c_2 x + d_2$$

The coordinates of the intersection point are

$$x = (d_2 - d_1)/(c_1 - c_2); \qquad y = (c_1 d_2 - c_2 d_1)/(c_1 - c_2)$$

We of course recognize that a line parallel to the y axis cannot be specified with a finite value of c_1 or c_2, and that lines for which c_1 is equal to c_2 either intersect at infinity or have an infinite number of intersections along their lengths. To write a robust code for the module, we would have to understand the data storage limitations of the computer so that we could treat these special cases. Data representation is discussed in Section 1.4; for our current problem, we shall simply exclude lines that might cause pathological conditions to occur.

The pseudocode for a module contains a **header** consisting of the module name and a list of **substitute** names, if any, for actual quantities. We use the generic term "substitute" instead of words like "argument" and "parameter," which mean different things in different languages. We shall first construct a module **without** substitute names.

Module Intersect:
    $x \leftarrow (d_2 - d_1)/(c_1 - c_2)$
    $y \leftarrow (c_1 d_2 - c_2 d_1)/(c_1 - c_2)$
End Module Intersect

The implied syntax here is that any other module that calls Intersect must supply data and receive data under the **same** names that are used within Intersect. Thus our code for finding the vertices of the triangle may be written as shown in Pseudocode 1–5.

Pseudocode 1–5   Code to use a module without substitute names

Program:

\ Program to find vertices of a triangle formed by
\ $y = a_1(x) + b_1$; $y = a_2(x) + b_2$; $y = a_3(x) + b_3$;
\ Module Intersect is called to find line intersections

Write: 'Constants a_1 and b_1 for line 1?'
Initialize: a_1, b_1
Write: 'Constants a_2 and b_2 for line 2?'
Initialize: a_2, b_2
Write: 'Constants a_3 and b_3 for line 3?'
Initialize: a_3, b_3

\ Intersection (x_{12}, y_{12}) of lines 1 and 2
$\qquad c_1 \leftarrow a_1$; $d_1 \leftarrow b_1$
$\qquad c_2 \leftarrow a_2$; $d_2 \leftarrow b_2$
Intersect
$\quad x_{12} \leftarrow x$
$\quad y_{12} \leftarrow y$

\ Intersection (x_{23}, y_{23}) of lines 2 and 3
$\qquad c_1 \leftarrow a_3$; $d_1 \leftarrow b_3$
Intersect
$\quad x_{23} \leftarrow x$
$\quad y_{23} \leftarrow y$

\ Intersection (x_{31}, y_{31}) of lines 3 and 1
$\qquad c_2 \leftarrow a_1$; $d_2 \leftarrow b_1$
Intersect
$\quad x_{31} \leftarrow x$
$\quad y_{31} \leftarrow y$

Write: 'Vertex 1-2: ', x_{12}, y_{12}
Write: 'Vertex 2-3: ', x_{23}, y_{23}
Write: 'Vertex 3-1: ', x_{31}, y_{31}

End Program

The quantities c_1, d_1, c_2, d_2, x, and y in the program and the module share the same memory locations under the same names; they are called **_global_** quantities. We are forced to reassign quantities to and from their natural names simply to meet the requirements for using global quantities. Because of the inconvenience caused by reassignment and the difficulty in accounting for several global quantities in large codes, we recommend that globals be eliminated from or reduced to a minimum in program modules.

A revised module that eliminates globals may be written with substitute names in the header as in the following pseudocode.

Module Intersect $(c_1, d_1, c_2, d_2, x, y)$: \ revised; with substitutes

$c \leftarrow c_1 - c_2$

$x \leftarrow (d_2 - d_1)/c$

$y \leftarrow (c_1 d_2 - c_2 d_1)/c$

End Module Intersect

The corresponding revised program is shown in Pseudocode 1–6.

The revised program is shorter than the original version because it can use its *own* names in the calls to the module. These names correspond in order and type to the names in the module header. The quantity c in the revised module is a *local* quantity, which can be accessed only by the module.

The names in the module header are treated as variables by the module. Quantities such as x and y whose values are returned to the program must have corresponding *variable* entities in the program to receive the values. The other quantities c_1, d_1, c_2, and d_2 are module inputs, which are not altered by the module; they may be associated either with variable entities (as in our example), or with expressions to be evaluated, or with directly given values in the program.

Some languages allow *recursive* modules, which are modules that can call themselves. Recursion refers to a process that is defined for some state in terms of the next

Pseudocode 1–6   Code to use a module with substitute names

Program: \ Revised

\ Program to find vertices of a triangle formed by

\ $y = a_1(x) + b_1; y = a_2(x) + b_2; y = a_3(x) + b_3;$

\ Module Intersect is called to find line intersections

Write: 'Constants a_1 and b_1 for line 1?'

Initialize: a_1, b_1

Write: 'Constants a_2 and b_2 for line 2?'

Initialize: a_2, b_2

Write: 'Constants a_3 and b_3 for line 2?'

Initialize: a_3, b_3

\ Intersection (x_{12}, y_{12}) of lines 1 and 2

Intersect$(a_1, b_1, a_2, b_2, x_{12}, y_{12})$

\ Intersection (x_{23}, y_{23}) of lines 2 and 3

Intersect$(a_2, b_2, a_3, b_3, x_{23}, y_{23})$

\ Intersection (x_{31}, y_{31}) of lines 3 and 1

Intersect$(a_3, b_3, a_1, b_1, x_{31}, y_{31})$

Write: 'Vertex 1-2: ', x_{12}, y_{12}

Write: 'Vertex 2-3: ', x_{23}, y_{23}

Write: 'Vertex 3-1: ', x_{31}, y_{31}

End Program

Pseudocode 1–7a   An iterative module for computing a mean

Module MeanI(*A, n*):   \ To find the mean *A* of *n* values of *x*

$A \leftarrow 0$

For [*j* = 1, 2, ..., *n*]
   Write: 'Give *x*(*j*) for *j* =', *j*
   Initialize: *x*
   $A \leftarrow A + x$ \ Accumulate sum of *x*
End For
$A \leftarrow A/n$

End Module MeanI

---

state until a well-defined **terminal** state is reached. Although recursive processes can generally be replaced by more computationally efficient iterative processes, recursion can significantly simplify coding in some cases.

An example of a recursive process is the one for computing the arithmetic mean of a set of values x_1 through x_n. The usual definition of the mean is

$$x_{mean} = (1/n) \sum_{j=1}^{n} (x_j)$$

An iterative module for computing x_{mean} under the name A is given in Pseudocode 1–7a.

To obtain a recursive model for the mean, we let A_j be the mean of the first j values x_1 through x_j. Then we can define A_j by the recursive formula

$$A_j = \begin{cases} x_j; & j = 1 \\ \{(j-1)A_{j-1} + x_j\}/j; & j = 2, 3, \ldots, n \end{cases}$$

and obtain x_{mean} as A_n. Pseudocode 1–7b is the recursive form of the module.

Pseudocode 1–7b   A recursive module for computing a mean

Module MeanR (*A, n*):   \ To find the mean *A* of *n* values of *x*

Write: 'Give *x*(*j*) for *j* =', *n*
Initialize: *x*

If [*n* = 1] **then**
   $A \leftarrow x$ \ Terminal state
Else
   MeanR(*A, n* − 1)   \ Recursive call
   $A \leftarrow \{(n-1)A + x\}/n$
End If

End Module MeanR

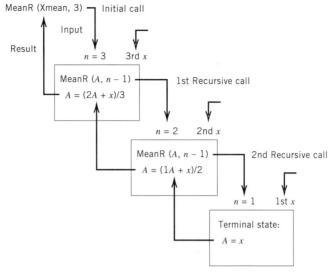

Figure 1–2 Illustration of recursive calls for finding the mean of three values.

We may use the module to compute the mean of three x values by employing a call statement such as

MeanR(Xmean, 3)

We note that the input for n is given as a value in the initial call and as an evaluated quantity in the **Else** clause of the module in recursive calls. We also note that Xmean is a variable to receive the result for A when n is equal to 3. The recursion process for computing the mean of three x values is illustrated in Fig. 1–2. We see that the computation of A (corresponding to A_3) in the initial call must await the return of A (corresponding to A_2) from the first recursive call, and that the computation of A (or A_2) in the first recursive call must in turn await the result for A (corresponding to the terminal state A_1) of the second recursive call.

1.4   DATA REPRESENTATION

We alluded earlier to the different ways in which fixed-point and floating-point numbers are stored in a computer's memory. We now consider the details of storage schemes (Ref. 1, 3) based on the binary number system.

1.4.1   Fixed-Point Numbers

We begin our discussion by considering the representation of fixed-point numbers or integers in an arbitrary number base b. An m–digit integer I_m is written as

$$[1.1] \qquad I_m = (d_{m-1} d_{m-2} \ldots d_1 d_0)_b \qquad d_j \in \{0, 1, \ldots, b-1\}$$

where the subscript b identifies the number base and the digits d_j of the number belong to the set of integers from 0 to $(b - 1)$. Thus digits may range from 0 to 1 for binary numbers and from 0 to 9 for decimal numbers. For hexadecimal numbers, with b equal to 16, the letters A through F are used as digits in place of the numbers 10 through 15. The decimal value of an integer described by Eq. [1.1] is obtained from

[1.2]
$$I_m = \sum_{j=0}^{m-1} \left(b^j d_j \right)$$

Examples of Eq. [1.2] for decimal and binary numbers are

$$(396)_{10} = \left(6 \times 10^0\right) + \left(9 \times 10^1\right) + \left(3 \times 10^2\right) = (396)_{10}$$
$$(11001)_2 = \left(1 \times 2^0\right) + \left(0 \times 2^1\right) + \left(0 \times 2^2\right) + \left(1 \times 2^3\right) + \left(1 \times 2^4\right) = (25)_{10}$$

The largest number that can be stored in m digits is one in which every digit is equal to $(b - 1)$. The decimal equivalent of this number is

[1.3]
$$(I_m)_{\max} = \sum_{j=0}^{m-1} \left\{ b^j (b - 1) \right\} = b^m - 1$$

The inversion of Eq. [1.2] to obtain the representation of a decimal number in the number base b is shown in Pseudocode 1–8.

The pseudocode introduces a new construction for declaring arrays. The process described for a positive integer is to replace I by the integer quotient of (I / b) while keeping track of the integer remainder as long as I remains nonzero. The digits of the converted integer in the number base b are the remainders in reverse order. For a negative number, the first position is filled with the character for the minus sign and I is replaced by its magnitude prior to conversion. An ***overflow*** (of the spaces reserved for storing the digits of the converted number) is indicated when elements 2 through k of the array c are filled before I is reduced to zero; j reaches a value of 1 at this point and causes the **While** loop to terminate. A terminated loop with a nonzero I value causes the overflow message to be displayed. The conversion process from decimal to binary is illustrated in Example 1.1.

Pseudocode 1–8   Conversion from decimal to number base b

```
\ Convert decimal integer to base-b number with at most k character positions.
\ Character array c with subscripts from 1 to k is used to store base-b form.
\ The first array element c(1) is reserved for a sign.
   Declare: c          \ Character array
   Initialize: b       \ Number base
```

Initialize: k \ Number of positions for storing converted value
Initialize: l \ Decimal integer to be converted
$c(1) \leftarrow \; ' \; '$ \ Assign blank character to $c(1)$
For $[j = 1, 2, \ldots, k]$
     $c(j) \leftarrow \; '0'$ \ Assign zero character to remaining elements of c
End For
     $j \leftarrow k$ \ Assign initial position counter
\ Processing for nonzero values; c already contains characters for $l = 0$
If $[l \neq 0]$ **then**
     If $[l < 0]$ **then**
         $c(1) \leftarrow \; ' - '$ \ Sign for negative l
             $l \leftarrow -l$ \ Replace l by its magnitude
     End If
     While $[(l \neq 0) \text{ and } (j > 1)]$
         $c(j) \leftarrow$ integer remainder of (l / b) \ j-th digit
         $j \leftarrow j - 1$ \ Reset j
         $l \leftarrow$ integer quotient of (l / b) \ Reduce l
     End While
End If
\ Output of converted number
If $[l > 0]$ **then**
     Write: 'Position overflow occurred'
Else
     Write: c \ In the order $c(1)$ through $c(k)$
End If

EXAMPLE 1.1

Convert $(39)_{10}$ to binary form.

Number base $b = 2$; reduction process of the pseudocode is as follows:

$$
\begin{array}{r|l}
2 & 39 \\ \hline
2 & 19 \quad + \quad \text{Remainder 1} \\ \hline
2 & 9 \quad + \quad \text{Remainder 1} \\ \hline
2 & 4 \quad + \quad \text{Remainder 1} \\ \hline
2 & 2 \quad + \quad \text{Remainder 0} \\ \hline
2 & 1 \quad + \quad \text{Remainder 0} \\ \hline
 & 0 \quad + \quad \text{Remainder 1}
\end{array}
$$

$\rightarrow (39)_{10} = (100111)_2$

The limitations of the statically allocated array structure in our pseudocode are similar to those of a memory cell for storing an integer in binary form. Memory cells usually consist of 8-bit or 1-byte multiples, with each position containing one binary digit or bit. Common cell lengths for integers are 16 bits (2 bytes) and 32 bits (4 bytes). In a k-bit cell, the first bit may be used for a sign, with 0 indicating a nonnegative number and 1 indicating a negative number. The remaining $(k - 1)$ bits are used to store the binary magnitude of the number. This type of storage uses what is called ***sign-magnitude*** notation; that is, a sign bit followed by the bits corresponding to the magnitude.

We use Eq. [1.3] with m equal to $(k - 1)$ to find that the decimal range of integers that may be stored in sign-magnitude notation in a k-bit cell is from $(-[2^{k-1} - 1])$ to $(+[2^{k-1} - 1])$. Numbers outside this range will cause overflow (for positive numbers) or underflow (for negative numbers). The limits for the commonly used 16-bit and 32-bit cells are shown as follows (with spaces for legibility):

$$\text{Maximum Integer} = (-[\text{Minimum Integer}]) = \begin{cases} (32\ 767)_{10}\ ; k = 16 \\ (2\ 147\ 483\ 647)_{10}\ ; k = 32 \end{cases}$$

The sign-magnitude convention is generally not used for integers because of the way in which additions and subtractions are performed by the computer's arithmetic/logic unit. The more common storage scheme involves the concept of ***two's complement*** notation. The two's complement of a negative integer I in a k-bit cell is given by

[1.4] $$\text{Two's complement of } I = 2^k + I$$

For example, the two's complement of $(-3)_{10}$ in a 3-bit cell is $(2^3 - 3)_{10}$ or $(101)_2$.

The storage scheme for storing an integer I in a k-bit cell with two's complement notation uses Eq. [1.4] for negative numbers and is described by

[1.5] $$\text{Stored value } C = \begin{cases} I\ ; & I \geq 0, \text{ first bit } = 0 \\ 2^k + I\ ; & I < 0 \end{cases}$$

The scheme of Eq. [1.5] is illustrated in Table 1–2; a 3-bit cell is used for simplicity.

We see in Table 1–2 that only two of the three bits are available for positive numbers and that the maximum positive value is therefore $(2^2 - 1)_{10}$ or $(3)_{10}$. A negative value is indicated by a 1 in the first bit position. The most negative number is stored as the 3-bit sequence $(100)_2$ which represents $(2^2 - 2^3)_{10}$ or $(2^2[1 - 2])_{10}$ or $(-4)_{10}$.

The general range of integers stored with two's complement notation in k-bit cells is from (-2^{k-1}) to $(2^{k-1} - 1)$. This range admits one more negative number

Table 1–2 Binary Storage of Integers with Two's Complement in a 3-Bit Cell

Binary Value Stored C	Binary Value Represented I	Decimal Equivalent of I
111	−001	−1
110	−010	−2
101	−011	−3
100	−100	−4
011	011	3
010	010	2
001	001	1
000	000	0

than can be represented in sign-magnitude notation. Limits for 16-bit and 32-bit cells with two's complement notation are

$$\text{Maximum Integer} = (-[\text{Minimum Integer} + 1]) = \begin{cases} (32\ 767)_{10} \,; k = 16 \\ (2\ 147\ 483\ 647)_{10} \,; k = 32 \end{cases}$$

Our discussion of fixed-point representation introduces elementary concepts of binary numbers and illustrates how integers are stored. The major aspects of which programmers should be aware are the underflow and overflow limits of the fixed-point data types available in their programming languages.

1.4.2 Floating-Point Numbers

Floating-point numbers or reals contain an integer part as represented in Eq. [1.1], and they also admit a fractional part. An n-digit fractional part P_n may be written in a number base b as

[1.6] $\quad P_n = (0.d_{-1}d_{-2}\dots d_{-n})_b \,; \qquad d_{-j} \in \{0, 1, \dots, b-1\}$

The point following the leading zero is called a ***radix*** point (where radix refers to the number base). The counterpart to Eq. [1.2] for obtaining the decimal value of P_n is

[1.7] $$P_n = \sum_{j=1}^{n} \left(b^{-j}d_{-j}\right)$$

The general floating-point number R_{mn} (equal to $[I_m + P_n]$) with m integer digits and n fractional digits is obtained by combining Eq. [1.1] and [1.6] to produce

[1.8] $\quad R_{mn} = (d_{m-1}\dots d_1 d_0.d_{-1}d_{-2}\dots d_{-n})_b \,; \qquad d_j \in \{0, 1, \dots, b-1\}$

Its decimal value from Eq. [1.2] and [1.7] is given by

[1.9]
$$R_{mn} = \sum_{j=(-n)}^{m-1} \left(b^j d_j\right)$$

For example, the binary number $(10.11)_2$ has a decimal value given by

$$(10.11)_2 = \left(1 \times 2^{-2}\right) + \left(1 \times 2^{-1}\right) + \left(0 \times 2^0\right) + \left(1 \times 2^1\right) = (2.75)_{10}$$

Conversion of a positive decimal fraction to a binary fraction is shown in the following procedure; application of the procedure is illustrated in Example 1.2.

■ **Procedure 1.1   Conversion of a positive decimal fraction D to a binary fraction.**

1. Set a digit counter j equal to zero.
2. Multiply the decimal fraction D by 2.
3. Decrease the counter j by 1.
4. Set the digit d_j equal to the integer part of $(2D)$ and reset D to the fractional part of $(2D)$.
5. If (the new value of) D is not zero, repeat the process from Step 2.   ■

EXAMPLE   1.2

Convert $(0.9)_{10}$ to a binary fraction.

Apply the conversion procedure as follows:

$$
\begin{array}{l}
0.9 \\
\times 2 \\
\hline
0.8 + \text{integer part 1 } (d_{-1}) \\
\times 2 \\
\hline
0.6 + \text{integer part 1 } (d_{-2}) \\
\times 2 \\
\hline
0.2 + \text{integer part 1 } (d_{-3}) \\
\times 2 \\
\hline
0.4 + \text{integer part 0 } (d_{-4}) \\
\times 2 \\
\hline
0.8 + \text{integer part 0 } (d_{-5})
\end{array}
$$

At this stage we see that a previous fractional part is repeated; continuation of the process will therefore cause the sequence of digits d_{-2} through d_{-5} to be repeated *ad infinitum*. The result of the process is therefore

$$(0.9)_{10} = (0.1\overline{1100})_2$$

Let us now consider storage of a general floating-point number such as $(-39.9)_{10}$. We obtain the binary representation of this number from Examples 1.1 and 1.2 as

$$(-39.9)_{10} = (-100111.1\overline{1100})_2$$

To store the binary form, we first **normalize** it; that is, we rewrite the number, in a binary equivalent of conventional scientific notation, by shifting the radix point immediately to the right of the number's first significant bit. We then multiply the result by (2^n) if the radix point was shifted n places to the left or by (2^{-n}) if the point was shifted n places to the right; the multiplication is necessary to recover the original value of the number. The normalized binary form of $(-39.9)_{10}$ is given by

$$(-39.9)_{10} = (-100111.1\overline{1100})_2 = (-1.001111\overline{1100})_2 \times (2^5)_{10}$$

A floating-point number can be written in the general normalized form

[1.10]   Floating-Point Number $= (-1)^s \times (1.f)_2 \times (2^t)_{10}$

The quantity s in Eq. [1.10] is the **sign bit**, whose value may be either 0 or 1; thus a negative number is indicated by s equal to 1, and a positive number is indicated by s equal to 0. The significant bits of the number are contained in the quantity $(1.f)$, which is known as the **mantissa**. The power t in the final piece of the representation is the **characteristic** or **exponent**.

A widely used storage scheme for the binary form is the **IEEE Standard for Binary Floating-Point Arithmetic**. This standard was defined by the Institute of Electrical and Electronic Engineers and adopted by the American National Standards Institute. The **single-precision** format employs 32 bits, of which the first or most significant bit is reserved for the sign bit s; the first bit for the number $(-39.9)_{10}$ is therefore equal to 1.

The next 8 bits are used to store a bit pattern to represent the exponent t. The binary value of t is not stored directly; rather, it is stored in **biased** or **offset** form as a nonnegative binary value c. The relation for the actual exponent t in terms of the stored value c and the bias b is

[1.11] $t = c - b$

The reason for using a bias is to make aligning of the radix point easier when two numbers are brought into registers to be operated on by the arithmetic/logic unit.

The 8-bit value of c ranges from $(0000\ 0000)_2$ to $(1111\ 1111)_2$ or from $(0)_{10}$ to $(255)_{10}$. The bias b has a value of $(0111\ 1111)_2$ or $(127)_{10}$. Again using our example of $(-39.9)_{10}$ for which t is equal to $(5)_{10}$, we obtain a c value of $(132)_{10}$ whose 8-bit form is $(1000\ 0100)_2$.

The remaining 23 bits of the 32-bit format are used for the mantissa. Only the part of the mantissa denoted by f is stored because the leading digit is always equal to 1 and is **understood** to be a part of the number. The f value for our example of $(-39.9)_{10}$ is an infinitely long binary sequence, which must be reduced to 23 bits for storage. The method of reduction dictated by the IEEE standard is **statistical rounding** in which we round up if the value of the bits beyond the 23rd bit of f exceeds (2^{-24}), round down if it is less than (2^{-24}), and round to either the nearest odd or the nearest even value if it is **exactly** (2^{-24}). The choice of odd or even in the last case is prescribed and must be used consistently.

The IEEE format for the decimal number $(-39.9)_{10}$ is shown in Fig. 1–3; the fractional part f of the mantissa is rounded up.

If we now use Eq. [1.9] and [1.10] to recover the decimal value of the stored number in Fig. 1–3, we obtain

$$\text{Decimal value of stored number} = (-1.0011\ 1111\ 0011\ 0011\ 0011\ 010)_2 \times (2^5)_{10}$$
$$= (-39.900\ 001\ 525\ 878\ 906\ 25)_{10}$$

Thus, even what we consider to be a relatively simple decimal value of $(-39.9)_{10}$ is represented with a small error in finite binary precision. Note that the pseudocode segment

$x \leftarrow -39.9$
Write: x

will **not** produce a 19-digit decimal equivalent of the stored value. To see why this is so, let us consider the relative error of a number stored in the 32-bit IEEE format.

The **relative error** ϵ_r between a value x and its approximation x_a is defined by

$$\epsilon_r = |x_a - x|/|x|$$

Figure 1–3   Storage of $(-39.9)_{10}$ or $(-100111.1\overline{1100})_2$ in IEEE 32-bit format.

The largest relative error for storage in the 32-bit format occurs when the mantissa is exactly equal to 1 and the actual error is $(\pm 2^{-24})$. The value of this error relative to 1 is equal to (2^{-24}) or approximately (6×10^{-8}); therefore, we can not rely on the accuracy of the floating-point representation of a decimal value to more than about 7 decimal digits. Because of this limitation, the decimal value that is reported for a number in storage is generally not computed beyond about 7 or 8 decimal digits when single precision is used.

We return now to the general storage scheme described by Eq. [1.10] and continue our discussion of the 32-bit format. The leading 1 in the mantissa $(1.f)$ is omitted in storage; therefore, the scheme as described so far is inadequate for storing a zero. A true zero is stored by setting each of the last 31 bits to zero, so that the scheme described by Eq. [1.10] is overridden in this special case; the sign bit may be either 0 or 1. The smallest nonzero magnitude that can be stored is limited by this exception and is (2^{-126}) or about (10^{-38}).

A similar exception exists when each of the last 31 bits is 1; the number represented in this case is infinite and is reported either as $(+\mathbf{Inf})$ or $(-\mathbf{Inf})$ depending on the sign bit. If each bit of the biased exponent c (described in Eq. [1.11]) is 1, and at least one bit of the stored part f of the mantissa is 0, the stored value is designated as (Not a Number) and is reported either as $(+\mathbf{NaN})$ or $(-\mathbf{NaN})$. The largest magnitude that can be stored according to Eq. [1.11] is therefore one whose biased exponent c is equal to $(1111\ 1110)_2$ and whose mantissa $(1.f)$ has a 1 in each bit position. The value of this magnitude is $(2^{128} - 2^{128-24})$ or about (3×10^{38}). An occurrence of **NaN** usually indicates an attempt to store a number with a larger than allowable magnitude (overflow) or an attempt to store a nonzero number with a smaller than allowable magnitude (underflow).

Double-precision storage uses 64 bits instead of 32 bits to store a value. The leading bit is again the sign bit, the next 11 bits are used to store the biased exponent, and the remaining 52 bits are used to store the fractional part f of the normalized mantissa. The bias b used in Eq. [1.11] is $(1023)_{10}$, and the maximum relative error for storing a number is (2^{-53}) or about (10^{-16}). Overflow occurs at about (10^{308}) and underflow occurs at about (10^{-308}).

We note in closing that variations in terminology and standards do exist. Some languages use 48-bit precision as normal, and others, which adhere to the 32-bit and 64-bit formats, may normalize the number so that the mantissa begins with a unit digit immediately *following* the radix point. The normal precision for supercomputers is 64 bits; the term used for the 32-bit format may be half-precision, and double-precision may refer to a 128-bit format on such computers. The 128-bit format may be found as extensions on general purpose computers and is often called quad-precision or extended-precision.

1.4.3 Some Consequences of Finite Data Representation

The most obvious things of which we should be aware are the overflow and underflow limits posed by finite data representation. Excessive magnitudes should be avoided with fixed-point data. For example, computations of factorials and conversion of floating-point numbers with large magnitudes to integers may cause integer overflow. The more often used floating-point arithmetic contains several traps for the unwary; some of these are computations of exponentials whose arguments have large magnitudes, logarithms and powers of quantities that are nearly zero, and tangents of angles near $(\pm \pi / 2)$ radians.

Examples 1.3 and 1.4 show how we can avoid overflow limits by recasting problems so that computed quantities are acceptable. The computational technique shown in Example 1.4 for building terms with powers and factorials is useful for summations of series.

EXAMPLE   1.3

Assume that a computer uses two's complement notation to store integers in a 32-bit format. How could one use the computer to determine the decimal value of the maximum integer $(2^{31} - 1)$?

We obviously cannot compute 2^{31} first and then subtract 1 because 2^{31} would cause an overflow. Recognize, however, that $(2^{31} - 1)$ is a binary sequence of 31 unit digits and use Eq. [1.2] to obtain

$$K = \left(2^{31} - 1\right) = \sum_{j=0}^{30} \left(2^{j}\right)$$

A pseudocode for computing the result K is as follows:

```
K ← 1        \ first term in the summation
P ← 1        \ j-th power of 2 for j = 0
For [j = 1, 2, …, 30]
     P ← 2P          \ Build j-th power of 2
     K ← K + P       \ Accumulate sum
End For
Write: K
```

EXAMPLE   1.4

Write a code to compute $15^{35}/(35!)$ on a machine whose overflow limit is 10^{38}. Note that (15^{35}) is approximately equal to (1.46×10^{41}) and that $(35!)$ is approximately equal to (1.03×10^{40}).

We use the relations

$$x^{n} = x\left(x^{n-1}\right); \qquad n! = n\left\{(n-1)!\right\}$$

to write

$$x^n/(n!) = [x/n][x^{n-1}/\{(n-1)!\}]$$

and start building the result z from the point at which n is equal to 1. The pseudocode is given by

```
z ← 15          \ At n equal to 1
For [n = 2, 3, ..., 35]
    z ← 15z/n         \ Build result (with x equal to 15)
End For
Write: z
```

The representation of floating-point numbers may contain errors that can propagate throughout a process and make a solution useless. Before discussing specific examples, we shall look at the errors in the results of simple operations. Let us denote the floating-point representation of a number x by $Fl(x)$. When we perform simple operations on two quantities x and y, we are really using $Fl(x)$ and $Fl(y)$ as the operands. The error in the result is **round-off error**, which has two sources; one is computational error due to the errors in the operands, and the other is the storage error that occurs when we store the computed result. For example, the result of the computation $(x + y)$ and its errors are

$$\text{Computed Result of } (x + y) = Fl[Fl(x) + Fl(y)];$$
$$\text{Computational Error} = [Fl(x) + Fl(y)] - (x + y);$$
$$\text{Storage Error} = Fl[Fl(x) + Fl(y)] - [Fl(x) + Fl(y)]$$

Let us define the errors ϵ_x in x and ϵ_y in y by

$$\epsilon_x = Fl(x) - x; \qquad \epsilon_y = Fl(y) - y$$

Then the computational errors ϵ_c for simple operations are given by

[1.12a]
$$Fl(x) \pm Fl(y) = x \pm y + \epsilon_x \pm \epsilon_y$$
$$\rightarrow \quad \epsilon_c(x \pm y) = \epsilon_x \pm \epsilon_y$$

[1.12b]
$$Fl(x) \times Fl(y) = xy + x\epsilon_y + y\epsilon_x + \epsilon_x\epsilon_y$$
$$\rightarrow \quad \epsilon_c(xy) = x\epsilon_y + y\epsilon_x + \epsilon_x\epsilon_y$$

[1.12c]
$$Fl(x) \div Fl(y) = [(x + \epsilon_x)/y]\left[1 + (1 + \epsilon_y/y)^{-1} - 1\right]$$
$$= x/y + \epsilon_x/(y + \epsilon_y) - (x/y)\epsilon_y/(y + \epsilon_y)$$
$$\rightarrow \quad \epsilon_c(x/y) = (\epsilon_x - [x/y]\epsilon_y)/(y + \epsilon_y)$$

The computational errors for multiplications and divisions typically have the same orders of magnitude as the storage errors. The computational error for division decreases as the divisor y increases in magnitude. Additions and multiplications may produce large relative errors depending on the sizes of the operands.

A large relative error may be produced when one number is subtracted from another that is almost equal to the first. Consider the approximation

$$f'(x) \cong D = [f(x + h) - f(x)]/h$$

that is sometimes used in numerical differentiation. We know that the true derivative is defined by D when h tends to zero. For numerical differentiation with a finite value of h, our mathematical intuition tells us to choose a small value of h if we wish D to be an accurate approximation to the true derivative $f'(x)$. We shall test this premise by using the expression for D to estimate $[d(\sin x)/dx]$ at x equal to 1 radian.

For simplicity, we shall simulate the computation of D on a decimal machine that can store only four significant digits of a decimal value. During the computation, the result of each intermediate operation is rounded to and stored in this four-digit form before it can be used in a subsequent operation. The results of the simulation for three values of h are shown in Table 1–3.

When we compare the values of D in Table 1–3 with the true result (cos 1) whose value to 4 significant digits is 0.5403, we see that the smallest value of h does **not** yield the most accurate result. The apparent contradiction of mathematical wisdom occurs because the three leading significant digits of $[\sin(x + h)]$ and $(\sin x)$ are "lost" in the subtraction even though the result is stored to 4 significant digits. This phenomenon is known as **subtractive cancellation** and can lead to significant errors. With the intermediate h value of 0.01, two leading digits are lost to cancellation, and a more accurate approximation is obtained. As we increase h to 0.1, only one leading digit is lost; however, the result for D is the least accurate among the three cases because the mathematical error in the model for the derivative is not sufficiently counteracted by the reduced cancellation error.

Further examination of our example shows one effect that occurs with addition. If we choose any (positive) h value less than 0.0005, the quantity $(x + h)$ would be rounded to 1.000 on our four-digit machine; in other words, the bits in too small a value of h would be dropped off the end of the larger number to which h is added.

Table 1–3   Simulation for Computation of a Derivative

Quantity	Value of Quantity with h equal to		
	0.001	0.010	0.100
$x + h$	1.001	1.010	1.100
$\sin(x + h)$	0.8420	0.8468	0.8912
$\sin(x)$	0.8415	0.8415	0.8415
$\sin(x + h) - \sin(x)$	0.0005000	0.005300	0.04970
D	0.5000	0.5300	0.4970

In these cases, the approximation D to the derivative of $(\sin x)$ at x equal to 1 would be *zero!*

When adding several numbers with dissimilar magnitudes, the effect of dropped bits can generally be reduced by adding in the direction from smallest to largest magnitude. The gain in accuracy is small in most cases; nevertheless, there are pathological cases in which the order of addition is important. Suppose, for example, we use our four-digit decimal machine to add one value of 1 to a thousand values of 0.0004 according to

$$S \;=\; 1.000 + 0.0004000 + \cdots; \qquad T \;=\; 0.0004000 + \cdots + 1.000$$

The computed value of S is 1.000 because all digits of the smaller numbers are dropped in each addition; on the other hand, T is computed as 1.400 because the smaller numbers accumulate during addition to a value that is significant when the last number 1.000 is added.

Propagation of round-off errors may generally be reduced by using efficient computational methods. Assume that n multiplications are required to compute (x^n) with an integer power greater than 1. Computation of the 4th degree polynomial

$$P(x) \;=\; a_4 x^4 + a_3 x^3 + a_2 x^2 + a_1 x + a_0$$

for a given value of x can be performed as it is written with 13 multiplications and four additions. A more efficient method is obtained by rewriting the polynomial with nested multiplications as

$$P(x) \;=\; a_0 + x\,(a_1 + x\,(a_2 + x\,(a_3 + x\,a_4)))$$

Evaluation of the nested form is called ***Horner's method*** and can be accomplished with 4 multiplications and 4 additions. For the general n-th degree polynomial

$$P_n(x) \;=\; \sum_{j=0}^{n} \left(a_j x^j \right)$$

Horner's method may be implemented as shown in Pseudocode 1–9.

Pseudocode 1–9   Horner's method for an n-th degree polynomial

```
Module Horner(P, a, x, n):
    P ← 0                    \ Initial P value
    For [j = n, (n − 1), . . . , 0]
        P ← xP + a(j)   \ Multiply and accumulate
    End For
End Module Horner
```

When small errors in parts of a quantity cause the computed value of the quantity to contain a relatively large error, the quantity is said to be ***ill conditioned***. A simple example of an ill-conditioned quantity on the four-digit decimal machine we used earlier is illustrated by the pseudocode segment

$$x \leftarrow 61.1/7; \qquad y \leftarrow \sqrt{(534 - 7x^2)/6}$$

Recalling that the result of an operation is rounded and stored to four significant digits, we compute x as 8.729 and obtain y from the sequence of operations

$$x^2 = 76.20; \qquad 7x^2 = 533.4; \qquad 534 - 7x^2 = 0.6000$$
$$(534 - 7x^2)/6 = 0.1000; \qquad y = 0.3162$$

The correct result for y (to 4 significant digits) is 0.3377; thus, an error of less than 0.005% in x causes an error of more than 6% in y.

The last topic that we shall discuss in this subsection is ***scaling***. Consider a quantity Q expressed by

$$Q = (uv + x^2)/(xy + u^2)$$

and let u, v, x, and y be quantities with magnitudes on the order of (10^{25}). An attempt to compute Q directly will cause overflows in the intermediate evaluations of (uv), (x^2), (xy), and (u^2) in single-precision arithmetic, even though Q itself has a magnitude on the order of 1. We can eliminate the overflow problem by reducing the magnitudes of u, v, x, and y to acceptable levels through the same scale factor s; then we rewrite Q as

$$Q = \left([u/s][v/s] + [x/s]^2\right) \Big/ \left([x/s][y/s] + [u/s]^2\right)$$

An appropriate choice for s can be made from a wide range of values; typically, we might consider some large power of 10 simply because we tend to think in the decimal number system. If we choose s in this manner, say as (10^{20}), we may introduce errors when the computer uses binary arithmetic to evaluate (u/s), (v/s), (x/s), and (y/s). A better choice for s is some large power of 2 such as (2^{70}). Such a number is represented with a mantissa that is exactly equal to $(1.0)_2$; thus, division by s will change only the biased exponents of u, v, x, and y but will not affect the significant bits in the mantissas.

1.5   CLOSURE

Numerical methods provide engineers with powerful tools for solving problems that are beyond the reach of ordinary mathematics. These methods and other computer aids have become so successful that they are often major developmental tools for new designs. Partial credit must be given to computing hardware and software for the success achieved by numerical techniques. As computers improve in speed and power, we are able to tackle larger problems and to reduce the costs of obtaining numerical solutions. We are also less hesitant about acquiring computing skills when the software is relatively easy to use.

Programming languages are the primary software tools for communicating with computers. Undergraduate students in technical programs are generally able to acquire fundamental programming skills in a high-level language in about 30–40 hours of instruction. Most continue to improve those skills as they achieve mathematical maturity and broaden their experiences in writing programs.

The popular programming languages in engineering are described in Section 1.3, and programming fundamentals are described in Section 1.4. We focus strongly in Section 1.4 on the relatively few major structures for controlling data, on the concept of modularity in the design of codes, on the representation of numerical data, and on errors that can arise when such data are stored with finite precision. Several examples of fundamental programming techniques are included in the discussions.

For generality, control structures are presented in pseudocode form rather than in the syntax of a particular language; these structures are also summarized in Appendix B. We shall use pseudocode structures extensively to explain the steps of various numerical procedures that are covered in the remaining chapters.

1.6 EXERCISES

1. A network of resistors $(R_1, R_2, \ldots)$ in a parallel arrangement has an equivalent resistance R_{eq} given by

$$1/R_{eq} = 1/R_1 + 1/R_2 + \cdots$$

Write a program to compute equivalent resistances for an arbitrary number of input data sets, each of which contains an arbitrary number of resistors (whose values in ohms are to be specified through keyboard input). *No arrays are to be used!* Execute the program for the following input data sets.

   Set 1: $R_1 = 5(10^3)$, $R_2 = 10^5$, $R_3 = 10^4$
   Set 2: $R_1 = 2(10^4)$, $R_2 = 10^3$, $R_3 = 10^4$, $R_4 = 5(10^4)$
   Set 3: $R_1 = 2(10^3)$, $R_2 = 10^3$, $R_3 = 10^5$

2. The geometric mean x_g of n quantities x_1 through x_n is defined by

$$(x_g)^n = \prod_{j=1}^{n}(x_j)$$

Write a program module to compute and return x_g. Input to the module is to be an array of x values and the number n of elements in the array. Use the module to compute x_g for an array whose elements are given by

$$x_j = e^{[10 + \cos(j/10)]}; \quad (j/10) \text{ in radians}, \quad j = 1, 2, \ldots, 100$$

3. The Poisson distribution $P(k)$ that is used in probabilistic models is defined by

$$P(k) = (\lambda^k e^{-\lambda})/k!$$

and represents the probability that k events will occur for a given value of λ. In a certain application in which λ is equal to 1.23, we wish to determine the smallest (integer) value of a quantity n such that

$$\sum_{k=0}^{n} P(k) \geq t$$

Write a program to determine n when a t value less than 1 is specified through keyboard input. Execute the program with an input value of 0.98 for t.

4. The height y and horizontal coordinate x of a point on a ramp are expressed in meters and are related through the quartic equation

$$y/5 = 1 - 6(x/15)^2 + 8(x/15)^3 - 3(x/15)^4; \qquad 0 \leq x \leq 15$$

An object descends the ramp under gravity, starting with a speed v_o (in m/s) from (x, y) equal to $(0,5)$ m. For negligible friction, the object's speed v (in m/s) at any point (x, y) on the ramp is found from

$$v^2 = (v_0)^2 + 2g(5 - y); \qquad g = 9.81 \text{ m/s}^2 = \text{gravitational acceleration}$$

The number of "gees" Γ experienced by the object (i.e., the ratio of the reaction force of the ramp on the object to the weight of the object) is given by

$$\Gamma = c(1 + v^2 c^2 [d^2 y/dx^2]/g); \qquad c^2 = 1/(1 + [dy/dx]^2)$$

Write a program **without arrays** to accept y_0 as input, and to compute estimates of the minimum and maximum Γ values and the x locations at which they occur. The estimates are to be computed by comparing Γ values at 150 x locations

$$x_j = (150 - j)/10; \qquad j = 0, 1, \ldots, 150$$

Execute the program with v_o equal to 4 m/s and with v_o equal to 1.5 m/s.

5. A shaft of length L (in inches) is fixed at one end and is subjected to a torque T (in lb·in) applied to the other end. The shaft has a rectangular cross section with sides of dimensions a and b (in inches). The maximum shear stress S_{max} (in psi) and the maximum angle of twist α_{max} (in degrees) are given by

$$S_{max} = c_1 T/(abq); \qquad \alpha_{max} = 180 T L/(\pi c_2 G a b q^2)$$

where G is the shear modulus (in psi), q is the smaller of a and b, and c_1 and c_2 are coefficients that depend on the larger of (a/b) and (b/a) as follows.

$r = \max[(a/b), (b/a)]$	c_1	c_2
1.0	4.81	0.141
1.5	4.33	0.196
2.0	4.06	0.229

Write a program to estimate S_{max} and α_{max} for steel shafts ($G = 12 \times 10^6$ psi) when a and b are in the range ($0.5 \leq a/b \leq 2.0$). The coefficients c_1 and c_2 are to be approximated by linear interpolation on ($1.0 \leq r \leq 1.5$) or on ($1.5 \leq r \leq 2.0$) as appropriate. Execute the program for the following input data sets.

Set 1: $T = 7500$ lb·in, $L = 12.0$ in, $a = 1.00$ in, $b = 0.75$ in
Set 2: $T = 8000$ lb·in, $L = 10.0$ in, $a = 0.75$ in, $b = 1.25$ in
Set 3: $T = 9000$ lb·in, $L = 15.0$ in, $a = 1.00$ in, $b = 1.25$ in

6. A large structure has a floor at elevation y equal to zero and a ceiling at elevation y equal to h. The coordinates of a sound source are (x_s, y_s), and those of a receiver are (x_r, y_r). If the floor and ceiling are assumed to be perfect reflectors and the side walls are assumed to be far away from the source and receiver, the sound energy density at the receiver is proportional to a quantity α given by

$$\alpha = \sum_{m=(-\infty)}^{\infty} 1 \Big/ \Big[(x_r - x_s)^2 + (mh + y_r - y_s + \beta)^2 \Big]$$

$$\beta = \begin{cases} 0; & m = \text{even} \\ h - 2y_r; & m \neq \text{even} \end{cases}.$$

(a) Write a program to estimate α and execute it for the input data set:

$$x_r = 8, y_r = 6, x_s = 0, y_s = 3, h = 10$$

(b) Simplify the expression for α for the special case ($y_r = y_s = h/2$).

7. The speed v (in ft/s) of a car in a coast-down test to determine its drag characterisics is sampled at intervals of 1 s. Eighteen data pairs for the variation of v with time t (in seconds) are as follows.

t [s]	v [ft/s]	t [s]	v [ft/s]
1	65.658	10	62.912
2	65.346	11	62.576
3	65.041	12	62.317
4	64.835	13	61.993
5	64.385	14	61.698
6	64.103	15	61.417
7	63.809	16	61.096
8	63.485	17	60.804
9	63.210	18	60.519

The data for such a test can be significantly affected by unsteady wind conditions and by a road surface which is only slightly uneven. The data may therefore be noisy and are filtered by averaging the original speeds at five times to give a smoothed speed v_s according to

$$v_s(t) = [v(t-2) + v(t-1) + v(t) + v(t+1) + v(t+2)]/5$$

Write and execute a program to determine v_s at t equal to $(3, 4, \ldots, 16)$, and to compute the (negative) accelerations at t equal to $(4, 5, \ldots, 15)$ from

$$a(t) = [v(t+1) - v(t-1)]/2$$
$$a_s(t) = [v_s(t+1) - v_s(t-1)]/2$$

Output t, $v(t)$, $v_s(t)$, $a(t)$, and $a_s(t)$ in tabular form.

8. The term ***machine epsilon*** refers to the smallest positive value ϵ such that the machine can distinguish between $(1 + \epsilon)$ and 1. This value depends on the hardware and on how the compiler for a programming language stores the mantissa of a floating-point number. The machine epsilon for a normalized mantissa $(1.f)$ in which f is stored in n- bit locations is equal to 2^{-n}. For an unnormalized mantissa $(.1f)$ which is stored in n-bit locations, the machine epsilon is 2^{1-n}. The error in storing a quantity x has an upper bound of $\epsilon|x|$. A pseudocode for computing ϵ is as follows.

```
ε ← 1.0
b ← 1.0 + ε
While [b > 1.0]    \ so that 1 and(1 + ε) are distinguishable
    ε ← ε/2.0
    b ← 1.0 + ε
End While
    ε ← 2.0(ε)    \ least value for which 1 and (1 + ε) are different
Write: ε
```

Determine the machine epsilon for the machine and language you will ordinarily use at the precision levels in which you expect to write your codes.

Caution: Follow the pseudocode as exactly as possible; some shorter versions will not give correct results for machines that use numerical coprocessors!

9. The machine epsilon (see Problem 8) allows us to estimate an upper bound for the error in an operation. Let the true result of an operation on x and y be $f(x, y)$ and let ϵ be the machine epsilon. The error E in computing and storing $f(x, y)$ is bounded by

$$E \le B = \epsilon|f(x, y)| + (1 + \epsilon)(|\epsilon x \partial f/\partial x| + |\epsilon y \partial f/\partial y|)$$

The terms with the partial derivatives account for the errors in the operation due to inexact representations of x and y; the multiplication by $(1 + \epsilon)$ and the addition of $\epsilon|f(x, y)|$ account for errors in storing the result.

Use a result from Problem 8 (or use a fictitious value of 10^{-7} for ϵ) to estimate the upper bound B for the error in z in the following operations.

$$
\begin{aligned}
&\textbf{(a)} \quad z = 1.897 + 9.635 \\
&\textbf{(b)} \quad z = 4.058 \times 13.27 \\
&\textbf{(c)} \quad z = 2.345 \div 19.28 \\
&\textbf{(d)} \quad z = 19.28 \div 2.345
\end{aligned}
$$

10. Use the relation $[(a^2 - b^2) = (a - b)(a + b)]$ to rewrite the quantity $(\sqrt{x} - \sqrt{y})$ so that it can be computed accurately when the values of x and y are very close to each other. Provide an example of how your method works.

11. A function $f(x)$ is described by

$$
f(x) = \left\{\tan^{-1}[(x + 2)/q] - \tan^{-1}[(x + 1)/q]\right\}\Big/ q
$$

$$
q^2 = 1.4 - x^2
$$

in which the arctangent produces a first-quadrant angle in radians. The derivative (df/dx) at x equal to 0.3 may be approximated by

$$
df/dx \cong [f(x + \Delta x) - f(x - \Delta x)]/[2\Delta x]
$$

Write and execute a program to estimate the derivative from this approximation with Δx values given by

$$
\Delta x = 1/2^k; \qquad k = 2, 3, \ldots, 50
$$

Explain your results and state (without actually performing a differentiation) the value of k that you believe produces the best approximation to df/dx.

12. A straight line $(y = x)$ extends from (0,0) to (1,1) as shown in Fig. 1–P12. A curve given by $(y = f(x))$ is symmetrical about $(x = 0.5)$ and has a maximum $(0.5, h)$. It passes through (0,0) and (1,1), and it intersects the line at point A, whose x coordinate is positive.

A particle starting anywhere on the x axis between $(x = 0)$ and $(x = 1)$ moves vertically to the curve and then moves horizontally to the line in a repetitive sequence. The figure shows the positions of the particle at steps k equal to 0 through 5. In general, the position of the particle at step k is given by

$$
\begin{aligned}
&x_k = x_{k-1}, \qquad y_k = f(x_{k-1}); \qquad k = \text{odd} \\
&x_k = y_k = y_{k-1}; \qquad k = \text{even}
\end{aligned}
$$

For certain values of the curve's maximum height h, the particle converges to the intersection point A (which is known as a simple attractor). The particle's motion

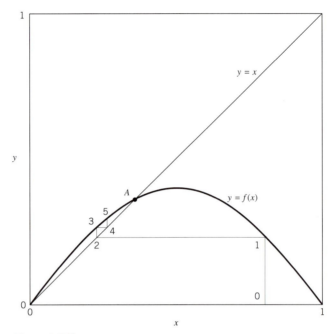

Figure 1–P12

is therefore a scheme for solving the equation $[x = f(x)]$. Write and execute a program to list (x_k, y_k) for k equal to $(0, 1, \ldots, 100)$ with x_0 equal to 0.8 and with h equal to 0.4 and to 0.8. The function to be used is

$$f(x) = h \sin(\pi x); \qquad (\pi x) \text{ in radians}$$

Describe the behavior of the motion when h is equal to 0.8.

2

Systems of Linear Algebraic Equations

A *linear algebraic equation* represents the linear dependence of a quantity ϕ on a set of variables x_1 through x_n and a set of constant coefficients α_1 through α_n; its form is

$$\alpha_1 x_1 + \alpha_2 x_2 + \cdots + \alpha_n x_n = \phi$$

If we replace ϕ by a singly subscripted quantity f_i and the coefficients α_j by a doubly subscripted quantity a_{ij}, we may write a *system* of n such equations in the form

$$[2.1] \quad \begin{cases} a_{11}x_1 & + & a_{12}x_2 & + & \cdots & + & a_{1j}x_j & + & \cdots & + & a_{1n}x_n & = & f_1 \\ a_{21}x_1 & + & a_{22}x_2 & + & \cdots & + & a_{2j}x_j & + & \cdots & + & a_{2n}x_n & = & f_2 \\ & & \vdots & & \vdots & & \vdots & & \vdots & & \vdots & & \vdots \\ a_{i1}x_1 & + & a_{i2}x_2 & + & \cdots & + & a_{ij}x_j & + & \cdots & + & a_{in}x_n & = & f_i \\ & & \vdots & & \vdots & & \vdots & & \vdots & & \vdots & & \vdots \\ a_{n1}x_1 & + & a_{n2}x_2 & + & \cdots & + & a_{nj}x_j & + & \cdots & + & a_{nn}x_n & = & f_n \end{cases}$$

The solution of such systems for the quantities x_j, when the coefficients a_{ij} and the values f_i are given, pervades engineering applications and computational methods. For example, systems of linear algebraic equations are used directly in mathematical models for electrical, structural, and pipe networks, and in some computational methods for fitting curves to data. In other cases, such as in finite difference and finite element solutions of partial differential equations, they represent approximations of mathematical equations, which cannot be solved by analytical methods. An ***analytical method*** is one that produces either exact or approximate solutions in closed form.

In general, almost any application that requires the solution of several quantities is likely to involve systems of linear equations; indeed, they are prerequisites for all but the more elementary material in the remaining chapters of the text. Any intent of going beyond the more elementary methods makes it worthwhile to consider linear algebraic systems at this early stage, even though the abstractions and computer implementations may be difficult concepts to deal with initially.

The form of Eq. [2.1] is cumbersome for large systems. A more compact notation is obtained from the concepts of matrices and vectors. Direct computational methods for solving the resulting matrix equations are in fact methods for calculating exact mathematical representations in an ordered and efficient manner. Because these representations are derived from algebraic manipulation, it is necessary to develop the fundamental concepts of linear algebra before discussing the numerical methods for solving systems of equations.

Fundamental linear algebra and its relation to systems of linear algebraic equations are presented in Section 2.1, and direct methods for solving the system of equations are treated in Section 2.2. Direct methods are inappropriate for certain systems because of excessive storage requirements, excessive build-up of computer round-off errors, or considerable computational inefficiency. Iterative methods, which yield approximate solutions, are attractive alternatives in those cases and are discussed in Section 2.3.

Several applied problems are used to illustrate the methods in Sections 2.2 and 2.3. These problems deal with a ***parabolic curve fit***, a ***spline curve fit***, the forces in the members of a ***plane truss***, the determination of currents in a ***resistor network***, and the stresses on a ***rectangular shaft in torsion***. A ***Hilbert matrix*** is also used to illustrate the effects of ill-conditioning.

2.1 FUNDAMENTALS OF LINEAR ALGEBRA

The concepts of linear algebra that form the foundation for the numerical methods of this chapter are presented in this section. We begin with the notation and definitions of matrices and vectors, and describe operations on these quantities. Thereafter, we focus on square matrices, some of their properties, and their relation to systems of linear equations.

2.1.1 Notation and Definitions

A *matrix* is a rectangular array of quantities. An example of a matrix is

$$[2.2] \qquad \mathbf{A} = [a_{ij}] = \begin{bmatrix} a_{11} & a_{12} & a_{13} & a_{14} \\ a_{21} & a_{22} & a_{23} & a_{24} \\ a_{31} & a_{32} & a_{33} & a_{34} \end{bmatrix}$$

In Eq. [2.2], the bold-faced $\mathbf{A}$ is the symbol for the matrix, the literal representation on the far right shows the arrangement of the *elements*, and the form $[a_{ij}]$ is a shorter abstraction of the literal form. The first subscript for each element denotes its row position (increasing from 1 in the top row as we move downward), and the second subscript denotes its column position (increasing from 1 in the leftmost column as we move to the right). The matrix has no particular meaning until we associate the elements with a given concept.

A matrix with m rows and n columns has *dimension* $(m \times n)$ and is referred to as an $(m \times n)$ matrix (in which "$\times$" is pronounced "by"). The matrix in Eq. [2.2] is a (3×4) matrix. A matrix with only one row is a special kind of matrix known as a *row vector*; similarly, a matrix with only one column is a *column vector*. The elements of a vector are called its *components*. A column vector $\mathbf{c}$ and a row vector $\mathbf{r}$ are shown in Eq. [2.3]:

$$[2.3] \qquad \mathbf{c} = [c_i] = \begin{bmatrix} c_1 \\ c_2 \\ c_3 \end{bmatrix}; \quad \mathbf{r} = [r_j] = \begin{bmatrix} r_1 & r_2 & r_3 & r_4 \end{bmatrix}$$

A simple quantity such as one we would use in ordinary arithmetic is called a *scalar*. A scalar should not be interpreted as a (1×1) matrix.

2.1.2 Operations

Addition and subtraction are defined for matrices (or vectors) which have the same dimensions; matrices with identical dimensions are said to be *conformable*. Addition and subtraction for conformable matrices $\mathbf{A}$ (equivalent to $[a_{ij}]$) and $\mathbf{B}$ (equivalent to $[b_{ij}]$) are summarized by

$$[2.4] \qquad \underset{(m \times n)}{\mathbf{A}} \quad \pm \quad \underset{(m \times n)}{\mathbf{B}} \quad = \quad \underset{(m \times n)}{\mathbf{C}} \quad = \quad [c_{ij}] \quad = \quad [a_{ij} \pm b_{ij}]$$

An example of applying Eq. [2.4] is as follows:

$$\begin{bmatrix} 3 & 4 & 7 \\ 2 & 6 & 8 \end{bmatrix} - \begin{bmatrix} 2 & 0 & 6 \\ 3 & 8 & 7 \end{bmatrix} = \begin{bmatrix} (3-2) & (4-0) & (7-6) \\ (2-3) & (6-8) & (8-7) \end{bmatrix} = \begin{bmatrix} 1 & 4 & 1 \\ -1 & -2 & 1 \end{bmatrix}$$

Matrix multiplication $\mathbf{A} \cdot \mathbf{B}$ is defined only if the number of columns in $\mathbf{A}$ is equal to the number of rows in $\mathbf{B}$. The "·" operator between the matrices is used deliberately throughout the text to emphasize matrix–matrix multiplication as different from scalar–scalar or scalar–matrix multiplication. The form $\mathbf{A} \cdot \mathbf{B}$ involves *premultiplication* by $\mathbf{A}$ (on the left) or *postmultiplication* by $\mathbf{B}$ (on the right).

In general, an $(m \times r)$ matrix $\mathbf{A}$ (equivalent to $[a_{ik}]$) postmultiplied by an $(r \times n)$ matrix $\mathbf{B}$ (equivalent to $[b_{kj}]$) has a product $\mathbf{A} \cdot \mathbf{B}$, which is an $(m \times n)$ matrix $\mathbf{P}$ such that

[2.5a]
$$\underset{(m \times r)}{\mathbf{A}} \cdot \underset{(r \times n)}{\mathbf{B}} = \underset{(m \times n)}{\mathbf{P}} = [p_{ij}]$$

[2.5b]
$$p_{ij} = \sum_{k=1}^{r} (a_{ik} \times b_{kj}) = a_{i1}b_{1j} + a_{i2}b_{2j} + \cdots + a_{ir}b_{rj}$$

For example, consider the multiplication

$$\begin{bmatrix} 1 & 2 \\ 3 & 4 \\ 5 & 6 \end{bmatrix} \cdot \begin{bmatrix} 9 & 8 & 7 & 6 \\ 5 & 4 & 3 & 2 \end{bmatrix} = \begin{bmatrix} 19 & 16 & 13 & 10 \\ 47 & 40 & 33 & 26 \\ 75 & 64 & 53 & 42 \end{bmatrix}$$

The (3×2) matrix premultiplies the (2×4) matrix to produce a (3×4) matrix. It is not possible to perform a multiplication if the positions of the two matrices on the left side of the equation are reversed. The result of the multiplication may be verified by applying Eq. [2.5b]. For example, the element 33 in row 2, column 3 of the result is obtained from Eq. [2.5b] with row 2 of the first matrix and column 3 of the second according to

$$33 = (3 \times 7) + (4 \times 3)$$

We see from Eq. [2.5a] and from the preceding example that the existence of $\mathbf{A} \cdot \mathbf{B}$ does not necessarily mean that $\mathbf{B} \cdot \mathbf{A}$ exists. Even in cases when both exist, matrix multiplication is in general *not commutative*; that is, $\mathbf{A} \cdot \mathbf{B}$ is not necessarily equal to $\mathbf{B} \cdot \mathbf{A}$, and we must be careful with the order of multiplication.

Matrix multiplication is both *distributive* and *associative*. If $\mathbf{A}$, $\mathbf{B}$, and $\mathbf{C}$ are matrices with appropriate dimensions to satisfy the addition/subtraction conditions of Eq. [2.4] and the multiplication conditions of Eq. [2.5a], the distributive and associative properties are expressed in Eq. [2.6] and Eq. [2.7], respectively, by

[2.6] $\mathbf{A} \cdot (\mathbf{B} + \mathbf{C}) = \mathbf{A} \cdot \mathbf{B} + \mathbf{A} \cdot \mathbf{C}; \quad (\mathbf{B} + \mathbf{C}) \cdot \mathbf{A} = \mathbf{B} \cdot \mathbf{A} + \mathbf{C} \cdot \mathbf{A}$

[2.7] $\mathbf{A} \cdot \mathbf{B} \cdot \mathbf{C} = \mathbf{A} \cdot (\mathbf{B} \cdot \mathbf{C}) = (\mathbf{A} \cdot \mathbf{B}) \cdot \mathbf{C}$

Multiplication of a matrix **A** (equivalent to $[a_{ij}]$) by a scalar q to form a product matrix **S** equal to $q\mathbf{A}$ is also a defined operation. The product in this case is given by

$$[2.8] \qquad q \quad \underset{(m \times n)}{\mathbf{A}} \quad = \quad \underset{(m \times n)}{\mathbf{S}} \quad = \quad [s_{ij}] \;=\; q[a_{ij}] \;=\; [qa_{ij}]$$

We see in the definition of Eq. [2.8] one of the differences between a scalar and a (1×1) matrix. It is possible to perform the scalar multiplication with a matrix of any dimension to produce a result **S** of the same dimension. The *matrix* multiplication $[q] \cdot \mathbf{A}$ is permitted only if **A** is a row vector.

2.1.3 Square Matrices

A *square* matrix is one that has the same number of rows as the number of columns; for example, a matrix with dimension $(n \times n)$. Conformable square matrices **A** and **B** may obviously be added or subtracted, and either multiplication $\mathbf{A} \cdot \mathbf{B}$ or $\mathbf{B} \cdot \mathbf{A}$ is defined (although we must keep the noncommutative nature of multiplication in mind). Special types of square matrices are described as follows.

A *diagonal* matrix **D** equal to $[d_{ij}]$ satisfies the condition

$$[2.9] \qquad\qquad d_{ij} = 0 \text{ for } i \neq j$$

Its name comes from the fact that nonzero elements may exist only on the diagonal; that is, in positions with the same row and column number. We may simplify our notation in this special case by using only one subscript for the diagonal elements since the others are implicitly understood to be zero valued. Then we may replace $[d_{ij}]$ by $[d_i]$, interpret it in the correct context so that it is not confused with the vector notation of Eq. [2.3], and express the diagonal matrix **D** in the form

$$[2.10] \qquad \mathbf{D} = [d_i] = \begin{bmatrix} d_1 & & & & & \\ & d_2 & & & \text{\Large 0} & \\ & & \ddots & & & \\ & & & d_i & & \\ & & & & \ddots & \\ & \text{\Large 0} & & & d_{n-1} & \\ & & & & & d_n \end{bmatrix}$$

The large zero symbols above and below the diagonal are used to indicate that the elements in those positions are all zero.

The *identity matrix* **I** is a diagonal matrix in which every diagonal element has a unit value; it is the matrix analog of the scalar unit value. The *zero* matrix **0** is one whose elements are all zero; it may be either a square matrix or a general rectangular matrix.

Multiplications involving the general diagonal matrix $\mathbf{D}$, the identity matrix $\mathbf{I}$, and the zero matrix $\mathbf{0}$ are summarized as follows; each of $\mathbf{D}$, $\mathbf{I}$, and $\mathbf{0}$ is assumed to be an $(n \times n)$ matrix.

$$[2.11a] \qquad \underset{(n \times n)}{\mathbf{D}} \cdot \underset{(n \times 1)}{\mathbf{c}} = \underset{(n \times 1)}{\mathbf{u}} = [u_i] = [d_i c_i]$$

$$[2.11b] \qquad \underset{(1 \times n)}{\mathbf{r}} \cdot \underset{(n \times n)}{\mathbf{D}} = \underset{(1 \times n)}{\mathbf{v}} = [v_j] = [r_j d_j]$$

$$[2.11c] \qquad \underset{(n \times n)}{\mathbf{D}} \cdot \underset{(n \times n)}{\mathbf{A}} = \underset{(n \times n)}{\mathbf{P}} = [p_{ij}] = [d_i a_{ij}]$$

$$[2.11d] \qquad \underset{(n \times n)}{\mathbf{A}} \cdot \underset{(n \times n)}{\mathbf{D}} = \underset{(n \times n)}{\mathbf{Q}} = [q_{ij}] = [d_j a_{ij}]$$

$$[2.12a] \qquad \underset{(n \times n)}{\mathbf{I}} \cdot \underset{(n \times 1)}{\mathbf{c}} = \underset{(n \times 1)}{\mathbf{c}}$$

$$[2.12b] \qquad \underset{(1 \times n)}{\mathbf{r}} \cdot \underset{(n \times n)}{\mathbf{I}} = \underset{(1 \times n)}{\mathbf{r}}$$

$$[2.12c] \qquad \underset{(n \times n)}{\mathbf{I}} \cdot \underset{(n \times n)}{\mathbf{A}} = \underset{(n \times n)}{\mathbf{A}} = \underset{(n \times n)}{\mathbf{A}} \cdot \underset{(n \times n)}{\mathbf{I}}$$

$$[2.13a] \qquad \underset{(n \times n)}{\mathbf{0}} \cdot \underset{(n \times 1)}{\mathbf{c}} = \underset{(n \times 1)}{\mathbf{0}}$$

$$[2.13b] \qquad \underset{(1 \times n)}{\mathbf{r}} \cdot \underset{(n \times n)}{\mathbf{0}} = \underset{(1 \times n)}{\mathbf{0}}$$

$$[2.13c] \qquad \underset{(n \times n)}{\mathbf{0}} \cdot \underset{(n \times n)}{\mathbf{A}} = \underset{(n \times n)}{\mathbf{0}} = \underset{(n \times n)}{\mathbf{A}} \cdot \underset{(n \times n)}{\mathbf{0}}$$

We have not discussed matrix division because it is not defined; however, a quantity that is analogous to the reciprocal of a scalar is the *inverse*. The inverse of a square matrix $\mathbf{A}$ is denoted by $\mathbf{A}^{-1}$ and is itself a square matrix with the same dimension as $\mathbf{A}$ such that

$$[2.14] \qquad \mathbf{A} \cdot \mathbf{A}^{-1} = \mathbf{A}^{-1} \cdot \mathbf{A} = \mathbf{I}$$

The analogous scalar relation is $(\alpha \alpha^{-1} = \alpha^{-1} \alpha = 1)$.

Two other types of square matrices that will be useful at a later stage are the *lower triangular* and *upper triangular* matrices. The lower triangular form **L** permits nonzero elements only on and below the diagonal; it may be represented by

[2.15]
$$
\mathbf{L} = [\ell_{ij}] =
\begin{bmatrix}
\ell_{11} & & & & & & & \\
\ell_{21} & \ell_{22} & & & & 0 & & \\
\cdot & & \cdot & & & & & \\
\cdot & & & \cdot & & & & \\
\cdot & & & & \cdot & & & \\
\cdot & & & & & \cdot & & \\
\cdot & & & & & & \cdot & \\
\cdot & & & & & & & \cdot \\
\ell_{n1} & \ell_{n2} & \cdot & \cdot & \cdot & \cdot & \cdot & \ell_{nn}
\end{bmatrix}
$$

The upper triangular form **U** permits nonzero elements only on and above the diagonal; it may be represented by

[2.16]
$$
\mathbf{U} = [u_{ij}] =
\begin{bmatrix}
u_{11} & u_{12} & \cdot & \cdot & \cdot & \cdot & \cdot & u_{1n} \\
 & u_{22} & & & & & & u_{2n} \\
 & & \cdot & & & & & \cdot \\
 & & & \cdot & & & & \cdot \\
 & & & & \cdot & & & \cdot \\
 & & & & & \cdot & & \cdot \\
 & 0 & & & & & \cdot & \cdot \\
 & & & & & & & u_{nn}
\end{bmatrix}
$$

2.1.4 The Determinant of a Square Matrix

The *determinant* of an $(n \times n)$ square matrix **A** (equivalent to $[a_{ij}]$) is written as $|\mathbf{A}|$ and is defined by either of

[2.17a]
$$
|\mathbf{A}| = \sum_{j=1}^{n} (a_{ij} C_{ij}) \text{ for any } \textbf{\textit{one}} \text{ value of } i
$$

or

[2.17b]
$$
|\mathbf{A}| = \sum_{i=1}^{n} (a_{ij} C_{ij}) \text{ for any } \textbf{\textit{one}} \text{ value of } j
$$

in which C_{ij} is known as the *cofactor* of the element a_{ij}.

The cofactor C_{ij} of an $(n \times n)$ square matrix $\mathbf{A}$ is obtained by first removing row i and column j to form an $((n - 1) \times (n - 1))$ matrix and then by performing the operation

[2.18] $C_{ij} = (-1)^{i+j} \times$ (determinant of $\mathbf{A}$ with row i and column j removed)

Since the cofactor involves a determinant, Eq. [2.17a] or [2.17b] (which are known as Laplace expansions) must be used as necessary in a recursive fashion until we reach a terminal case of the (1×1) matrix $\mathbf{Q}$ whose determinant is

[2.19] $$|\mathbf{Q}| = |[q]| = q$$

Examples of using Eq. [2.17a] through [2.19] to obtain the determinants of a (2×2) and a (3×3) matrix are as follows. For the (2×2) case, we have

$$\begin{vmatrix} a_{11} & a_{12} \\ a_{21} & a_{22} \end{vmatrix} = a_{11}C_{11} + a_{21}C_{21}$$

$$= a_{11}(-1)^2|[a_{22}]| + a_{21}(-1)^3|[a_{12}]|$$

$$= a_{11}a_{22} - a_{21}a_{12}$$

The result is shown in a mnemonic form with products of diagonally aligned elements:

[2.20] $$\begin{vmatrix} a_{11} & a_{12} \\ a_{21} & a_{22} \end{vmatrix} = a_{11}a_{22} - a_{21}a_{12}$$

The (3×3) result is obtained with the help of Eq. [2.20] as follows:

$$\begin{vmatrix} a_{11} & a_{12} & a_{13} \\ a_{21} & a_{22} & a_{23} \\ a_{31} & a_{32} & a_{33} \end{vmatrix}$$

$$= a_{11}C_{11} + a_{12}C_{12} + a_{13}C_{13}$$

$$= a_{11}(-1)^2 \begin{vmatrix} a_{22} & a_{23} \\ a_{32} & a_{33} \end{vmatrix} + a_{12}(-1)^3 \begin{vmatrix} a_{21} & a_{23} \\ a_{31} & a_{33} \end{vmatrix} + a_{13}(-1)^4 \begin{vmatrix} a_{21} & a_{22} \\ a_{31} & a_{32} \end{vmatrix}$$

$$= a_{11}(a_{22}a_{33} - a_{32}a_{23}) - a_{12}(a_{21}a_{33} - a_{31}a_{23}) + a_{13}(a_{32}a_{21} - a_{31}a_{22})$$

A mnemonic form using diagonally aligned elements is shown below for the (3×3) matrix:

[2.21]

$$\begin{vmatrix} a_{11} & a_{12} & a_{13} \\ a_{21} & a_{22} & a_{23} \\ a_{31} & a_{32} & a_{33} \end{vmatrix} \begin{matrix} a_{11} & a_{12} \\ a_{21} & a_{22} \\ a_{31} & a_{32} \end{matrix} = \begin{matrix} a_{11}a_{22}a_{33} + a_{12}a_{23}a_{31} + a_{13}a_{21}a_{32} \\ -a_{31}a_{22}a_{13} - a_{32}a_{23}a_{11} - a_{33}a_{21}a_{12} \end{matrix}$$

It should be obvious from the two examples that evaluation of a determinant for a large matrix is a time-consuming task. In the special cases of diagonal and triangular matrices, the process is simplified considerably because of the zero elements in those matrices. For the diagonal matrix **D** described in Eq. [2.10] and the triangular matrices **L** and **U** described in Eq. [2.15] and [2.16], the determinants are simply the products of the diagonal elements as expressed by

[2.22]
$$|\mathbf{D}| = \prod_{i=1}^{n}(d_i); \quad |\mathbf{L}| = \prod_{i=1}^{n}(\ell_{ii}); \quad |\mathbf{U}| = \prod_{i=1}^{n}(u_{ii})$$

2.1.5 The Matrix Equation for Linear Algebraic Systems

the concept of matrix multiplication expressed in Eq. [2.5a] and [2.5b] allows us to represent the system of linear algebraic equations given by Eq. [2.1] in the form

[2.23a]
$$\begin{bmatrix} a_{11} & a_{12} & \cdots & \cdots & a_{1j} & \cdots & a_{1n} \\ a_{21} & a_{22} & \cdots & \cdots & a_{2j} & \cdots & a_{2n} \\ \vdots & \vdots & & & \vdots & & \vdots \\ a_{i1} & a_{i2} & \cdots & \cdots & a_{ij} & \cdots & a_{in} \\ \vdots & \vdots & & & \vdots & & \vdots \\ a_{n1} & a_{n2} & \cdots & \cdots & a_{nj} & \cdots & a_{nn} \end{bmatrix} \cdot \begin{bmatrix} x_1 \\ x_2 \\ \vdots \\ \vdots \\ \vdots \\ x_n \end{bmatrix} = \begin{bmatrix} f_1 \\ f_2 \\ \vdots \\ f_i \\ \vdots \\ f_n \end{bmatrix}$$

or, more compactly, in the form

[2.23b]
$$\mathbf{A} \cdot \mathbf{x} = \mathbf{f}$$

Here, **A** is the *coefficient matrix* formed by the coefficients of the linear system, the original right-hand sides of the system are now the components of the column vector **f**, and the original unknown quantities are now the components of the column vector **x**. The original equations may be recovered individually by performing the operations between row i of **A** and the vector **x** to produce the component f_i.

A unique solution for **x** requires **A** to be a square matrix (so that there are as many equations as unknown components of **x**), and it requires the system of equations to be *linearly independent*. No matrix row or equation in a linearly independent system can be derived by adding multiples of the remaining rows or equations. The determinant $|\mathbf{A}|$ is nonzero for a linearly independent system and is zero for a *linearly dependent* one.

For example, consider the system

$$\begin{bmatrix} 1 & 1 & 3 \\ 3 & -2 & 2 \\ 5 & -5 & 1 \end{bmatrix} \begin{bmatrix} x_1 \\ x_2 \\ x_3 \end{bmatrix} = \begin{bmatrix} 8 \\ 5 \\ 2 \end{bmatrix}$$

We see that the system is linearly dependent because the third row may be obtained from $(2 \times [\text{row}2] + (-1) \times [\text{row}1])$. The solution **x** belongs to the family given by

$$x_1 = \text{arbitrary}; \quad x_2 = (1 + 7x_1)/8; \quad x_3 = 2 - 5(x_1 - x_2)$$

and we may verify from Eq. [2.21] that the coefficient matrix has a zero determinant.

A coefficient matrix with a zero determinant is *singular*; a unique solution for **x** requires a *nonsingular* matrix. An $(n \times n)$ nonsingular matrix is said to be of *rank n*. It is impractical or inefficient to compute a determinant for large systems merely to establish the linear independence of a system. We instead look for equivalent indicators of a singular matrix during the solution process. We shall consider these indicators more fully in the methods of Sections 2.2 and 2.3.

The solution of Eq. [2.23b] may be derived with the help of the associative law from Eq. [2.7], the concept of the inverse from Eq. [2.14], and operations with the identity matrix from Eq. [2.12a] as follows:

[2.24] $$\mathbf{x} = \mathbf{I} \cdot \mathbf{x} = (\mathbf{A}^{-1} \cdot \mathbf{A}) \cdot \mathbf{x} = \mathbf{A}^{-1} \cdot (\mathbf{A} \cdot \mathbf{x}) = \mathbf{A}^{-1} \cdot \mathbf{f}$$

We observe that computing the inverse of an $(n \times n)$ matrix **A** from $(\mathbf{A} \cdot \mathbf{A}^{-1} = \mathbf{I})$ as given in Eq. [2.14] is equivalent to solving an equation of the type in Eq. [2.23b] for each of the n columns of the inverse. The objective of computational methods for determining **x** is to avoid this ludicrous situation of solving n systems for the inverse when we must use the same methods to solve the single, original system.

Among the additional tools used in methods for solving matrix equations are row and column exchanges. Some consequences of these operations are now considered. When two rows or two columns of a matrix **A** are exchanged to form a modified matrix $\mathbf{A_e}$, the determinant of the new matrix is given by

[2.25] $$|\mathbf{A_e}| = -|\mathbf{A}|; \quad \mathbf{A_e} = \mathbf{A} \text{ with 2 rows/columns exchanged}$$

Determinant relations after row and column exchanges are shown in the example below:

$$\mathbf{A} = \begin{bmatrix} 1 & 2 & 3 \\ 4 & 5 & 6 \\ 7 & 8 & 9 \end{bmatrix} \rightarrow \mathbf{B} = \begin{bmatrix} 4 & 5 & 6 \\ 1 & 2 & 3 \\ 7 & 8 & 9 \end{bmatrix} \rightarrow \mathbf{C} = \begin{bmatrix} 6 & 5 & 4 \\ 3 & 2 & 1 \\ 9 & 8 & 7 \end{bmatrix} \rightarrow |\mathbf{A}| = -|\mathbf{B}| = |\mathbf{C}|$$

A row exchange in the coefficient matrix **A** of Eq. [2.23b] is simply a position exchange of two equations. The sequence of the components of **x** is preserved, but those of the right-hand side vector **f** must be exchanged to correspond to the new positions. A column exchange in the coefficient matrix does not alter the sequence of equations, but it requires a corresponding exchange in the components of **x** so that the components are multiplied by the appropriate coefficients. Illustrations of these concepts are as follows.

Exchange of rows i and k:

[2.26a]
$$\begin{bmatrix} \vdots & & \vdots \\ a_{i1} & \cdots & a_{in} \\ \vdots & & \vdots \\ a_{k1} & \cdots & a_{kn} \\ \vdots & & \vdots \end{bmatrix} \cdot \mathbf{x} = \begin{bmatrix} \vdots \\ f_i \\ \vdots \\ f_k \\ \vdots \end{bmatrix}$$

Exchange of columns j and k:

[2.26b]
$$\begin{bmatrix} \cdots & a_{1j} & \cdots & a_{1k} & \cdots \\ & \vdots & & \vdots & \\ \cdots & a_{nj} & \cdots & a_{nk} & \cdots \end{bmatrix} \cdot \begin{bmatrix} \vdots \\ x_j \\ \vdots \\ x_k \\ \vdots \end{bmatrix} = \mathbf{f}$$

2.1.6 Computational Techniques for Basic Operations

We close Section 2.1 with a presentation of pseudocodes (described in Chapter 1 and Appendix B) for the basic operations of sums, differences, and products of matrices. Computation of the determinant directly from the definitions in Eq. [2.17a] through [2.19] is an inefficient process. More efficient methods for computing determinants will be discussed at a later stage.

Pseudocode 2–1 for addition/subtraction, and multiplication follows. It is assumed that the operands are of appropriate dimensions and that their element values are accessible.

Pseudocode 2–1   Segments for basic matrix operations

```
\ Addition or subtraction pseudocode segment:
\ A = [aᵢⱼ]; B = [bᵢⱼ]; C = [cᵢⱼ] = A ± B; refer to Eq. [2.4]
  For [j = 1, 2, ..., n]                \ n = number of columns
      For [i = 1, 2, ..., m]            \ m = number of rows
          cᵢⱼ ← aᵢⱼ ± bᵢⱼ
      End For              \ with row counter i
  End For                  \ with column counter j

\ Multiplication pseudocode segment:
\ A = [aᵢₖ]; B = [bₖⱼ]; P = [pᵢⱼ] = A · B; refer to Eq. [2.5a] and [2.5b]
  For [j = 1, 2, ..., n]                \ n = number of columns of B and P
      For [i = 1, 2, ..., m]            \ m = number of rows of A and P
          pᵢⱼ ← 0                       \ initialize element prior to summation
          For [k = 1, 2, ..., r]        \ r = columns of A = rows of B
              pᵢⱼ ← pᵢⱼ + aᵢₖbₖⱼ
          End For          \ with k counter
      End For              \ with row counter i
  End For                  \ with column counter j
```

2.2 DIRECT METHODS FOR LINEAR SYSTEMS

A *direct* method refers to a procedure for computing a solution from a form that is mathematically exact. We shall begin with simple cases involving triangular coefficient matrices and Cramer's rule with determinants. We shall then continue with the Gauss elimination method and its variants, methods for symmetric and tridiagonal coefficient matrices, and methods for block matrix representations.

2.2.1 Methods for Triangular Matrices

Some of the later methods of Section 2.2 involve reduction of a matrix equation to one of the forms

[2.27a] $\mathbf{L} \cdot \mathbf{g} = \mathbf{b}$; $\mathbf{L}$ = Lower triangular matrix of Eq. [2.15]

[2.27b] $\mathbf{U} \cdot \mathbf{h} = \mathbf{c}$; $\mathbf{U}$ = Upper triangular matrix of Eq. [2.16]

The unknowns to be determined in these equations are the vectors $\mathbf{g}$ and $\mathbf{h}$.

The expanded form of Eq. [2.27a] is the system of equations

$$
\begin{aligned}
\ell_{11}g_1 & & = b_1 \\
\ell_{21}g_1 + \ell_{22}g_2 & & = b_2 \\
\vdots \qquad \vdots & & \vdots \\
\ell_{n1}g_1 + \ell_{n2}g_2 + \cdots + \ell_{nn}g_n & & = b_n
\end{aligned}
$$

The component g_1 is found from the first equation and substituted into the second to obtain g_2. In general, the component g_i is found from the i-th equation of the system after the previously determined values g_1 through g_{i-1} have been substituted. The general solution for g_i is obtained from Eq. [2.28a] in the prescribed sequence.

[2.28a]
$$
\begin{cases}
g_1 = b_1/\ell_{11} \\
g_i = (1/\ell_{ii})\left[b_i - \displaystyle\sum_{j=1}^{i-1}(\ell_{ij}g_j)\right]; \quad i = 2, 3, \ldots, n
\end{cases}
$$

The expanded system of equations for Eq. [2.27b] is

$$
\begin{aligned}
u_{11}h_1 + u_{12}h_2 + \cdots + u_{1n}h_n &= c_1 \\
u_{22}h_2 + \cdots + u_{2n}h_n &= c_2 \\
\vdots \qquad \vdots & \\
u_{nn}h_n &= c_n
\end{aligned}
$$

The sequence of computations in this case begins with obtaining h_n; then we work backward through the system as described in Eq. [2.28b].

[2.28b]
$$
\begin{cases}
h_n = c_n/u_{nn} \\
h_i = (1/u_{ii})\left[c_i - \displaystyle\sum_{j=i+1}^{n}(u_{ij}h_i)\right]; \quad i = (n-1), (n-2), \ldots, 1
\end{cases}
$$

If we consider multiplications and divisions as the major operations in Eq. [2.28a] and [2.28b], the *operational count* Ω is the same in each case as summarized:

Division by ℓ_{ii} or u_{ii}: n

Multiplication by ℓ_{ij} or u_{ij} in accumulation loop for Σ: $1 + 2 + \cdots + (n-1) = n(n-1)/2$

The summation formula for the multiplication count is given in Appendix C. The total operational count for solutions of triangular systems is

[2.29] $$\Omega_{triangular} = n(n+1)/2$$

2.2.2 Cramer's Rule

Cramer's rule for solving the nonsingular system $(A \cdot x = f)$ of Eq. [2.23b] gives the components x_j of x in terms of determinants according to

[2.30] $$x_j = |A_j| / |A| ; \quad A_j = (A \text{ with } f \text{ replacing column } j)$$

Because evaluation of a determinant for an $(n \times n)$ matrix by a Laplace expansion is inefficient, Cramer's rule is rarely used for systems with more than about three or four equations.

EXAMPLE 2.1

Solve the system $A \cdot c = f$ given by

$$\begin{bmatrix} 1 & 2 & 4 \\ 1 & 3 & 9 \\ 1 & 4 & 16 \end{bmatrix} \begin{bmatrix} c_1 \\ c_2 \\ c_3 \end{bmatrix} = \begin{bmatrix} 2 \\ 4 \\ 7 \end{bmatrix}$$

to determine the coefficients c_i of the parabola $(y = c_1 + c_2 x + c_3 x^2)$ that passes through points with (x, y) coordinates (2,2), (3,4), and (4,7).

From Eq. [2.21]: $|A| = (1)(3)(16) + (2)(9)(1) + (4)(1)(4)$
$$- (1)(3)(4) - (4)(9)(1) - (16)(1)(2) = 2$$

The matrices A_j in Eq. [2.30] are:

$$A_1 = \begin{bmatrix} 2 & 2 & 4 \\ 4 & 3 & 9 \\ 7 & 4 & 16 \end{bmatrix}; \quad A_2 = \begin{bmatrix} 1 & 2 & 4 \\ 1 & 4 & 9 \\ 1 & 7 & 16 \end{bmatrix}; \quad A_3 = \begin{bmatrix} 1 & 2 & 2 \\ 1 & 3 & 4 \\ 1 & 4 & 7 \end{bmatrix}$$

We again use Eq. [2.21] to obtain the determinants:

$$|A_1| = 2; \quad |A_2| = -1 \quad |A_3| = 1$$

The solutions for c_j from Eq. [2.30] are therefore:

$$c_1 = |A_1| / |A| = 1.0; \quad c_2 = |A_2| / |A| = -0.5; \quad c_3 = |A_3| / |A| = 0.5$$

2.2.3 Gauss Elimination with Row Pivoting

The goal of the *Gauss elimination* method for solving a system of the type ($\mathbf{A} \cdot \mathbf{x} = \mathbf{f}$) in Eq. [2.23b] is to reduce it to the upper triangular form of Eq. [2.27b]; we may then use the back substitution scheme of Eq. [2.28b] to obtain the components of $\mathbf{x}$. The basic reduction step is to use one equation of the system to eliminate one of the $\mathbf{x}$ components from each of the remaining equations. The elimination process should be performed in an orderly fashion for efficient computer implementation. One such process is to use equation k with successive k values equal to $(1, 2, ..., [n-1])$ to eliminate x_k from equation $(k+1)$ through equation n.

To illustrate the method, we consider the problem of fitting a curve $y(z)$ to a set of data. The data for the problem are as follows.

$$y = 1.5 \text{ and } y'(= dy/dz) = -0.2 \text{ when } z \text{ is equal to } 1$$
$$y = 1.2 \text{ and } y'(= dy/dz) = +0.8 \text{ when } z \text{ is equal to } 2$$

A curve that is commonly used for data of this kind is a cubic spline for which y and y' are expressed by

$$y = x_1 + x_2 z + x_3 z^2 + x_4 z^3; \quad x_1, x_2, x_3, \text{ and } x_4 = \text{coefficients of the cubic}$$
$$y' = x_2 + 2x_3 z + 3x_4 z^2$$

We use these expressions and the given data to generate a system of linear algebraic equations for the unknown coefficients x_1 through x_4.

The linear equations for the curve-fitting problem are obtained as follows.

$$y' = -0.2 \text{ at } z = 1 \rightarrow x_2 + 2x_3 + 3x_4 \qquad = -0.2$$
$$y' = +0.8 \text{ at } z = 2 \rightarrow x_2 + 4x_3 + 12x_4 \qquad = +0.8$$
$$y = \quad 1.5 \text{ at } z = 1 \rightarrow x_1 + x_2 + x_3 + x_4 \qquad = 1.5$$
$$y = \quad 1.2 \text{ at } z = 2 \rightarrow x_1 + 2x_2 + 4x_3 + 8x_4 = 1.2$$

The equivalent matrix equation for the four linear equations is

[2.31]
$$\begin{bmatrix} 0 & 1 & 2 & 3 \\ 0 & 1 & 4 & 12 \\ 1 & 1 & 1 & 1 \\ 1 & 2 & 4 & 8 \end{bmatrix} \cdot \begin{bmatrix} x_1 \\ x_2 \\ x_3 \\ x_4 \end{bmatrix} = \begin{bmatrix} -0.2 \\ 0.8 \\ 1.5 \\ 1.2 \end{bmatrix}$$

Following the elimination sequence given earlier, we wish to begin the solution process by using the first equation to eliminate x_1 from the remaining equations. This is equivalent to subtracting an appropriate multiple of row 1 from each of the remaining rows so that the coefficients in column 1 of rows 2, 3, and 4 become zero.

Corresponding operations must be performed on the right-hand side. The presence of a zero in column 1 of row 1 does not permit us to accomplish this task because no finite multiple of zero can be subtracted from the unit values in column 1 of rows 3 and 4 to reduce them to zero.

The row k whose multiples are to be subtracted from the other rows is called the *pivot row*. The *pivot element* is the element a_{kk}, which, as we have just seen, cannot be zero. To solve the difficulty posed by a zero value for the pivot element, we may employ a technique known as *row pivoting*, or sometimes as *partial pivoting*. The technique consists of the following steps:

1. Search rows k through n for the element in column k with the largest magnitude; denote the row containing this element by m.

2. If m is different from k, exchange rows k and m, and exchange f_k and f_m.

The row and right-hand side exchanges, if necessary, do not change the system of equations; they simply change the sequence of equations as illustrated in Eq. [2.26a].

Before continuing, we observe that a row exchange is accompanied by an exchange of corresponding right-hand side components. Similarly, subtraction of a multiple of the pivot row k from another row i in the elimination step is accompanied by subtraction of the same multiple of f_k from f_i. To make treatment of the right-hand side components more convenient, we introduce at this stage the concept of an *augmented matrix*.

The augmented matrix $\mathbf{A}^+$ is formed by combining the original matrix $\mathbf{A}$ and the right-hand vector $\mathbf{f}$ to form an $(n \times p)$ matrix, where p is equal to $(n + 1)$. The first n columns of $\mathbf{A}^+$ are identical to those of $\mathbf{A}$; the extra column, column p, is the same as the vector $\mathbf{f}$. In general, we write $\mathbf{A}^+$ in the form

$$[2.32] \qquad \mathbf{A}^+ = \begin{bmatrix} a_{11} & \cdots & a_{1n} & \vdots & a_{1p} \\ \vdots & & \vdots & & \vdots \\ a_{n1} & \cdots & a_{nn} & \vdots & a_{np} \end{bmatrix} = \begin{bmatrix} a_{11} & \cdots & a_{1n} & \vdots & f_1 \\ \vdots & & \vdots & & \vdots \\ a_{n1} & \cdots & a_{nn} & \vdots & f_n \end{bmatrix}$$

We shall retain the notation a_{ij} for an element of $\mathbf{A}^+$; recognize, however, that the component f_i is now denoted by a_{ip}. For our cubic spline problem, the initial augmented matrix for Eq. [2.31] is

$$\mathbf{A}^+ = \begin{bmatrix} 0 & 1 & 2 & 3 : & -0.2 \\ 0 & 1 & 4 & 12 : & 0.8 \\ 1 & 1 & 1 & 1 : & 1.5 \\ 1 & 2 & 4 & 8 : & 1.2 \end{bmatrix}$$

We return now to the description of the Gauss elimination method and perform the row pivoting step for pivot row k equal to 1. The element in column 1 with the largest magnitude is either a_{31} or a_{41}. In cases where duplication exists, we use the first occurrence of the maximum magnitude; that is, the row number m for the row pivoting step is 3. The condition $(m \neq k)$ is true, and we therefore exchange rows 1 and 3 to produce a modified $\mathbf{A}^+$ matrix

$$(\mathbf{A}^+)_{\text{1st pivot}} = \begin{bmatrix} 1 & 1 & 1 & 1 : & 1.5 \\ 0 & 1 & 4 & 12 : & 0.8 \\ 0 & 1 & 2 & 3 : & -0.2 \\ 1 & 2 & 4 & 8 : & 1.2 \end{bmatrix} \begin{matrix} \text{new row}_1 = \text{old row}_3 \\ \\ \text{new row}_3 = \text{old row}_1 \\ \end{matrix}$$

The exchange serves the obvious purpose of making the pivot element a_{11} nonzero; it also helps to reduce round-off errors in the overall method (as we shall soon see) and is therefore performed even if the original pivot element is nonzero.

The next step of the method is to perform the elimination pass with pivot row k equal to 1. The operations for the pass are to subtract (a_{21}/a_{11}) times row 1 from row 2, (a_{31}/a_{11}) times row 1 from row 3, and (a_{41}/a_{11}) times row 1 from row 4. The general elimination pass is summarized by

[2.33]
$$\left. \begin{cases} (a_{ij})_{\text{new}} = a_{ij} - (a_{ik}/a_{kk})a_{kj}; & \forall \ j > k \\ (a_{ik})_{\text{new}} = 0 \end{cases} \right\}; \quad \forall \ i > k$$

In practice, we drop the subscript "new" because the new values replace the old ones in the same computer memory locations, and we continue to use the notation a_{ij} to refer to the current element in row i and column j of the augmented matrix. The multipliers of a_{kj} in Eq. [2.33] contain a divisor a_{kk}. We see from Eq. [1.12c] of Chapter 1 that the magnitude of round-off error in division is reduced by increasing the magnitude of the divisor. Row pivoting places the element with the largest magnitude allowed by the scheme in the pivot element position and thus retards the growth of round-off errors.

Application of Eq. [2.33] to the cubic spline problem with the pivot row number k equal to 1 completes the first elimination pass and yields the new augmented matrix

$$(\mathbf{A}^+)_{\text{1st pass}} = \begin{bmatrix} 1 & 1 & 1 & 1 : & 1.5 \\ 0 & 1 & 4 & 12 : & 0.8 \\ 0 & 1 & 2 & 3 : & -0.2 \\ 0 & 1 & 3 & 7 : & -0.3 \end{bmatrix} \begin{matrix} \\ \text{old row}_2 - (0/1) \times \text{pivot row}_1 \\ \text{old row}_3 - (0/1) \times \text{pivot row}_1 \\ \text{old row}_4 - (1/1) \times \text{pivot row}_1 \end{matrix}$$

The pivot element a_{22} for the second elimination pass is already the one with the largest magnitude in column 2 of rows 2, 3, and 4; a row exchange is therefore unnecessary, and we proceed immediately to the application of Eq. [2.33] with k equal to 2. The augmented matrix after the second pass is

$$(\mathbf{A}^+)_{\text{2nd pass}} = \begin{bmatrix} 1 & 1 & 1 & 1 : & 1.5 \\ 0 & 1 & 4 & 12 : & 0.8 \\ 0 & 0 & -2 & -9 : & -1.0 \\ 0 & 0 & -1 & -5 : & -1.1 \end{bmatrix} \begin{matrix} \\ \\ \text{old row}_3 - (1/1) \times \text{pivot row}_2 \\ \text{old row}_4 - (1/1) \times \text{pivot row}_2 \end{matrix}$$

A row exchange is again unnecessary prior to the final pass with the pivot row number k equal to 3. The final application of Eq. [2.33] yields

$$(\mathbf{A}^+)_{3\text{rd pass}} = \begin{bmatrix} 1 & 1 & 1 & 1: & 1.5 \\ 0 & 1 & 4 & 12: & 0.8 \\ 0 & 0 & -2 & -9: & -1.0 \\ 0 & 0 & 0 & -0.5: & -0.6 \end{bmatrix} \text{old row}_4 - (-1)/(-2) \times \text{pivot row}_3$$

The first four columns of the final matrix form an upper triangular matrix corresponding to $\mathbf{U}$ in Eq. [2.27b], the last column corresponds to the vector $\mathbf{c}$ on the right-hand side of Eq. [2.27b], and the required unknown vector $\mathbf{x}$ now corresponds to the vector $\mathbf{h}$ in Eq. [2.27b]. The procedure in Eq. [2.28b] for solving an upper triangular system gives the solution vector $\mathbf{x}$ for our problem. The components of $\mathbf{x}$ are

$$x_1 = -0.8; \quad x_2 = 6.0; \quad x_3 = -4.9; \quad x_4 = 1.2$$

The transformation of the augmented matrix by Gauss elimination is expressed by

$$[\mathbf{A} : \mathbf{f}] \rightarrow [\mathbf{U} : \mathbf{h}]$$

The determinant $|\mathbf{A}|$ is related to the determinant $|\mathbf{U}|$ through

$$|\mathbf{A}| = |\mathbf{U}| \times (-1)^{\text{number of row interchanges}}$$

For our cubic spline problem, which involved only one row interchange, we obtain

$$|\mathbf{A}| = -|\mathbf{U}| = \begin{vmatrix} 1 & 1 & 1 & 1.0 \\ 0 & 1 & 4 & 12.0 \\ 0 & 0 & -2 & -9.0 \\ 0 & 0 & 0 & -0.5 \end{vmatrix} = -1$$

The relation between $|\mathbf{A}|$ and $|\mathbf{U}|$ gives us a simple way to determine, during the Gauss elimination process, if the coefficient matrix $\mathbf{A}$ is singular. A singular matrix $\mathbf{A}$, for which there is no unique solution $\mathbf{x}$ to $(\mathbf{A} \cdot \mathbf{x} = \mathbf{f})$, is indicated by a zero-valued pivot element a_{kk} after pivoting. Because of finite machine precision, we should not test for a singular matrix $\mathbf{A}$ by looking for a pivot element that is exactly zero; rather, we should look for an occurrence of $|a_{kk}|$ that is less than some very small, positive value ϵ. Although such an occurrence does not necessarily mean that $\mathbf{A}$ is singular, it does mean that $|\mathbf{A}|$ is too close to zero for an accurate determination of the solution $\mathbf{x}$ to be likely. A recommended value of ϵ is easily computed from

[2.34] $$\epsilon = \left\{ \max \left| a_{ij} \right| \right\} / 10^4; \quad i, j = 1, 2, \ldots, n$$

We now consider the computer implementation of Gauss elimination with partial pivoting. The three major parts are the elimination pass, the row pivoting step, and the solution by back substitution. The module for the elimination pass is given in Pseudocode 2–2.

Pseudocode 2–2   Program module for Gauss elimination passes

Module GaussElim($\mathbf{A}^+$, ϵ, n, p, $\mathbf{x}$, rowx, err):

       rowx ← 0   \ initial counter for row exchanges
       err ← 0   \ indicates nonsingular matrix
       k ← 1   \ initial pivot row number

While [err $=$ 0 **and** $k < n$]   \ pivot row loop

    Pivot($\mathbf{A}^+$, n, k, rowx)   \ call partial pivoting module

    If [$|a_{kk}| < \epsilon$] **then**
        Write: 'Matrix may be singular; execution discontinued'
        err ← 1   \ indicates (nearly) singular matrix
    Else
        For [$i = (k + 1), (k + 2), \ldots, n$]   \ row loop
            $c \leftarrow a_{ik} / a_{kk}$ \ row multiplier
            $a_{ik} \leftarrow 0$ \ column k value (Eq. [2.33])
            For [$j = (k + 1), (k + 2), \ldots, p$]   \ column loop
                $a_{ij} \leftarrow a_{ij} - c(a_{kj})$ \ column j value (Eq. [2.33])
            End For   \ with column counter j
        End For   \ with row counter i
    End If

    $k \leftarrow k + 1$ \ increment pivot row counter
End While

If [err $=$ 0] **then**

    If [$|a_{nn}| < \epsilon$] **then**   \ last check for a singular matrix
        Write: 'Matrix may be singular; execution discontinued'
        err ← 1   \ indicates (nearly) singular matrix
    Else
        BackSub($\mathbf{A}^+$, n, $\mathbf{x}$)   \ call back substitution module
    End If

End If
End Module GaussElim

The initial augmented matrix $\mathbf{A}^+$, the indicator ϵ, the number of rows n, and the number of columns p are supplied to the elimination module. This module in turn calls the row pivoting and back substitution modules. If the possibility of a singular matrix is detected, the error code with a value (err = 1) is returned to the calling module; otherwise, the value (err = 0) is returned along with the solution $\mathbf{x}$ and the number (rowx) of row exchanges that occurred during pivoting. In either case, the matrix $\mathbf{A}^+$ is returned in its modified form. The value of "rowx" and the modified augmented matrix $\mathbf{A}^+$ may be used to compute $|\mathbf{A}|$ when $\mathbf{A}$ is nonsingular.

The modules for the row pivoting and the back substitution steps are given in Pseudocode 2–3.

Pseudocode 2–3   Modules for row pivoting and back substitution

---

\ Row pivoting module
   Module Pivot($\mathbf{A}^+, n, k$, rowx) :
        amax $\leftarrow |a_{kk}|$ \ begin search for element with largest magnitude
         $m \leftarrow k$

For $[i = (k + 1), (k + 2), \dots, n]$ \ row search loop

   If $[|a_{ik}| >$ amax$]$ **then**
            amax $\leftarrow |a_{ik}|; m \leftarrow i$ \ find maximum and row location
   End If

End For   \ with row counter i

If $[$not$(m = k)]$ **then**   \ row exchange necessary

        rowx $\leftarrow$ rowx $+ 1$ \ update exchange counter
        For $[j = k, (k + 1), \dots, p]$ \ column loop
            temp $\leftarrow a_{kj}; a_{kj} \leftarrow a_{mj}; a_{mj} \leftarrow$ temp
        End For   \ with column counter j

End If

End Module Pivot

\ Back substitution module (refer to Eq. [2.28b])
   Module BackSub($\mathbf{A}^+, n, \mathbf{x}$) :
   $x_n \leftarrow a_{np}/a_{nn}$ \ begin upper triangular solution
   For $[k = (n - 1), (n - 2), \dots, 1]$ \ component loop

        $z \leftarrow a_{kp}$ \ initialize $z = a_{kk}x_k$
        For $[j = (k + 1), (k + 2), \dots, n]$ \ accumulation loop
            $z \leftarrow z - x_j a_{kj}$ \ accumulate z value
        End For   \ with column counter j
        $x_k \leftarrow z/a_{kk}$

   End For   \ with component counter k

   End Module BackSub

The three pseudocode modules for Gauss elimination with partial pivoting are written to correspond with our earlier fundamental description. We now examine some modifications to improve efficiency and accuracy.

The modifications of the fundamental method center on the row pivoting step. In our description, row pivoting was accomplished by an actual exchange of rows. A more efficient alternative is to leave the rows in place while keeping track of the pivot row through an order vector **v**. The order is initially

$$\mathbf{v} = [v_i] = \begin{bmatrix} 1 & 2 & 3 & \cdots & n \end{bmatrix}$$

When an exchange of rows m and k is indicated in the pivoting step, we no longer perform a physical row exchange; instead, we exchange elements m and k of the order vector. For example, the row exchange of rows 1 and 3 prior to the first elimination pass in our case study of the cubic spline problem is replaced by changing the initial order vector to

$$\mathbf{v} = \begin{bmatrix} 3 & 2 & 1 & 4 \end{bmatrix}$$

The first pivot row is now identified by the component v_1, and the coefficients of x_1 in rows v_2, v_3, and v_4 are reduced to zero in the elimination pass. In general, we use pivot row v_k in pass k to reduce the coefficients of x_k in rows v_{k+1} through v_n to zero. A similar process is used to solve for **x** in the back substitution process. We solve for x_n through x_1 by working backward from row v_n through row v_1. Use of the order vector for a process that is similar to Gauss elimination is shown in more detail in §2.2.5.

The two purposes of the partial pivoting step are to ensure that the pivot element is not zero and to help in the reduction of round-off errors. In some cases where the elements of the augmented matrix have widely differing magnitudes, row pivoting can actually **aggravate** the growth of round-off errors. A recommended step for such cases is to **scale** each row of the augmented matrix.

Scaling, also known as **equilibration**, is accomplished by dividing each row of a matrix and the corresponding right-hand vector by the element with the largest magnitude in that row. Although we have not used scaling in our discussions, scaling should generally be carried out before attempting to solve the system of equations.

Because operations are required to perform the pivoting, we should be aware of conditions for which pivoting is unnecessary. One such easily recognized condition is diagonal dominance. A **diagonally dominant** matrix **A** is one in which the magnitude $|a_{ii}|$ of each element on the diagonal of the matrix is greater than the sum of the magnitudes of all other elements in that row or column.

We note finally that full pivoting may be used to achieve slightly more accuracy in the solution. Full pivoting generally requires both a row and a column exchange to bring the element with the largest magnitude to the pivot element position. There is increased program complexity because column exchanges require the elements of the solution vector **x** to be reordered. The improvement in accuracy is not considered significant enough to warrant the increased complexity; therefore, full pivoting is rarely used.

The operational count for multiplications and divisions in Gauss elimination may be obtained by looking at the loops in the pseudocodes. This count is based on "major" operations and is at best a crude estimate of the efficiency of a method; it ignores the time required for tasks such as testing of conditions, storage of values, overhead in incrementing loop variables, and address calculations for access to the matrix elements. Counting for the Gauss elimination process may be conveniently separated into counts for triangularization of the matrix, for processing of the right-hand side vector **f** (i.e., column p of the augmented matrix), and for the solution of the final upper triangular system to obtain **x**.

The triangularization requires $(n - k)$ divisions by a_{kk} to form the pivot row multipliers c, equal to (a_{ik}/a_{kk}), for each value of k equal to $[1, 2, \ldots, (n - 1)]$. With the formulas of Appendix C, we find that

$$\text{Number of divisions by } a_{kk} = 1 + 2 + \cdots + (n - 1) = n(n - 1)/2$$

The remaining operations of the triangularization are multiplications of a_{kj} by c in the computation of a_{ij} values. These are performed $(n - k)^2$ times for each value of k equal to $[1, 2, \ldots, (n - 1)]$. The counts for the multiplications and overall triangularization are therefore

$$\text{Number of multiplications } [c(a_{kj})] = 1^2 + 2^2 + \cdots + (n - 1)^2$$
$$= n(n - 1)(2n - 1)/6$$

$$\text{Triangularization count} = n(n - 1)/2 + n(n - 1)(2n - 1)/6 = (n^3 - n)/3$$

Operations on **f** are the multiplications of c by a_{kj} as above, but only for j equal to $(n + 1)$. The total of these operations are

$$\text{Number of operations on } \mathbf{f} = 1 + 2 + \cdots + (n - 1) = n(n - 1)/2$$

Finally, from Eq. [2.29], the operations for solving the upper triangular system are

$$\text{Number of upper triangular system operations} = n(n + 1)/2$$

The total number of operations for Gauss elimination is then found to be

[2.35]
$$\Omega_{\text{ge}} = n^3/3 + n^2 - n/3$$

If the same matrix **A** is to be used with s different right-hand vectors to solve s systems of equations ($\mathbf{A} \cdot \mathbf{x_1} = \mathbf{f_1}, \mathbf{A} \cdot \mathbf{x_2} = \mathbf{f_2}, \ldots, \mathbf{A} \cdot \mathbf{x_s} = \mathbf{f_s}$), the augmented matrix is modified to contain $(n + s)$ columns with column $(n + t)$ corresponding to the vector $\mathbf{f_t}$. The number of triangularization operations is the same as before, but the operational count for processing the right-hand sides and for the back substitution increases by a factor of s. The count for s systems of equations with a common coefficient matrix $\mathbf{A}$ is then

[2.36] $$(\Omega_{\text{ge}})_s = n^3/3 + sn^2 - n/3$$

2.2.4   The Gauss–Jordan Method

The **_Gauss–Jordan_** method is essentially a variation of Gauss elimination; as such, §2.2.3 on Gauss elimination is a prerequisite for the current discussion. The goal of the Gauss–Jordan method is to reduce the original matrix to a diagonal form. There are two basic ways to achieve the reduction. One way is to perform the triangularization step of the Gauss elimination method, and to follow that step by another set of elimination passes with rows $[n, (n - 1), \ldots, 2]$ as the sequence of pivot rows. The reverse elimination process and the subsequent solution of $\mathbf{x}$ replace the back substitution step of the Gauss elimination method. This approach is not appealing because there is neither a reduction in programming complexity nor increased efficiency; in fact, the Gauss–Jordan method is somewhat less efficient than the Gauss elimination method.

The second approach is to include rows above the pivot row k in the elimination passes, which are carried out for k equal to $[1, 2, \ldots, n]$. We also normalize the pivot row with respect to the pivot element a_{kk}. The pivot row normalization and the Gauss–Jordan counterpart to the elimination step of Eq. [2.33] are expressed by

[2.37a] $$(a_{kj})_{\text{new}} = a_{kj}/a_{kk}, \quad \forall\, j > k; \quad (a_{kk})_{\text{new}} = 1$$

[2.37b] $$\left\{ \begin{array}{l} (a_{ij})_{\text{new}} = a_{ij} - a_{ik}(a_{kj})_{\text{new}}; \quad \forall\, j > k \\ (a_{ik})_{\text{new}} = 0 \end{array} \right\}; \quad \forall\, i \neq k$$

This approach yields a transformation of the augmented matrix given by

$$[\mathbf{A} : \mathbf{f}] \rightarrow [\mathbf{I} : \mathbf{x}]$$

Despite the relative inefficiency of the Gauss–Jordan method compared with Gauss elimination, the final form of the augmented matrix is attractive because it contains the solution vector $\mathbf{x}$. The final form makes the Gauss–Jordan method particularly well suited to computing the inverse of a matrix through the transformation

$$[\mathbf{A} : \mathbf{I}] \rightarrow [\mathbf{I} : \mathbf{A}^{-1}]$$

To illustrate the Gauss–Jordan method relative to the Gauss elimination method, we shall again solve Eq. [2.31] of §2.2.3. We begin with the augmented matrix following the first row pivoting step; this matrix is repeated below for convenience.

$$(\mathbf{A}^+)_{\text{1st pivot}} = \begin{bmatrix} 1 & 1 & 1 & 1 : & 1.5 \\ 0 & 1 & 4 & 12 : & 0.8 \\ 0 & 1 & 2 & 3 : & -0.2 \\ 1 & 2 & 4 & 8 : & 1.2 \end{bmatrix} \begin{matrix} \text{new row}_1 = \text{old row}_3 \\ \\ \text{new row}_3 = \text{old row}_1 \\ \\ \end{matrix}$$

The pivot row with k equal to 1 is already normalized with respect to a_{11}; we may therefore proceed with the application of Eq. [2.37b] to obtain

$$(\mathbf{A}^+)_{\text{1st pass}} = \begin{bmatrix} 1 & 1 & 1 & 1 : & 1.5 \\ 0 & 1 & 4 & 12 : & 0.8 \\ 0 & 1 & 2 & 3 : & -0.2 \\ 0 & 1 & 3 & 7 : & -0.3 \end{bmatrix} \begin{matrix} \\ \text{old row}_2 - (0) \times \text{pivot row}_1 \\ \text{old row}_3 - (0) \times \text{pivot row}_1 \\ \text{old row}_4 - (1) \times \text{pivot row}_1 \end{matrix}$$

Every element a_{i2} for i equal to 2, 3, or 4 has a unit value. The row pivoting and normalization steps are therefore unnecessary prior to the second pass. The immediate application of Eq. [2.37b] yields the following augmented matrix after the second elimination pass.

$$(\mathbf{A}^+)_{\text{2nd pass}} = \begin{bmatrix} 1 & 0 & -3 & -11 : & 0.7 \\ 0 & 1 & 4 & 12 : & 0.8 \\ 0 & 0 & -2 & -9 : & -1.0 \\ 0 & 0 & -1 & -5 : & -1.1 \end{bmatrix} \begin{matrix} \text{old row}_1 - (1) \times \text{pivot row}_2 \\ \\ \text{old row}_3 - (1) \times \text{pivot row}_2 \\ \text{old row}_4 - (1) \times \text{pivot row}_2 \end{matrix}$$

Row pivoting is again unnecessary for the third pass; however, the pivot row 3 must be normalized with respect to (-2). The matrix after normalization is

$$(\mathbf{A}^+)_{\text{row 3 normalized}} = \begin{bmatrix} 1 & 0 & -3 & -11.0 : & 0.7 \\ 0 & 1 & 4 & 12.0 : & 0.8 \\ 0 & 0 & 1 & 4.5 : & 0.5 \\ 0 & 0 & -1 & -5.0 : & -1.1 \end{bmatrix}$$

and the matrix after the third elimination pass is

$$(\mathbf{A}^+)_{\text{3rd pass}} = \begin{bmatrix} 1 & 0 & 0 & 2.5 : & 2.2 \\ 0 & 1 & 0 & -6.0 : & -1.2 \\ 0 & 0 & 1 & 4.5 : & 0.5 \\ 0 & 0 & 0 & -0.5 : & -0.6 \end{bmatrix} \begin{matrix} \text{old row}_1 - (-3) \times \text{pivot row}_3 \\ \text{old row}_2 - (4) \times \text{pivot row}_3 \\ \\ \text{old row}_4 - (-1) \times \text{pivot row}_3 \end{matrix}$$

A fourth pass is required in the Gauss–Jordan method. We first normalize row 4 with respect to (-0.5) to obtain

$$(\mathbf{A}^+)_{\text{row 4 normalized}} = \begin{bmatrix} 1 & 0 & 0 & 2.5 : & 2.2 \\ 0 & 1 & 0 & -6.0 : & -1.2 \\ 0 & 0 & 1 & 4.5 : & 0.5 \\ 0 & 0 & 0 & 1.0 : & 1.2 \end{bmatrix}$$

and apply Eq. [2.37b] once more to obtain the final augmented matrix

$$(\mathbf{A}^+)_{\text{final}} = \begin{bmatrix} 1 & 0 & 0 & 0 : & -0.8 \\ 0 & 1 & 0 & 0 : & 6.0 \\ 0 & 0 & 1 & 0 : & -4.9 \\ 0 & 0 & 0 & 1 : & 1.2 \end{bmatrix} \begin{array}{l} \text{old row}_1 - (2.5) \times \text{pivot row}_4 \\ \text{old row}_2 - (-6.0) \times \text{pivot row}_4 \\ \text{old row}_3 - (4.5) \times \text{pivot row}_4 \\ \end{array}$$

The last column of the final augmented matrix is the solution vector **x**.

The application of the Gauss–Jordan method to matrix inversion follows the same sequence of steps that we have just described. The initial augmented matrix in this case has $(2n)$ columns with the identity matrix **I** in the last n columns. On completion of the passes, the last n columns contain the inverse. Application of the Gauss–Jordan method to inversion of the matrix in Eq. [2.31] yields the result

$$\begin{bmatrix} 0 & 1 & 2 & 3 \\ 0 & 1 & 4 & 12 \\ 1 & 1 & 1 & 1 \\ 1 & 2 & 4 & 8 \end{bmatrix}^{-1} = \begin{bmatrix} -4 & -2 & -4 & 5 \\ 8 & 5 & 12 & -12 \\ -5 & -4 & -9 & 9 \\ 1 & 1 & 2 & -2 \end{bmatrix}$$

A module for the Gauss–Jordan elimination passes is given in Pseudocode 2–4; it is written to be compatible with the row pivoting module given in §2.2.3. The major features of the Gauss–Jordan method are that all n rows are used in the elimination passes, the pivot row is normalized with respect to the pivot element, and coefficients are eliminated from all rows except the pivot row. The solution vector (or solution matrix for multiple right-hand side vectors) is contained in the final form of the augmented matrix.

The major operations for the Gauss–Jordan method are the divisions in the normalization step and the multiplications in the row operations. For each k, we must perform $(n - k)$ divisions (through column n) for the diagonalization process. The number of divisions is therefore

$$\text{Number of divisions (through column } n) = (n - 1) + (n - 2) + \cdots + 1 + 0$$
$$= n(n - 1)/2$$

Pseudocode 2–4   Program module for Gauss–Jordan elimination passes

Module GaussJordan(A$^+$, ϵ, n, p, err) :

   err $\leftarrow$ 0   \ indicates nonsingular matrix
   $k \leftarrow$ 1   \ initial pivot row number

While [err = 0 **and** $k \le n$]   \ pivot row loop

  If [$k < n$] **then**
     Pivot(A$^+$, n, k, rowx)   \ call partial pivoting module with
                \ rowx as a dummy (unused) variable
  End If

  If [$|a_{kk}| < \epsilon$] **then**
    Write: 'Matrix may be singular; execution discontinued'
    err $\leftarrow$ 1   \ indicates (nearly) singular matrix
  Else
    For [$j = (k + 1), (k + 2), \ldots, p$]   \ column loop
      $a_{kj} \leftarrow a_{kj} / a_{kk}$ \ pivot row normalization
    End For   \ with column counter j
    $a_{kk} \leftarrow$ 1
    For [$i = 1, 2, \ldots, n$]   \ row loop
      If [**not** ($i = k$)] **then**
        $a_{ik} \leftarrow$ 0   \ column k value (Eq. [2.37b])
        For [$j = (k + 1), (k + 2), \ldots, p$] \ column loop
          $a_{ij} \leftarrow a_{ij} - a_{ik}(a_{kj})$ \ column j value (Eq. [2.37b])
        End For   \ with column counter j
      End If   \ to exclude row k from the elimination pass
    End For   \ with row counter i
  End If

  $k \leftarrow k + 1$ \ increment pivot row counter
End While

End Module GaussJordan

The multiplications (through column n) are performed $[(n - 1)(n - k)]$ times for each k value from 1 through n. The count for these operations is

$$\text{Multiplication count (through column } n) = (n - 1)[n(n - 1)/2] = n(n - 1)^2/2$$

and the total number of operations for reducing **A** to the identity matrix is

$$\text{Operations for reduction to } \mathbf{I} = n^2(n - 1)/2$$

For a single right-hand side vector $\mathbf{f}$, the operations on column $(n + 1)$ are the $(n - 1)$ multiplications and the single normalization division for each value of k. We then obtain

$$\text{Number of operations on } \mathbf{f} = n^2$$

The total count for the Gauss–Jordan method with one right-hand vector $\mathbf{f}$ is therefore

[2.38]
$$\Omega_{\text{gj}} = (n^3 + n^2)/2$$

With s right-hand vectors, the right-hand-side operations are increased by a factor of s so that the total operational count becomes

[2.39]
$$(\Omega_{\text{gj}})_s = (n^3 + [2s - 1]n^2)/2$$

2.2.5 LU Decomposition

A third method for the solution of general systems of linear algebraic equations is the *LU decomposition* method. The objective of this method is to find a lower triangular factor $\mathbf{L}$ and an upper triangular factor $\mathbf{U}$ such that the system of equations can be transformed according to

[2.40]
$$\mathbf{A} \cdot \mathbf{x} = \mathbf{f} \rightarrow (\mathbf{L} \cdot \mathbf{U}) \cdot \mathbf{x} = \mathbf{A}^* \cdot \mathbf{x} = \mathbf{f}^*$$

The matrix $\mathbf{A}^*$ in Eq. [2.40] is the matrix $\mathbf{A}$ after row exchanges have been made to allow the factors $\mathbf{L}$ and $\mathbf{U}$ to be computed accurately; the vector $\mathbf{f}^*$ is the vector $\mathbf{f}$ after an identical set of row exchanges.

A decomposition in which each diagonal element ℓ_{ii} of $\mathbf{L}$ has a unit value is known as the *Doolittle method*; one in which each diagonal element u_{ii} of $\mathbf{U}$ has a unit value is known as the *Crout method*. Another method in which corresponding diagonal elements ℓ_{ii} and u_{ii} are equal to each other is known as the *Cholesky method*. Regardless of which method is used to obtain the factors $\mathbf{L}$ and $\mathbf{U}$ of $\mathbf{A}^*$, the methods described in §2.2.1 for triangular matrices are used to obtain $\mathbf{x}$ by solving

[2.41]
$$\mathbf{L} \cdot \mathbf{g} = \mathbf{f}^*; \quad \mathbf{U} \cdot \mathbf{x} = \mathbf{g}$$

The Doolittle method for obtaining the factors $\mathbf{L}$ and $\mathbf{U}$ is contained in the Gauss elimination method described in §2.2.3. The elements of $\mathbf{U}$ are the same as those for the upper triangular matrix found by Gauss elimination. The elements ℓ_{ik} of $\mathbf{L}$, except for the diagonal elements which have unit value, are the same as the multipliers (a_{ik}/a_{kk}) used in Eq. [2.33] for the elimination process. In practice, the original elements of the matrix **A** are overwritten by the elements of **L** and **U** except for the diagonal elements of **L**, which are not stored.

Pseudocodes for the Doolittle decomposition are now given. In these pseudocodes, we shall use the order vector **v** described in §2.2.3 instead of performing physical row exchanges in the pivoting step. The decomposition module given in Pseudocode 2–5 is similar to the one for the Gauss elimination passes.

Pseudocode 2–5   Program module for Doolittle decomposition

Module Doolittle(A, ϵ, n, v, err):

    err $\leftarrow$ 0   \ indicates nonsingular matrix
    k $\leftarrow$ 1   \ initial pivot row number
For [$i = 1, 2, \ldots, n$]   \ initialize order vector
    $v_i \leftarrow i$
End For
While [err = 0 **and** $k < n$]   \ pivot row loop
    NewPivot(A, n, k, v)   \ call partial pivoting module
    $r \leftarrow v_k$ \ pivot row
    If [$|a_{rk}| < \epsilon$] **then**
        Write: 'Matrix may be singular; execution discontinued'
        err $\leftarrow$ 1   \ indicates (nearly) singular matrix
    Else
        For [$z = (k + 1), (k + 2), \ldots, n$]   \ row loop
            $i \leftarrow v_z$
            $a_{ik} \leftarrow a_{ik}/a_{rk}$ \ element of **L**
            For [$j = (k + 1), (k + 2), \ldots, p$]   \ column loop
                $a_{ij} \leftarrow a_{ij} - a_{ik}(a_{rj})$ \ element of **U**
            End For   \ with column counter j
        End For   \ with row counter z
    End If
    $k \leftarrow k + 1$ \ increment pivot row counter
End While
If [err = 0] **then**
    $r \leftarrow v_n$
    If [$|a_{rn}| < \epsilon$] **then**   \ last check for a singular matrix
        Write: 'Matrix may be singular; execution discontinued'
        err $\leftarrow$ 1   \ indicates (nearly) singular matrix
    End If
End If
End Module Doolittle

Pseudocode 2–6   Row pivoting module with order vector

Module NewPivot(A, n, k, **v)** :

      $r \leftarrow v_k$

    amax $\leftarrow |a_{rk}|$ \ begin search for element with largest magnitude

      $m \leftarrow r$

For $[z = (k + 1), (k + 2), \ldots, n]$ \ row search loop

    $i \leftarrow v_z$

    If$[|a_{ik}| >$ amax] **then**

        amax $\leftarrow |a_{ik}|; m \leftarrow i$ \ find maximum and row location

    End If

End For   \ with row counter z

If [**not**$(m = r)$] **then**   \ row reordering necessary

    $v_k \leftarrow v_m; v_m \leftarrow r$ \ update order vector

End If

End Module NewPivot

    The module for the row pivoting step with the order vector is shown in Pseudocode 2–6. A comparison with the corresponding module in §2.2.3 shows how the pivoting is implemented through the order vector.

    The Doolittle decomposition begins with the original coefficient matrix **A**. On completion of the decomposition process, the same physical computer locations that were used for the elements of **A** are used to store the elements of **L** and **U** except for the unit diagonal elements of **L**. The final order vector **v** may be used to decode the storage scheme to produce **L**, **U**, and the reordered matrix **A*** of Eq. [2.40]. To see how the decoding is actually accomplished, we consider the following truss problem.

    A *truss* is a structure consisting of bars or members whose ends are pinned together to form joints. The basic form of the truss consists of three members and three joints; the members are sides of a triangle and the joints are the vertices. The truss is then built up by adding one joint and two members at a time (to produce an additional triangle). The number of joints J and the number of members M are related by

$$M = 2J - 3$$

    The diagram in Fig. 2–1 shows a planar truss with four joints P, Q, R, and S. The truss symmetrically supports a sign of weight W by cables from joints P and S at an angle p to the horizontal. The truss itself is held in place next to a fixed vertical wall by a pin support at joint R and a roller support at joint S. The problem

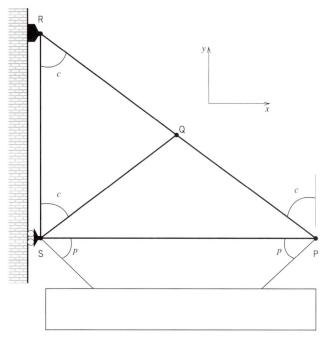

Figure 2–1   Diagram for a truss supporting a sign.

is to find the (internal) forces in the members and the (external) forces due to the supports at joints R and S.

The equilibrium conditions for a two-dimensional structure are met by

$$\begin{cases} \text{Sum of forces in the } x \text{ direction} = 0 \\ \text{Sum of forces in the } y \text{ direction} = 0 \end{cases}$$

The vertical component (see Appendix D) of the tension T in each of the cables supporting the sign is $(T \sin p)$. The sum of the forces on the sign in the y direction is therefore set to zero to give

$$2T \sin p - W = 0 \rightarrow T = W/(2 \sin p)$$

The tension T is the known load applied to the truss at joints P and S.

The unknown forces are those in the members and the supports. The equilibrium conditions allow us to write two equations at each joint. We therefore have enough equations to solve for the member forces plus three additional forces. The three additional forces are the forces at R and S. The pin support at R can have a two-

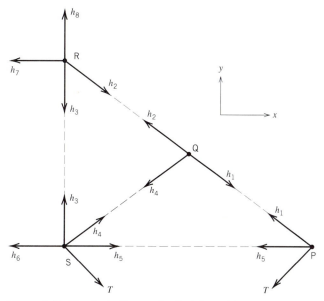

Figure 2–2 Free-body diagram for the joints of the truss.

component force, and the roller support at S can have a horizontal force. The member forces can be tensions (which pull on the joints at each end of the member) or compressions (which push on the joints at the ends of the member). We shall assume that all member forces are tensions (acting away from the joints) and that the external support forces also act away from the joints. The actual directions of these forces will be indicated by the solution to the problem; a negative force means that the actual direction is opposite to the assumed direction.

A free-body diagram for each joint is shown in Fig. 2–2. The external components at R are h_7 and h_8, and the single external force at S is h_6; forces h_1 through h_5 are member forces. The two equilibrium conditions at each joint are written below.

Joint P :
$$\begin{cases} x \text{ component:} & -T \cos p - h_1 \sin c - h_5 = 0 \\ y \text{ component:} & -T \sin p + h_1 \cos c \qquad\quad = 0 \end{cases}$$

Joint Q :
$$\begin{cases} x \text{ component:} & h_1 \sin c - h_2 \sin c - h_4 \sin c = 0 \\ y \text{ component:} & -h_1 \cos c + h_2 \cos c - h_4 \cos c = 0 \end{cases}$$

Joint R :
$$\begin{cases} x \text{ component:} & h_2 \sin c - h_7 \qquad\quad = 0 \\ y \text{ component:} & -h_2 \cos c - h_3 + h_8 = 0 \end{cases}$$

Joint S :
$$\begin{cases} x \text{ component:} & T \cos p + h_4 \sin c + h_5 - h_6 = 0 \\ y \text{ component:} & -T \sin p + h_4 \cos c + h_3 \qquad\quad = 0 \end{cases}$$

These conditions may be rewritten in the form

[2.42a]
$$
\mathbf{A} \cdot
\begin{bmatrix}
h_1 \\
h_2 \\
h_3 \\
h_4 \\
h_5 \\
h_6 \\
h_7 \\
h_8
\end{bmatrix}
=
\begin{bmatrix}
T \cos p \\
T \sin p \\
0 \\
0 \\
0 \\
0 \\
-T \cos p \\
T \sin p
\end{bmatrix}
$$

The coefficient matrix $\mathbf{A}$ in Eq. [2.42a] depends on the angle c in Fig. 2–1. If we choose c so that $(\cos c)$ is equal to 0.6 and $(\sin c)$ is equal to 0.8, we obtain $\mathbf{A}$ as

[2.42b]
$$
\mathbf{A} =
\begin{bmatrix}
-4/5 & 0 & 0 & 0 & -1 & 0 & 0 & 0 \\
3/5 & 0 & 0 & 0 & 0 & 0 & 0 & 0 \\
4/5 & -4/5 & 0 & -4/5 & 0 & 0 & 0 & 0 \\
-3/5 & 3/5 & 0 & -3/5 & 0 & 0 & 0 & 0 \\
0 & 4/5 & 0 & 0 & 0 & 0 & -1 & 0 \\
0 & -3/5 & -1 & 0 & 0 & 0 & 0 & 1 \\
0 & 0 & 0 & 4/5 & 1 & -1 & 0 & 0 \\
0 & 0 & 1 & 3/5 & 0 & 0 & 0 & 0
\end{bmatrix}
$$

After the decomposition is performed, the stored matrix (occupying the same memory locations used for $\mathbf{A}$) is $\mathbf{A_s}$. The stored matrix and the order vector $\mathbf{v}$ are given by

$$
\mathbf{A_s} =
\begin{bmatrix}
-4/5 & 0 & 0 & 0 & -1 & 0 & 0 & 0 \\
-3/4 & 0 & 0 & 0 & 3/4 & 0 & 3/4 & 0 \\
-1 & -4/5 & 0 & -4/5 & -1 & 0 & 0 & 0 \\
3/4 & -3/4 & 0 & -6/5 & 0 & 0 & 0 & 0 \\
0 & -1 & 0 & 2/3 & -1 & 0 & -1 & 0 \\
0 & 3/4 & -1 & 3/5 & 3/4 & 0 & 0 & 1 \\
0 & 0 & 0 & -2/3 & -1 & -1 & -1 & 0 \\
0 & 0 & -1 & -1 & -3/4 & 0 & -1 & 1
\end{bmatrix}
$$

$$
\mathbf{v} = [1\ 3\ 6\ 4\ 5\ 7\ 2\ 8]
$$

The matrix $\mathbf{A}^*$ that is actually factored is obtained from $\mathbf{A}$ in Eq. [2.42b] by writing its rows in the sequence given by the order vector. The matrix $\mathbf{A}^*$ is

$$\mathbf{A}^* = \begin{bmatrix} -4/5 & 0 & 0 & 0 & -1 & 0 & 0 & 0 \\ 4/5 & -4/5 & 0 & -4/5 & 0 & 0 & 0 & 0 \\ 0 & -3/5 & -1 & 0 & 0 & 0 & 0 & 1 \\ -3/5 & 3/5 & 0 & -3/5 & 0 & 0 & 0 & 0 \\ 0 & 4/5 & 0 & 0 & 0 & 0 & -1 & 0 \\ 0 & 0 & 0 & 4/5 & 1 & -1 & 0 & 0 \\ 3/5 & 0 & 0 & 0 & 0 & 0 & 0 & 0 \\ 0 & 0 & 1 & 3/5 & 0 & 0 & 0 & 0 \end{bmatrix} \begin{matrix} \text{row 1 of } \mathbf{A} \\ \text{row 3 of } \mathbf{A} \\ \text{row 6 of } \mathbf{A} \\ \text{row 4 of } \mathbf{A} \\ \text{row 5 of } \mathbf{A} \\ \text{row 7 of } \mathbf{A} \\ \text{row 2 of } \mathbf{A} \\ \text{row 8 of } \mathbf{A} \end{matrix}$$

The $\mathbf{L}$ and $\mathbf{U}$ factors of $\mathbf{A}^*$ are obtained by reordering the matrix $\mathbf{A}_s$ in the row sequence prescribed by the order vector, writing the upper triangular part of the new matrix as $\mathbf{U}$, and writing the remaining part plus the identity matrix $\mathbf{I}$ as $\mathbf{L}$. The results of these operations are

$$\mathbf{L} = \begin{bmatrix} 1 & 0 & 0 & 0 & 0 & 0 & 0 & 0 \\ -1 & 1 & 0 & 0 & 0 & 0 & 0 & 0 \\ 0 & 3/4 & 1 & 0 & 0 & 0 & 0 & 0 \\ 3/4 & -3/4 & 0 & 1 & 0 & 0 & 0 & 0 \\ 0 & -1 & 0 & 2/3 & 1 & 0 & 0 & 0 \\ 0 & 0 & 0 & -2/3 & -1 & 1 & 0 & 0 \\ -3/4 & 0 & 0 & 0 & 3/4 & 0 & 1 & 0 \\ 0 & 0 & -1 & -1 & -3/4 & 0 & -1 & 1 \end{bmatrix}$$

$$\mathbf{U} = \begin{bmatrix} -4/5 & 0 & 0 & 0 & -1 & 0 & 0 & 0 \\ 0 & -4/5 & 0 & -4/5 & -1 & 0 & 0 & 0 \\ 0 & 0 & -1 & 3/5 & 3/4 & 0 & 0 & 1 \\ 0 & 0 & 0 & -6/5 & 0 & 0 & 0 & 0 \\ 0 & 0 & 0 & 0 & -1 & 0 & -1 & 0 \\ 0 & 0 & 0 & 0 & 0 & -1 & -1 & 0 \\ 0 & 0 & 0 & 0 & 0 & 0 & 3/4 & 0 \\ 0 & 0 & 0 & 0 & 0 & 0 & 0 & 1 \end{bmatrix}$$

It is easy to verify that $\mathbf{L}$ and $\mathbf{U}$ are factors of $\mathbf{A}^*$ by performing the multiplication $\mathbf{L} \cdot \mathbf{U}$.

We, of course, do not perform an actual decoding to obtain $\mathbf{L}$, $\mathbf{U}$, and $\mathbf{A}^*$. Instead, we continue to use the order vector when we apply the methods for triangular matrices to complete the solution process. The elements g_k of the vector $\mathbf{g}$ in Eq. [2.41] are obtained in the sequence g_1 through g_n by using element v_k of $\mathbf{f}$ and the elements of $\mathbf{L}$ in row v_k of the stored matrix $\mathbf{A_s}$. Then the elements x_k of the solution vector $\mathbf{x}$ are found in the sequence x_n through x_1 by using element v_k of $\mathbf{g}$ and the elements of $\mathbf{U}$ in row v_k of $\mathbf{A_s}$.

The operational count for the LU decomposition method is the same as for the Gauss elimination method for a set of s right-hand vectors $\mathbf{f}_1$ through $\mathbf{f_s}$ when they are known *a priori*. In situations where a right-hand vector $\mathbf{f_t}$ depends on the solution with a previous right-hand vector $\mathbf{f_{t-1}}$, the LU method is superior to Gauss elimination. With the LU method, the triangular systems of Eq. [2.41] must be solved anew for each right-hand vector; however, the factors $\mathbf{L}$ and $\mathbf{U}$ are computed only once. With Gauss elimination, the method must be completely reexecuted for each new right-hand vector because the row operations that are needed to process the right-hand sides are not saved.

2.2.6 Applicability of General Direct Methods

Gauss elimination, the Gauss–Jordan method, and LU decomposition are methods for solving general systems of linear algebraic equations. As we have seen in the discussion of these methods, the number of operations for an $(n \times n)$ coefficient matrix and a single right-hand vector is on the order of n^3. These operations produce round-off errors and thus limit the size of the system that can be solved to about 20 equations if arithmetic is performed in single precision. Larger systems can be solved with adequate accuracy if the coefficient matrix is well behaved or if greater precision is used; on the other hand, severe errors in the solution can occur for systems with only a few equations. Before we consider the conditions that produce significant errors, we shall first look at how we might choose an appropriate method.

The Gauss elimination and LU decomposition methods are equally efficient for a single right-hand vector or for multiple right-hand vectors that are prescribed. In such cases, the Gauss elimination method is preferred because its implementation is somewhat less complex than that of the LU decomposition. If the same coefficient matrix is used for several right-hand sides that are given one at a time, the LU decomposition method is vastly superior to Gauss elimination because the latter must be completely reexecuted for each new right-hand vector.

The Gauss–Jordan method has the least complex implementation, and it is comparable in efficiency to Gauss elimination and LU decomposition for very small systems of equations. For large systems, the Gauss–Jordan method requires nearly 50% more computational effort than Gauss elimination or LU decomposition for a single right-hand vector. The major use of the Gauss–Jordan method is in computing the inverse of an $(n \times n)$ matrix, which is equivalent to solving systems with n right-hand vectors. The effort for obtaining the inverse of a large matrix by the Gauss–Jordan method is only about 10% more than by Gauss elimination or

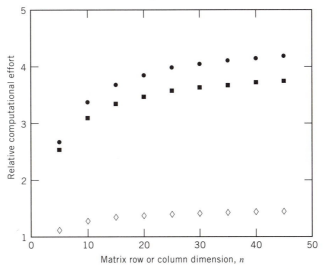

• Gauss–Jordan (*n* simultaneous right-hand vectors)
■ LU decomposition (*n* right-hand vectors)
◇ Gauss–Jordan (1 right-hand vector)

Figure 2–3 Computational effort for direct solvers relative to Gauss elimination with a single right-hand vector.

LU decomposition, and the form of the final augmented matrix is a very attractive feature of the Gauss–Jordan method.

The computational effort relative to Gauss elimination with a single right-hand vector is shown in Fig. 2–3 for different sizes of the $(n \times n)$ coefficient matrix. The locations of the plot symbols in the figure are obtained from Eq. [2.35], [2.36], [2.38], and [2.39]. The figure shows the effort for LU decomposition with n right-hand vectors given simultaneously or one at a time, and for the Gauss–Jordan method with one and with n simultaneous right-hand vectors. The relative effort for Gauss elimination and for the Gauss–Jordan method with n right-hand vectors given one at a time is n times the effort for a single right-hand vector.

Regardless of which method is used, unacceptable errors in the solution may occur if the coefficient matrix is ***ill-conditioned.*** To discuss ill-conditioned matrices, we introduce the concepts of ***vector norms*** and ***matrix norms*** (Ref. 4,5). Let $\mathbf{x}$ be a vector in a linear space $\mathbb{R}$; then the norm of $\mathbf{x}$ is written as $\|\mathbf{x}\|$ and may be thought of as a measure of the vector's magnitude. A vector norm satisfies the following conditions:

a. $\|\mathbf{x}\| \geq 0, \forall\, \mathbf{x} \in \mathbb{R}; \|\mathbf{x}\| = 0$ if and only if $\mathbf{x}$ is the zero vector.

b. $\|\alpha \mathbf{x}\| = |\alpha| \times \|\mathbf{x}\|; \forall$ scalars α and $\forall\, \mathbf{x} \in \mathbb{R}$.

c. $\|\mathbf{x} + \mathbf{y}\| \leq \|\mathbf{x}\| + \|\mathbf{y}\|; \forall\, \mathbf{x}, \mathbf{y} \in \mathbb{R}$

The *p-norm* of an *n*-dimensional vector **x** is defined by

[2.43]
$$\|\mathbf{x}\|_p = \left[\sum_{j=1}^{n} |x_j|^p \right]^{(1/p)} ; \quad p \geq 1$$

Two special cases are $\|\mathbf{x}\|_1$, which is the sum of the absolute values of the components of **x**, and $\|\mathbf{x}\|_\infty$, which is the maximum absolute value of the components of **x**. The norm $\|\mathbf{x}\|_2$ is known as the Euclidean norm because it is analogous to the distance formula in geometry.

The norm $\|\mathbf{A}\|$ of a matrix **A** must satisfy the same types of conditions as a vector norm. It must also satisfy an additional condition

$$\|\mathbf{A} \cdot \mathbf{B}\| \leq \|\mathbf{A}\| \times \|\mathbf{B}\|$$

If **x** is a column vector such that the multiplication $\mathbf{A} \cdot \mathbf{x}$ is defined, the norm of **A** is given by the supremum (or maximum)

[2.44a]
$$\|\mathbf{A}\| = \sup\{ \|\mathbf{A} \cdot \mathbf{x}\| / \|\mathbf{x}\| \} \text{ for } \mathbf{x} \neq \mathbf{0}$$

An alternative definition is

[2.44b]
$$\|\mathbf{A}\| = \max\{ \|\mathbf{A} \cdot \mathbf{x}\| \} , \ \|\mathbf{x}\| = 1$$

The norms $\|\mathbf{A}\|_1$ and $\|\mathbf{A}\|_\infty$ for an $(n \times n)$ matrix **A** that are compatible with the vector norms $\|\mathbf{x}\|_1$ and $\|\mathbf{x}\|_\infty$ are given by

[2.45a]
$$\|\mathbf{A}\|_1 = \max_j \left[\sum_{i=1}^{n} |a_{ij}| \right] = \text{maximum absolute column sum}$$

[2.45b]
$$\|\mathbf{A}\|_\infty = \max_i \left[\sum_{j=1}^{n} |a_{ij}| \right] = \text{maximum absolute row sum}$$

Now suppose that a process for solving $(\mathbf{A} \cdot \mathbf{x} = \mathbf{f})$ causes perturbations $\mathbf{\Delta A}$ in **A** and $\mathbf{\Delta f}$ in **f** so that the solution obtained is $(\mathbf{x} + \mathbf{\Delta x})$, which satisfies

$$(\mathbf{A} + \mathbf{\Delta A}) \cdot (\mathbf{x} + \mathbf{\Delta x}) = \mathbf{f} + \mathbf{\Delta f}$$

If we assume that the perturbation $\mathbf{\Delta A}$ is small enough so that $\|\mathbf{\Delta A}\|$ is less than $1/\|\mathbf{A}^{-1}\|$, we can show that

[2.46] $\|\mathbf{\Delta x}\| / \|\mathbf{x}\| \leq \left(\kappa(\mathbf{A})/\{1 - \kappa(\mathbf{A})\,\|\mathbf{\Delta A}\| / \|\mathbf{A}\|\}\right)\left(\|\mathbf{\Delta f}\| / \|\mathbf{f}\| + \|\mathbf{\Delta A}\| / \|\mathbf{A}\|\right)$

where $\kappa(\mathbf{A})$ is the **condition number** of the matrix $\mathbf{A}$ and is given by

[2.47] $$\kappa(\mathbf{A}) = \|\mathbf{A}\| \cdot \|\mathbf{A}^{-1}\|$$

Small relative changes in $\mathbf{A}$ and $\mathbf{f}$ may therefore cause large relative changes in $\mathbf{x}$ if the condition number is large (on the order of 1000 for arithmetic in single precision).

An illustration of the effects of an ill-conditioned matrix is provided by the **Hilbert matrix** $\mathbf{H_n}$ whose elements h_{ij} are given by

[2.48] $$h_{ij} = 1/(i + j - 1); \quad i, j = 1, 2, \ldots, n$$

The equation $(\mathbf{H_7} \cdot \mathbf{x} = \mathbf{f})$, in which $\mathbf{f}$ is chosen so that $\mathbf{x}$ has elements x_i equal to i, is solved in single precision by Gauss elimination with partial pivoting. The results are

$$
\begin{bmatrix} x_1 \\ x_2 \\ x_3 \\ x_4 \\ x_5 \\ x_6 \\ x_7 \end{bmatrix}_{\text{computed}}
=
\begin{bmatrix}
0.9964314 \times 10^0 \\
0.2126550 \times 10^1 \\
0.1882656 \times 10^1 \\
0.8057573 \times 10^1 \\
-0.2039473 \times 10^1 \\
0.1181091 \times 10^2 \\
0.5164301 \times 10^1
\end{bmatrix}
$$

Use of double precision (approximately 16 decimal digits) gives considerably more accurate results with errors appearing in the tenth significant digit.

The errors in approximately the last six significant digits are due to the large condition number of the Hilbert matrix $\mathbf{H_7}$, whose inverse is

$$
(\mathbf{H_7})^{-1} =
\begin{bmatrix}
49 & -1176 & 8820 & -29400 & 48510 & -38808 & 12012 \\
-1176 & 37632 & -317520 & 1128960 & -1940400 & 1596672 & -504504 \\
8820 & -317520 & 2857680 & -10584000 & 18711000 & -15717240 & 5045040 \\
-29400 & 1128960 & -10584000 & 40320000 & -72765000 & 62092800 & -20180160 \\
48510 & -1940400 & 18711000 & -72765000 & 133402500 & -115259760 & 37837800 \\
-38808 & 1596672 & -15717240 & 62092800 & -115259760 & 100590336 & -33297264 \\
12012 & -504504 & 5045040 & -20180160 & 37837800 & -33297264 & 11099088
\end{bmatrix}
$$

Either of the definitions in Eq. [2.45a] and [2.45b] yields the norms

$$\|\mathbf{H_7}\| = 363/140; \|(\mathbf{H_7})^{-1}\| = 379\,964\,970$$

The condition number for $\mathbf{H_7}$ is therefore found from Eq. [2.47] to be

$$\kappa(\mathbf{H_7}) \cong 9.852 \times 10^8$$

EXAMPLE   2.2

Does the multiplication $\mathbf{H_7} \cdot \mathbf{x}_{computed}$ provide a check on the accuracy of the computed solution in the preceding discussion on the Hilbert matrix?

The right-hand vector $\mathbf{f}$ in the equation $(\mathbf{H_7} \cdot \mathbf{x} = \mathbf{f})$ that produces a solution in which x_i is equal to i has components given by

$$f_i = \sum_{j=1}^{7} (j h_{ij}); \quad i = 1, 2, \ldots, 7$$

The elements of $\mathbf{f}$ are:

$$f_1 = 7; \quad f_2 = 1479/280; \quad f_3 = 5471/1260; \quad f_4 = 3119/840;$$
$$f_5 = 22549/6930; \quad f_6 = 16081/5544; \quad f_7 = 157309/60060$$

The multiplication $\mathbf{H_7} \cdot \mathbf{x}_{computed}$ with the computed elements of $\mathbf{x}$ as in the preceding discussion yields a vector whose elements match those of $\mathbf{f}$ to seven significant digits!

The results of the multiplication are compatible with Eq. [2.46]; that is, large errors in $\mathbf{x}$ may occur with small errors in $\mathbf{f}$ for an ill-conditioned coefficient matrix.

The fact that $\mathbf{H_7} \cdot \mathbf{x}_{computed}$ is very close to the original right-hand vector $\mathbf{f}$ does not necessarily indicate that $\mathbf{x}_{computed}$ is an accurate approximation of $\mathbf{x}$.

In solving systems of linear equations of the form $(\mathbf{A} \cdot \mathbf{x} = \mathbf{f})$ by one of the fundamental direct methods, we avoid an actual inversion of the coefficient matrix. We cannot therefore predict the likelihood of obtaining an inaccurate solution by computing the condition number of the matrix. The point of Example 2.2 is that comparing $\mathbf{f}$ with the product $\mathbf{A} \cdot \mathbf{x}_{computed}$ is also not an indicator of accuracy.

A simple test for ill-conditioning that does not require computation of the inverse is to solve the system for an additional right-hand vector that differs slightly from $\mathbf{f}$. From Eq. [2.46], we see that an ill-conditioned coefficient matrix will produce solutions for $\mathbf{f}$ and the additional right-hand vector that are significantly different from each other.

2.2.7 Cholesky Decomposition for Symmetric Matrices

Cholesky decomposition, which we mentioned in §2.2.5, is particularly useful for symmetric coefficient matrices. A few additional definitions are necessary before we proceed with the method.

The *transpose* of an ($m \times n$) matrix **A** (or of a vector) is the ($n \times m$) matrix $\mathbf{A}^T$ whose elements are related to the elements of **A** by

[2.49]
$$(a^T)_{ji} = a_{ij}; \qquad \begin{cases} i = 1, 2, \ldots, m \\ j = 1, 2, \ldots, n \end{cases}$$

A *symmetric* matrix is one which is equal to its transpose, and must therefore be a square matrix. Any symmetric matrix **S** can be factored in the form

[2.50]
$$\mathbf{S} = \mathbf{\Psi}^T \cdot \mathbf{\Psi}$$

The matrix **Ψ** is generally not unique, and it may contain elements that are complex numbers (i.e., numbers with real and imaginary parts).

An ($n \times n$) symmetric matrix **S** has n real *eigenvalues* λ_1 through λ_n (which are discussed in Chapter 8). Any eigenvalue λ of **S** must satisfy the equation

$$\left| \mathbf{S} - \lambda \mathbf{I} \right| = 0$$

If every eigenvalue of **S** is positive, **S** is said to be a *positive definite* matrix. Such matrices have particularly simple factors.

A factor **Ψ** for a symmetric, positive definite matrix **S** is the upper triangular matrix **U** described in Eq. [2.16]; its transpose is a lower triangular matrix. By rewriting Eq. [2.50] in the form

$$\mathbf{S} = \mathbf{U}^T \cdot \mathbf{U}$$

and performing the matrix multiplication for the elements of **S**, we obtain the elements of the first row of **S** as

$$s_{1j} = (u_{11})(u_{1j}); \quad j = 1, 2, \ldots, n$$

The elements in the first row of **U** are therefore found to be

[2.51a]
$$u_{11} = \sqrt{s_{11}}$$

[2.51b] $u_{1j} = s_{1j}/u_{11}; \quad j = 2, 3, \ldots, n$

For rows 2 through n in succession, the matrix multiplication gives

$$s_{ii} = \sum_{k=1}^{i} (u_{ki})^2$$

$$s_{ij} = \sum_{k=1}^{i} (u_{ki})(u_{kj}); \quad j = i + 1, i + 2, \ldots, n$$

The elements in row i of **U** are therefore

[2.52a] $u_{ii} = \left[s_{ii} - \sum_{k=1}^{i-1} (u_{ki})^2 \right]^{(1/2)} ; \quad i = 2, 3, \ldots, n$

[2.52b] $u_{ij} = (1/u_{ii}) \left[s_{ij} - \sum_{k=1}^{i-1} (u_{ki})(u_{kj}) \right]; \quad j = i + 1, i + 2, \ldots, n$

We note several points about the application of Eq. [2.51a, b] and [2.52a, b]. First, the factor **U** that is produced is not unique because either positive or negative values may be assigned to the square roots in Eq. [2.51a] and [2.52a]. For simplicity, we shall use the positive square root interpretation.

Second, Eq. [2.51a] requires the element s_{11} to be positive. If a system contains a symmetric coefficient matrix with a negative element s_{11}, we simply change the signs of all elements of the matrix and all components of any right-hand vector. The result is an equivalent system which also meets the requirement that s_{11} be positive.

Third, the derivations of Eq. [2.51a, b] and [2.52a, b] are based on the symmetry of **S** and on the triangular forms of the factor **U** and its transpose. Pivoting *cannot* be used since it will destroy these properties.

Last, the decomposition is applicable to positive definite matrices; however, we have no general way of establishing positive definiteness without computation of the eigenvalues. A sufficient *but not necessary* indicator of positive definiteness is *diagonal dominance*. The matrix **S** is diagonally dominant if the magnitude $|s_{ii}|$ of every diagonal element is greater than the sum of the magnitudes of every other element in row i or column i.

Even if a symmetric matrix is not diagonally dominant, it may still be positive definite. A practical test for positive definiteness is contained in Eq. [2.52a]. The square root operation in that equation requires a positive argument; a negative argument at any stage of the decomposition indicates that the matrix is not positive definite. This concept is illustrated in the following example.

EXAMPLE 2.3

Find the upper triangular factor **U** with real elements for

$$\mathbf{S} = \begin{bmatrix} 1 & c \\ c & 2 \end{bmatrix}$$

and establish the bounds on c for which the factorization is possible.

From Eq. [2.51a,b]: $\qquad u_{11} = 1$ and $u_{12} = c$

From Eq. [2.52a]: $\qquad u_{22} = \sqrt{2 - c^2}$

Note that Eq. [2.52b] is never invoked for a (2×2) matrix.
 Allowable values of c must satisfy $(2 - c^2 > 0)$
 Thus c is bounded by

$$-\sqrt{2} < c < \sqrt{2}$$

Diagonally dominant, and hence positive definite, coefficient matrices arise naturally in many physical applications. One instance is in the solution of the electrical resistance network of Fig. 2–4. The network has resistors R_1 through R_8 and two

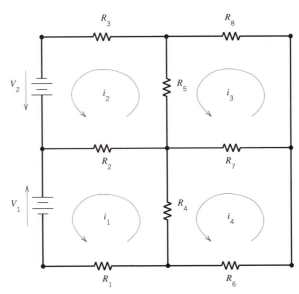

Figure 2–4 A resistor network with two voltage sources.

voltage sources V_1 and V_2 in the directions shown. Currents in the four closed loops of the network are i_1 through i_4, which are drawn in assumed counterclockwise directions.

The equations for the currents in terms of known resistances and voltages are derived from Ohm's Law and Kirchhoff's Voltage Law (see Appendix D). These are:

Ohm's Law: The voltage drop across a resistance R is (iR) in the direction of the current i.

Kirchhoff's Voltage Law: The net voltage drop in a closed loop is zero.

Four equations are obtained by applying Kirchhoff's Law to the four loops indicated by the currents i_1 through i_4. Note that when we express the voltage drop across a resistor such as R_2 in writing the equation for Loop 1, the current across R_2 is $(i_1 - i_2)$ in the counterclockwise sense.

The equations from Kirchhoff's Law are obtained for each of the four loops by expressing the voltage drops across resistors according to Ohm's Law. The equations are

Loop 1: $\quad i_1 R_1 + (i_1 - i_4)R_4 + (i_1 - i_2)R_2 + V_1 = 0$

Loop 2: $\quad (i_2 - i_1)R_2 + (i_2 - i_3)R_5 + i_2 R_3 - V_2 = 0$

Loop 3: $\quad (i_3 - i_4)R_7 + i_3 R_8 + (i_3 - i_2)R_5 = 0$

Loop 4: $\quad i_4 R_6 + (i_4 - i_3)R_7 + (i_4 - i_1)R_4 = 0$

The corresponding matrix equation is

[2.53a]
$$\mathbf{S} \cdot \begin{bmatrix} i_1 \\ i_2 \\ i_3 \\ i_4 \end{bmatrix} = \mathbf{f} = \begin{bmatrix} -V_1 \\ V_2 \\ 0 \\ 0 \end{bmatrix}$$

in which $\mathbf{S}$ is a symmetric, diagonally dominant matrix given by

[2.53b] $\quad \mathbf{S} = \begin{bmatrix} R_1 + R_2 + R_4 & -R_2 & 0 & -R_4 \\ -R_2 & R_2 + R_3 + R_5 & -R_5 & 0 \\ 0 & -R_5 & R_5 + R_7 + R_8 & -R_7 \\ -R_4 & 0 & -R_7 & R_4 + R_6 + R_7 \end{bmatrix}$

The elements of the current vector are not to be confused with the subscript that we have generally used to indicate row number.

Let us now consider the case for which

$$R_k = \begin{cases} 5; & k = 1, 2, 3, \text{ and } 4 \\ 2; & k = 5, 6, 7, \text{ and } 8 \end{cases}$$

The matrix S for this case is

$$S = \begin{bmatrix} 15 & -5 & 0 & -5 \\ -5 & 12 & -2 & 0 \\ 0 & -2 & 6 & -2 \\ -5 & 0 & -2 & 9 \end{bmatrix}$$

The elements of the first row of the factor U are found from Eq. [2.51a, b] (to six decimal places) as

$$u_{11} = 3.872983; \quad u_{12} = u_{14} = -1.290994; \quad u_{13} = 0$$

We then apply Eq. [2.52a, b] to find the elements of row 2 of U as

$$u_{22} = 3.214550; \quad u_{23} = -0.622171; \quad u_{24} = -0.518476$$

and again to find the elements of row 3 of U as

$$u_{33} = 2.369157; \quad u_{34} = -0.980341$$

The last element to be determined is u_{44}, which is found from Eq. [2.52a] to be

$$u_{44} = 2.470516$$

The solution for the current vector i may then be determined for given values of the voltages V_1 and V_2 by solving triangular systems as in Eq. [2.41]; that is, by solving

$$U^T \cdot g = f; \quad U \cdot i = g$$

Computer implementation of the Cholesky decomposition does not require separate storage for the matrices S and U. The elements of U may overwrite the corresponding elements of S because those elements (of S) are never used again.

The operational count for the Cholesky decomposition of an ($n \times n$) matrix S includes n square root operations to obtain the elements u_{ii}. We also have (with the help of the formulas in Appendix C) the remaining counts associated with the decomposition:

Divisions in Eq.[2.51b] and [2.52b]: $\quad 1 + 2 + \cdots + (n-1) = n(n-1)/2$

Multiplications in Eq.[2.52a]: $\quad 1 + 2 + \cdots + (n-1) = n(n-1)/2$

Multiplications in Eq.[2.52b]: $\quad 1(n-2) + 2(n-3) + \cdots$
$$+ (n-2)(n - [n-1])$$
$$= n(n-1)(n-2)/6$$

Finally, the number of operations for solving the two triangular systems with the factors of S is found from Eq. [2.29] to be $[n(n + 1)]$. The total operational count for the Cholesky method for symmetric matrices is therefore

[2.54]
$$\Omega_c = (n^3/6 + 3n^2/2 + n/3) + n \text{ square roots}$$

If the square root operation is equivalent to k multiplications, a comparison of Eq. [2.54] and [2.35] shows that the Cholesky method becomes more efficient than either the Doolittle LU decomposition or Gauss elimination when n exceeds 9 for k equal to 10, when n exceeds 12 for k equal to 20, and when n exceeds 15 for k equal to 30.

2.2.8 The Thomas Algorithm for Tridiagonal Matrices

A *tridiagonal or Jacobi matrix* T contains zero-valued elements t_{ij} whenever $|i - j|$ exceeds 1. Such matrices occur frequently, especially in numerical methods for differential equations. We may store the nonzero elements of T more efficiently if we use single subscripts as shown in the notation

[2.55]
$$T = \begin{bmatrix} a_1 & c_1 & & & & \\ b_2 & a_2 & c_2 & & \mathbf{0} & \\ & \cdot & \cdot & \cdot & & \\ & & \cdot & \cdot & \cdot & \\ & & & \cdot & \cdot & \cdot \\ \mathbf{0} & & & \cdot & \cdot & c_{n-1} \\ & & & & b_n & a_n \end{bmatrix}$$

The matrix T may be decomposed into a lower bidiagonal factor L and an upper bidiagonal factor U whose forms are given by

[2.56a]
$$L = \begin{bmatrix} 1 & & & & & \\ \beta_2 & 1 & & & \mathbf{0} & \\ & \cdot & \cdot & & & \\ & & \cdot & \cdot & & \\ & & & \cdot & \cdot & \\ \mathbf{0} & & & & \cdot & \cdot \\ & & & & & \beta_n & 1 \end{bmatrix}$$

$$\mathbf{U} = \begin{bmatrix} \alpha_1 & c_1 & & & & & \\ & \alpha_2 & c_2 & & & \mathbf{0} & \\ & & \cdot & \cdot & & & \\ & & & \cdot & \cdot & & \\ & & & & \cdot & \cdot & \\ & \mathbf{0} & & & & \cdot & c_{n-1} \\ & & & & & & \alpha_n \end{bmatrix}$$

[2.56b]

The **Thomas algorithm** is used to determine the elements β_i and α_i in Eq. [2.56a,b], and it may be regarded as a special case of the Doolittle LU decomposition **without** pivoting. The equations for α_i and β_i are

[2.57a] $$\alpha_1 = a_1$$

[2.57b] $$\beta_i = b_i/\alpha_{i-1}, \quad \alpha_i = a_i - \beta_i c_{i-1}; \qquad i = 2, 3, \ldots, n$$

The values for α_i and β_i overwrite the corresponding values a_i and b_i in the computer's memory. They are then used, along with the elements c_i, to solve the system $(\mathbf{T} \cdot \mathbf{x} = \mathbf{f})$ by solving the two bidiagonal systems

$$\mathbf{L} \cdot \mathbf{g} = \mathbf{f}; \qquad \mathbf{U} \cdot \mathbf{x} = \mathbf{g}$$

The continuation of the Thomas algorithm for solving the bidiagonal systems is given by

[2.58a] $$g_1 = f_1$$

[2.58b] $$g_i = f_i - \beta_i g_{i-1}; \qquad i = 2, 3, \ldots, n$$

followed by

[2.59a] $$x_n = g_n/\alpha_n$$

[2.59b] $$x_i = (g_i - c_i x_{i+1})/\alpha_i; \qquad i = (n-1), (n-2), \ldots, 1$$

The Thomas algorithm is remarkably efficient for a system of n equations. There are $(n-1)$ divisions and $(n-1)$ multiplications in Eq. [2.57b], $(n-1)$ multiplications in Eq. [2.58b], n divisions in Eq. [2.59a,b], and $(n - 1)$ multiplications in Eq. [2.59b]. The total count for the tridiagonal system is therefore

[2.60]
$$\Omega_t = 5n - 4$$

2.2.9   Block Matrix Methods

We close our treatment of direct methods with an illustration of **block matrix** formulations. These are useful in reducing computational effort when dealing with systems that contain **sparse** coefficient matrices. A sparse matrix is one that contains a large percentage of zero-valued elements; the matrix in Eq. [2.42b] is an example of a sparse matrix.

To illustrate the use of the block matrix, we revisit the problem of the truss in Fig. 2–1 and Fig. 2–2. Two equilibrium conditions were written at each joint of the truss. These conditions are reproduced here in a different sequence:

Joint P :
$$\begin{cases} x \text{ component:} & -T \cos p - h_1 \sin c - h_5 = 0 \\ y \text{ component:} & -T \sin p + h_1 \cos c \quad\quad = 0 \end{cases}$$

Joint Q :
$$\begin{cases} x \text{ component:} & h_1 \sin c - h_2 \sin c - h_4 \sin c = 0 \\ y \text{ component:} & -h_1 \cos c + h_2 \cos c - h_4 \cos c = 0 \end{cases}$$

Joint S :
$$\begin{cases} x \text{ component:} & T \cos p + h_4 \sin c + h_5 - h_6 = 0 \\ y \text{ component:} & -T \sin p + h_4 \cos c + h_3 \quad\quad = 0 \end{cases}$$

Joint R :
$$\begin{cases} x \text{ component:} & h_2 \sin c - h_7 \quad\quad = 0 \\ y \text{ component:} & -h_2 \cos c - h_3 + h_8 = 0 \end{cases}$$

With this sequence, at most two unknown elements of $\mathbf{h}$ are introduced with each pair of equations—h_1 and h_5 for joint P, h_2 and h_4 for joint Q, h_3 and h_6 for joint S, and h_7 and h_8 for joint R.

The introduction of and solution for two unknown forces at a time is known as the method of joints for truss problems in statics. We may write the counterpart of Eq. [2.42a] by listing the elements of $\mathbf{h}$ in the order of their introduction as

[2.61a]
$$\mathbf{A} \cdot \begin{bmatrix} h_1 \\ h_5 \\ \cdots \\ h_2 \\ h_4 \\ \cdots \\ h_3 \\ h_6 \\ \cdots \\ h_7 \\ h_8 \end{bmatrix} = \begin{bmatrix} T \cos p \\ T \sin p \\ \cdots \\ 0 \\ 0 \\ \cdots \\ -T \cos p \\ T \sin p \\ \cdots \\ 0 \\ 0 \end{bmatrix}$$

The matrix **A** with (cos c) equal to 0.6 and (sin c) equal to 0.8 is now

$$
[2.61b] \quad \mathbf{A} = \begin{bmatrix}
-4/5 & -1 & : & 0 & 0 & : & 0 & 0 & : & 0 & 0 \\
3/5 & 0 & : & 0 & 0 & : & 0 & 0 & : & 0 & 0 \\
\cdots & \cdots & \cdots & \cdots & \cdots & \cdots & \cdots & \cdots & \cdots & \cdots & \cdots \\
4/5 & 0 & : & -4/5 & -4/5 & : & 0 & 0 & : & 0 & 0 \\
-3/5 & 0 & : & 3/5 & -3/5 & : & 0 & 0 & : & 0 & 0 \\
\cdots & \cdots & \cdots & \cdots & \cdots & \cdots & \cdots & \cdots & \cdots & \cdots & \cdots \\
0 & 1 & : & 0 & 4/5 & : & 0 & -1 & : & 0 & 0 \\
0 & 0 & : & 0 & 3/5 & : & 1 & 0 & : & 0 & 0 \\
\cdots & \cdots & \cdots & \cdots & \cdots & \cdots & \cdots & \cdots & \cdots & \cdots & \cdots \\
0 & 0 & : & 4/5 & 0 & : & 0 & 0 & : & -1 & 0 \\
0 & 0 & : & -3/5 & 0 & : & -1 & 0 & : & 0 & 1
\end{bmatrix}
$$

the partitions in Eq. [2.61a] separate the vectors into four subvectors with two components each. Similarly, the matrix **A** is separated into a (4×4) square array of (2×2) submatrices or block matrices. We may represent Eq. [2.61a,b] in the shorter symbolic form

$$
[2.61c] \quad \begin{bmatrix}
\mathbf{A}_{11} & \mathbf{0} & \mathbf{0} & \mathbf{0} \\
\mathbf{A}_{21} & \mathbf{A}_{22} & \mathbf{0} & \mathbf{0} \\
\mathbf{A}_{31} & \mathbf{A}_{32} & \mathbf{A}_{33} & \mathbf{0} \\
\mathbf{A}_{41} & \mathbf{A}_{42} & \mathbf{A}_{43} & \mathbf{A}_{44}
\end{bmatrix} \cdot \begin{bmatrix}
\boldsymbol{\eta}_1 \\
\boldsymbol{\eta}_2 \\
\boldsymbol{\eta}_3 \\
\boldsymbol{\eta}_4
\end{bmatrix} = \begin{bmatrix}
\mathbf{f}_1 \\
\mathbf{f}_2 \\
\mathbf{f}_3 \\
\mathbf{f}_4
\end{bmatrix}
$$

Each $\mathbf{A}_{ij}$ is a (2×2) matrix, each $\boldsymbol{\eta}_i$ is a vector with two of the **h** components, and each $\mathbf{f}_i$ is a vector with two of the right-hand components.

The solution of Eq. [2.61c] is completely analogous to that of the lower triangular system in Eq. [2.27a], except that submatrices and subvectors are used in place of scalars. The block counterpart of Eq. [2.28a] for lower triangular systems is

$$
[2.62] \quad \begin{cases}
\boldsymbol{\eta}_1 = (\mathbf{A}_{11})^{-1} \cdot \mathbf{f}_1 \\
\boldsymbol{\eta}_i = (\mathbf{A}_{ii})^{-1} \cdot \left[\mathbf{f}_i - \sum_{j=1}^{i-1} (\mathbf{A}_{ij} \cdot \boldsymbol{\eta}_j) \right]; \quad i = 2, 3, 4
\end{cases}
$$

In Eq. [2.62], premultiplications by inverses replace the scalar divisions in Eq. [2.28a]. The forms of the $\boldsymbol{\eta}$ vectors in Eq. [2.62] do not imply that we must compute inverses of the matrices $\mathbf{A}_{ii}$. These vectors can be obtained by solving the equivalent system with any appropriate method. For example, we can determine $\boldsymbol{\eta}_1$ by using LU decomposition to solve the system $(\mathbf{A}_{11} \cdot \boldsymbol{\eta}_1 = \mathbf{f}_1)$.

Let us now look at the reduction in computational effort that can be achieved by recasting the general system of Eq. [2.42a,b] in the block form of Eq. [2.61a, b,c]. The solution from Eq. [2.62] requires four solutions of two-equation systems to compute the $\boldsymbol{\eta}_i$ vectors. If we use Gauss elimination or LU decomposition to solve these systems, the operational count from Eq. [2.35] is 6 for each system. In addition, each of the six matrix–vector multiplications in Eq. [2.62] has a scalar operational count of 4. The total operational count for the block matrix formulation is therefore equal to 48. In contrast, the general solution of the original system of eight linear equations by either Gauss elimination or LU decomposition requires 232 operations!

Our illustration shows the dramatic reduction in computational effort that can be realized when sparse systems are recast in an appropriate block form, even for a relatively small system. The block approach is generally very effective if the resulting block form is simple (diagonal, triangular, or tridiagonal) and if a significant number of operations on zero elements is eliminated as a result of the reformulation.

In seeking to reduce the computational effort for sparse systems, we should not attempt to extract more than the system will allow. To demonstrate this point, we perform a set of row and column exchanges on the matrix in Eq. [2.61b] to obtain

$$
\mathbf{A} =
\begin{bmatrix}
-1 & 0 & : & 0 & -4/5 & : & 0 & 0 & : & 0 & 0 \\
0 & 0 & : & 0 & 3/5 & : & 0 & 0 & : & 0 & 0 \\
\cdots & \cdots & \cdots & \cdots & \cdots & \cdots & \cdots & \cdots & \cdots & \cdots & \cdots \\
1 & -1 & : & 4/5 & 0 & : & 0 & 0 & : & 0 & 0 \\
0 & 0 & : & 3/5 & 0 & : & 1 & 0 & : & 0 & 0 \\
\cdots & \cdots & \cdots & \cdots & \cdots & \cdots & \cdots & \cdots & \cdots & \cdots & \cdots \\
0 & 0 & : & -4/5 & 4/5 & : & 0 & -4/5 & : & 0 & 0 \\
0 & 0 & : & -3/5 & -3/5 & : & 0 & 3/5 & : & 0 & 0 \\
\cdots & \cdots & \cdots & \cdots & \cdots & \cdots & \cdots & \cdots & \cdots & \cdots & \cdots \\
0 & 0 & : & 0 & 0 & : & 0 & 4/5 & : & -1 & 0 \\
0 & 0 & : & 0 & 0 & : & -1 & -3/5 & : & 0 & 1
\end{bmatrix}
$$

which has a block tridiagonal form. The tridiagonal form is more efficient than the lower triangular form we employed earlier, and we might be tempted to solve the system with the block counterpart of the Thomas algorithm. We would quickly find, however, that the Thomas algorithm is not applicable in this case because the required inverse of the submatrix in the top-left corner (corresponding to $1/\alpha_{i-1}$ for i equal to 2 in Eq. [2.57b]) does not exist.

2.3 ITERATIVE METHODS FOR LINEAR SYSTEMS

Iterative methods for systems of the form $(\mathbf{A} \cdot \mathbf{x} = \mathbf{f})$ may be used to refine a solution that is obtained by a direct method or as complete methods for solving the system. We shall begin our discussion with a method to improve the solution from a direct method, and continue with the Jacobi and Gauss–Seidel methods, and follow with the use of successive overrelaxation to accelerate the convergence of the Gauss–Seidel method. We shall then close our discussion with a treatment of the conjugate gradient method.

2.3.1 The Method of Residual Correction

Direct methods for solving systems of the form $(\mathbf{A} \cdot \mathbf{x} = \mathbf{f})$ usually involve a reduction of the coefficient matrix $\mathbf{A}$ to a simpler form. Let us assume that some form of LU decomposition (as in §2.2.5, §2.2.7, or §2.2.8) is used and rewrite the equation for the system as

$$\mathbf{A} \cdot \mathbf{x} = (\mathbf{L} \cdot \mathbf{U}) \cdot \mathbf{x} = \mathbf{f}$$

The process for determining the factors $\mathbf{L}$ and $\mathbf{U}$ introduces machine round-off error, so that the computed factors are only approximations.

Let the computed approximation of $\mathbf{L}$ be $\mathbf{L}'$, let the computed approximation of $\mathbf{U}$ be $\mathbf{U}'$, and let the resulting computed approximation of $\mathbf{x}$ be $\mathbf{x}^{(0)}$. The *residual vector* $\mathbf{r}^{(0)}$ that is associated with $\mathbf{x}^{(0)}$ is given by

[2.63]
$$\mathbf{r}^{(0)} = \mathbf{f} - \mathbf{A} \cdot \mathbf{x}^{(0)}$$

The residual vector may differ significantly from the zero vector when large systems of equations are solved, even if the coefficient matrix is well-conditioned. The *method of residual correction* (Ref. 4) is an iterative process for modifying the solution vector $\mathbf{x}^{(0)}$ so that the residual approaches the zero vector. The starting point is the computation of $\mathbf{r}^{(0)}$ from Eq. [2.63].

The improvement $\mathbf{x}^{(k)}$ at iteration k is obtained from

[2.64]
$$\mathbf{x}^{(k)} = \mathbf{x}^{(k-1)} + \boldsymbol{\delta}\mathbf{x}^{(k-1)}$$

in which the vector change $\boldsymbol{\delta}\mathbf{x}^{(k-1)}$ satisfies

[2.65]
$$(\mathbf{L}' \cdot \mathbf{U}') \cdot \boldsymbol{\delta}\mathbf{x}^{(k-1)} = \mathbf{r}^{(k-1)}$$

The same procedures that are used to compute the initial solution $\mathbf{x}^{(0)}$ are used to compute $\boldsymbol{\delta}\mathbf{x}^{(k-1)}$.

In general, only one iteration is used to accomplish the correction. Because the factors $\mathbf{L}'$ and $\mathbf{U}'$ already exist from the initial solution, the cost of the correction is relatively small.

Residual correction is generally not applicable for a system with an ill-conditioned coefficient matrix if the same level of precision is used for the correction as for the initial solution. We have shown in Example 2.2 that ill-conditioned systems can cause erroneous results that produce residual vectors whose components are at the "noise" level for the precision that is used. The vector change $\boldsymbol{\delta}\mathbf{x}^{(k-1)}$ from Eq. [2.65] may correct the solution at some stage; however, it is impossible to establish when the appropriate correction is achieved, and subsequent iterations may again cause an erroneous solution. If residual correction is to be effective for ill-conditioned systems, the correction should be performed at a higher level of precision than was used in acquiring the initial solution.

2.3.2 Jacobi and Gauss–Seidel Iterations

One of the most popular iterative methods for systems of linear algebraic equations is the successive overrelaxation method of §2.3.3. The bases for that method are the *Jacobi* and *Gauss–Seidel* iterative methods, which are presented here. To illustrate these methods, we shall use the system given by Eq. [2.53a, b] for the resistor network in §2.2.7. The equation with V_1 equal to 3.45, V_2 equal to 9.96, R_1 through R_4 equal to 5, and R_5 through R_8 equal to 2 is expressed in the form ($\mathbf{A} \cdot \mathbf{x} = \mathbf{f}$) by

[2.66a]
$$
\begin{bmatrix}
15 & -5 & 0 & -5 \\
-5 & 12 & -2 & 0 \\
0 & -2 & 6 & -2 \\
-5 & 0 & -2 & 9
\end{bmatrix}
\cdot
\begin{bmatrix}
x_1 \\ x_2 \\ x_3 \\ x_4
\end{bmatrix}
=
\begin{bmatrix}
-3.45 \\ 9.96 \\ 0.00 \\ 0.00
\end{bmatrix}
$$

The solution for Eq. [2.66a] is

[2.66b]
$$
\mathbf{x} =
\begin{bmatrix}
x_1 \\ x_2 \\ x_3 \\ x_4
\end{bmatrix}
=
\begin{bmatrix}
0.14 \\ 0.95 \\ 0.37 \\ 0.16
\end{bmatrix}
$$

The coefficient matrix $\mathbf{A}$ for the Jacobi and Gauss–Seidel methods is split into three parts $\mathbf{D}$, $\mathbf{V}$, and $\mathbf{W}$. $\mathbf{D}$ is the diagonal part of $\mathbf{A}$ with

[2.67a] $\qquad d_{ij} = a_{ij}, i = j; \qquad d_{ij} = 0, i \neq j$

$\mathbf{V}$ is the part of $\mathbf{A}$ below the diagonal, and $\mathbf{W}$ is the part of $\mathbf{A}$ above the diagonal. The elements of $\mathbf{V}$ and $\mathbf{W}$ are given by

[2.67b] $\qquad v_{ij} = a_{ij}, j < i; \qquad v_{ij} = 0, j \geq i$

[2.67c] $\qquad w_{ij} = a_{ij}, j > i; \qquad w_{ij} = 0, j \leq i$

The matrix $\mathbf{A}$ may then be written as

[2.67d] $\qquad\qquad\qquad\qquad \mathbf{A} = \mathbf{V} + \mathbf{D} + \mathbf{W}$

The matrices $\mathbf{D}$, $\mathbf{V}$, and $\mathbf{W}$ for Eq. [2.66a] are

$$
\mathbf{D} =
\begin{bmatrix}
15 & 0 & 0 & 0 \\
0 & 12 & 0 & 0 \\
0 & 0 & 6 & 0 \\
0 & 0 & 0 & 9
\end{bmatrix}
$$

$$\mathbf{V} = \begin{bmatrix} 0 & 0 & 0 & 0 \\ -5 & 0 & 0 & 0 \\ 0 & -2 & 0 & 0 \\ -5 & 0 & -2 & 0 \end{bmatrix}; \quad \mathbf{W} = \begin{bmatrix} 0 & -5 & 0 & -5 \\ 0 & 0 & -2 & 0 \\ 0 & 0 & 0 & -2 \\ 0 & 0 & 0 & 0 \end{bmatrix}$$

The methods begin with an initial guess or estimate $\mathbf{x}^{(0)}$, which is updated by an iterative process until convergence to a solution is achieved. To derive the iterative steps, we first use Eq. [2.67d] to recast the original equation in the form

[2.68] $$(\mathbf{V} + \mathbf{D} + \mathbf{W}) \cdot \mathbf{x} = \mathbf{f}$$

The general scheme for Jacobi iterations is obtained by rewriting Eq. [2.68] as

$$\mathbf{D} \cdot \mathbf{x} = -(\mathbf{V} + \mathbf{W}) \cdot \mathbf{x} + \mathbf{f}$$

The updated solution vector $\mathbf{x}^{(k)}$ replaces $\mathbf{x}$ on the left side, and the previous vector $\mathbf{x}^{(k-1)}$ replaces $\mathbf{x}$ on the right side. The result is the iterative process

[2.69a] $$\mathbf{x}^{(k)} = \mathbf{J} \cdot \mathbf{x}^{(k-1)} + \mathbf{D}^{-1} \cdot \mathbf{f}$$

in which the Jacobi iteration matrix $\mathbf{J}$ is given by

[2.69b] $$\mathbf{J} = -\mathbf{D}^{-1} \cdot (\mathbf{V} + \mathbf{W})$$

The matrix form of the Jacobi method may seem daunting; in reality, we are merely finding new elements x_i at iteration k by solving the equation corresponding to row i of $\mathbf{A}$ with all other elements of $\mathbf{x}$ having values at iteration $(k - 1)$. The scalar form of Eq. [2.69a] for an n-dimensional vector $\mathbf{x}$ is found with the help of Eq. [2.67a,b,c] and the inverse of $\mathbf{D}$ to be

$$(x_i)^{(k)} = (1/a_{ii}) \left[f_i - \sum_{\substack{j=1 \\ j \neq i}}^{n} a_{ij}(x_j)^{(k-1)} \right]; \quad i = 1, 2, \ldots, n$$

A more convenient form of this result is

[2.69c] $$(x_i)^{(k)} = (x_i)^{(k-1)} + (1/a_{ii}) \left[f_i - \sum_{j=1}^{n} a_{ij}(x_j)^{(k-1)} \right]; \quad i = 1, 2, \ldots, n$$

The application of Eq. [2.69c] to our example in Eq. [2.66a] yields four scalar equations for the elements of $\mathbf{x}$; these are

$$(x_1)^{(k)} = \left\{ -3.45 + 5(x_2)^{(k-1)} + 5(x_4)^{(k-1)} \right\} \Big/ 15$$

$$(x_2)^{(k)} = \left\{ 9.96 + 5(x_1)^{(k-1)} + 2(x_3)^{(k-1)} \right\} \Big/ 12$$

$$(x_3)^{(k)} = \left\{ 2(x_2)^{(k-1)} + 2(x_4)^{(k-1)} \right\} \Big/ 6$$

$$(x_4)^{(k)} = \left\{ 5(x_1)^{(k-1)} + 2(x_3)^{(k-1)} \right\} \Big/ 9$$

We see that every updated element of $\mathbf{x}$ depends only on values of the elements at the previous iteration. For this reason, the Jacobi method is also known as the method of *simultaneous iterations*.

The Gauss–Seidel method differs from the Jacobi method by incorporating the updated elements of $\mathbf{x}$ *immediately*. The rewritten form of Eq. [2.68] is now

$$(\mathbf{V} + \mathbf{D}) \cdot \mathbf{x} = -\mathbf{W} \cdot \mathbf{x} + \mathbf{f}$$

The Gauss–Seidel iterative process is then given by

[2.70a] $$\mathbf{x}^{(k)} = \mathbf{G} \cdot \mathbf{x}^{(k-1)} + (\mathbf{V} + \mathbf{D})^{-1} \cdot \mathbf{f}$$

in which the Gauss–Seidel iteration matrix $\mathbf{G}$ is given by

[2.70b] $$\mathbf{G} = -(\mathbf{V} + \mathbf{D})^{-1} \cdot \mathbf{W}$$

The new elements x_i at iteration k are found from the equation corresponding to row i of $\mathbf{A}$. In the Gauss–Seidel method, however, other elements of $\mathbf{x}$ from iteration k are used immediately. The scalar form of Eq. [2.70a] for an n-dimensional vector $\mathbf{x}$ is best found by using Eq. [2.28a] for solving lower triangular systems. This form is

[2.70c] $$(x_i)^{(k)} = (x_i)^{(k-1)}$$
$$+ (1/a_{ii}) \left[f_i - \sum_{j=1}^{i-1} a_{ij}(x_j)^{(k)} - \sum_{j=i}^{n} a_{ij}(x_j)^{(k-1)} \right]; \qquad i = 1, 2, \ldots, n$$

Unlike the Jacobi method of Eq. [2.69c], which requires us to maintain separate vectors for iterations $(k-1)$ and k, the Gauss–Seidel method of Eq. [2.70c] allows us to overwrite the previous elements of $\mathbf{x}$ with the updated values as soon as they are obtained.

The application of Eq. [2.70c] to our example in Eq. [2.66a] is used to update x_1, x_2, x_3, and x_4 in sequence. The sequence is important because the right-hand side of Eq. [2.70c] contains elements of **x** at both iterations $(k-1)$ and k. The four scalar equations for our example are

$$(x_1)^{(k)} = \left\{ -3.45 + 5(x_2)^{(k-1)} + 5(x_4)^{(k-1)} \right\} \Big/ 15$$

$$(x_2)^{(k)} = \left\{ 9.96 + 5(x_1)^{(k)} + 2(x_3)^{(k-1)} \right\} \Big/ 12$$

$$(x_3)^{(k)} = \left\{ 2(x_2)^{(k)} + 2(x_4)^{(k-1)} \right\} \Big/ 6$$

$$(x_4)^{(k)} = \left\{ 5(x_1)^{(k)} + 2(x_3)^{(k)} \right\} \Big/ 9$$

The Jacobi and Gauss–Seidel methods require the iteration matrices **J** and **G** given by Eq. [2.69b] and [2.70b] to be convergent. The eigenvalues of a square matrix are discussed briefly in §2.2.7 and in more detail in Chapter 8. The largest magnitude of the eigenvalues is called the ***spectral radius*** (see Appendix C) and is denoted by ρ. A ***convergent*** matrix is one whose spectral radius is smaller than 1.

The spectral radii $\rho(\mathbf{J})$ and $\rho(\mathbf{G})$ for the Jacobi and Gauss–Seidel iteration matrices associated with Eq. [2.66a] are

[2.71] $$\rho(\mathbf{J}) = \sqrt{49/108} = 0.673575; \qquad \rho(\mathbf{G}) = 0.467095$$

To see how these spectral radii are related to the convergence of the methods, we solve Eq. [2.66a] and, at each iteration k, monitor the norms defined by

[2.72a] $$E_k = \left\| \mathbf{x}^{(k)} - \mathbf{x} \right\|; \qquad k = 0, 1, \ldots$$

[2.72b] $$\delta_k = \left\| \mathbf{x}^{(k)} - \mathbf{x}^{(k-1)} \right\|; \qquad k = 1, 2, \ldots$$

E_k is the norm of the error vector at iteration k, and δ_k is the norm of the solution vector change from iteration $(k-1)$ to iteration k. The vector **x** in Eq. [2.72a] is provided by Eq. [2.66b]. A convergent process drives the values of E_k and δ_k to zero as k increases.

The solution for the Jacobi method is started with $\mathbf{x}^{(0)}$ equal to the zero vector. Table 2–1 shows the elements of $\mathbf{x}^{(k)}$ for the first 16 iterations, and Table 2–2 shows the behavior of the norms E_k and δ_k which are computed as 1-norms as defined in Eq. [2.43]. The convergence of $\mathbf{x}^{(k)}$ to **x** is evident in Table 2–1, and the corresponding decrease in E_k and δ_k are obvious from Table 2–2.

An especially important set of values in Table 2–2 are the ratios (E_k/E_{k-1}) and (δ_k/δ_{k-1}), which are seen to be approximately equal to $\rho(\mathbf{J})$. Indeed, the value $\rho(\mathbf{J})$

Table 2–1   Jacobi Solutions of Eq. [2.66a] at Iteration k

k	$(x_1)^{(k)}$	$(x_2)^{(k)}$	$(x_3)^{(k)}$	$(x_4)^{(k)}$
1	−0.230000	0.830000	0.000000	0.000000
2	0.046667	0.734167	0.276667	−0.127778
3	−0.027870	0.895556	0.202130	0.087407
4	0.097654	0.852076	0.327654	0.029434
5	0.063837	0.925298	0.293837	0.127064
6	0.120788	0.905571	0.350788	0.100762
7	0.105444	0.938793	0.335444	0.145057
8	0.131283	0.929843	0.361283	0.133123
9	0.124322	0.944915	0.354322	0.153220
10	0.136045	0.940854	0.366045	0.147806
11	0.132887	0.947693	0.362887	0.156924
12	0.138206	0.945851	0.368206	0.154468
13	0.136773	0.948953	0.366773	0.158604
14	0.139186	0.948117	0.369186	0.157490
15	0.138536	0.949525	0.368536	0.159367
16	0.139631	0.949146	0.369631	0.158861

approximates the reduction factor per iteration of the error and vector change norms as the iteration number k becomes sufficiently large.

Similar sets of results for the Gauss–Seidel method are given in Tables 2–3 and 2–4 for the first 10 iterations. The convergence is again evident, and the ratios of the norms are approximately equal to $\rho(\mathbf{G})$ in this case. We also note that Gauss–Seidel

Table 2–2   Behavior of the Vector Norms for Jacobi Iterations

k	E_k	E_k/E_{k-1}	δ_k	δ_k/δ_{k-1}
1	1.02000	0.629630	1.06000	—
2	$6.90278(10^{-1})$	0.676743	$7.76944(10^{-1})$	0.732966
3	$4.62778(10^{-1})$	0.670423	$5.25648(10^{-1})$	0.676558
4	$3.13182(10^{-1})$	0.676743	$3.52503(10^{-1})$	0.670606
5	$2.09964(10^{-1})$	0.670423	$2.38489(10^{-1})$	0.676558
6	$1.42092(10^{-1})$	0.676743	$1.59932(10^{-1})$	0.670606
7	$9.52614(10^{-2})$	0.670423	$1.08203(10^{-1})$	0.676558
8	$6.44675(10^{-2})$	0.676743	$7.25616(10^{-2})$	0.670606
9	$4.32205(10^{-2})$	0.670422	$4.90922(10^{-2})$	0.676558
10	$2.92491(10^{-2})$	0.676743	$3.29215(10^{-2})$	0.670606
11	$1.96093(10^{-2})$	0.670422	$2.22733(10^{-2})$	0.676558
12	$1.32704(10^{-2})$	0.676743	$1.49366(10^{-2})$	0.670606
13	$8.89680(10^{-3})$	0.670422	$1.01055(10^{-2})$	0.676558
14	$6.02084(10^{-3})$	0.676743	$6.77679(10^{-3})$	0.670606
15	$4.03650(10^{-3})$	0.670422	$4.58489(10^{-3})$	0.676558
16	$2.73167(10^{-3})$	0.676742	$3.07465(10^{-3})$	0.670606

Table 2–3 Gauss–Seidel Solutions of Eq. [2.66a] at Iteration k

k	$(x_1)^{(k)}$	$(x_2)^{(k)}$	$(x_3)^{(k)}$	$(x_4)^{(k)}$
1	−0.230000	0.734167	0.244722	−0.073395
2	−0.009743	0.866728	0.264444	0.053353
3	0.076693	0.906030	0.319794	0.113673
4	0.109901	0.929091	0.347588	0.138298
5	0.125796	0.940346	0.359548	0.149786
6	0.133378	0.945499	0.365095	0.155231
7	0.136910	0.947895	0.367709	0.157774
8	0.138556	0.949017	0.368930	0.158960
9	0.139326	0.949541	0.369500	0.159514
10	0.139685	0.949785	0.369767	0.159773

solution converges more quickly than the Jacobi solution. In general, we may state the following:

> **If the Jacobi method is convergent, the Gauss–Seidel method is also convergent, and it converges more quickly.**

We have observed that the spectral radius ρ for either the Jacobi or Gauss–Seidel method is related to the error and vector change norms by

$$[2.73] \qquad \rho \simeq E_k/E_{k-1} \simeq \delta_k/\delta_{k-1}; \qquad \text{for large } k$$

We note that the norm E_k cannot be computed if the solution we are seeking is initially unknown. However, we can compute the norm δ_k and use it to establish convergence. The usual approach is to end the iterative process when δ_k becomes smaller than some small, positive quantity ϵ. The choice of ϵ depends on the precision level that is used, the definition of the norm, and the magnitudes of the elements of the solution vector.

Table 2–4 Behavior of the Vector Norms for Gauss–Seidel Iterations

k	E_k	E_k/E_{k-1}	δ_k	δ_k/δ_{k-1}
1	$9.44506(10^{-1})$	0.583028	1.28228	—
2	$4.45218(10^{-1})$	0.471377	$4.99288(10^{-1})$	0.389374
3	$2.03810(10^{-1})$	0.457775	$2.41408(10^{-1})$	0.483505
4	$9.51224(10^{-2})$	0.466721	$1.08688(10^{-1})$	0.450223
5	$4.45228(10^{-2})$	0.468058	$5.05996(10^{-2})$	0.465551
6	$2.07978(10^{-2})$	0.467126	$2.37251(10^{-2})$	0.468878
7	$9.71251(10^{-3})$	0.466998	$1.10853(10^{-2})$	0.467239
8	$4.53664(10^{-3})$	0.467092	$5.17588(10^{-3})$	0.466915
9	$2.11908(10^{-3})$	0.467104	$2.41756(10^{-3})$	0.467082
10	$9.89807(10^{-4})$	0.467093	$1.12927(10^{-3})$	0.467113

For example, the solution of Eq. [2.66a] involves the computation of four vector elements. The 1-norm definition therefore requires us to add four values to compute δ_k. If we use single precision with approximately seven decimal digits and we assume that the magnitudes of the elements of the solution are on the order of 1, the smallest δ_k value that we can expect to compute is about (4×10^{-6}). If we choose ϵ smaller than this value, we may not be able to detect convergence.

Another important factor that dictates how quickly an accurate solution can be obtained is the choice of the initial solution vector $\mathbf{x}^{(0)}$. Many physical problems allow us to estimate the initial vector so that the initial error norm E_0 is small. It is clear from Eq. [2.72] that smaller values of E_0 will result in fewer iterations to an accurate solution.

2.3.3 Successive Overrelaxation

We have seen in §2.3.2 that the Gauss–Seidel iterative method uses updated information immediately and converges more quickly than the Jacobi method. In some large systems of equations, the spectral radius $\rho(\mathbf{G})$ may be so close to 1 that even the Gauss–Seidel method converges at an agonizingly slow rate. One popular process for accelerating convergence of the Gauss–Seidel method is ***successive overrelaxation*** or ***SOR*** (Ref. 6).

A qualitative look at the meaning of overrelaxation is provided by the following situation. Suppose we have a starting value S of a quantity. Suppose further that we wish to approach a target value T by some process. Let an application of the process change the value of the quantity from S to S'. If S' is between S and T, we may get to a value S'', which is even closer to T by magnifying the change $(S' - S)$. To do so, we use an amplification factor ω greater than 1 and obtain

$$S'' - S = \omega(S' - S) \rightarrow S'' = \omega S' + (1 - \omega)S$$

This amplification process is an extrapolation and is an example of overrelaxation. If the intermediate result S' tends to overshoot the target T, we may wish to use a value of ω that is smaller than 1; this case is called underrelaxation.

The application of the overrelaxation concept to the Gauss–Seidel method is perhaps easiest to see from the scalar form of Eq. [2.70c]. Let us replace the left side of that equation by $(\xi_i)^{(k)}$ so that we obtain

[2.74a] $(\xi_i)^{(k)} = (x_i)^{(k-1)}$

$$+ (1/a_{ii}) \left[f_i - \sum_{j=1}^{i-1} a_{ij}(x_j)^{(k)} - \sum_{j=i}^{n} a_{ij}(x_j)^{(k-1)} \right]; \quad i = 1, 2, \ldots, n$$

We use $(\xi_i)^{(k)}$ as the intermediate value of the overrelaxation process and compute $(x_i)^{(k)}$ from

[2.74b] $$(x_i)^{(k)} = \omega(\xi_i)^{(k)} + (1 - \omega)(x_i)^{(k-1)}$$

In short, the Gauss–Seidel process of Eq. [2.70c] is used successively for each element of **x** to provide an intermediate value, and overrelaxation is used to obtain the updated value. The difficulty in the SOR method described by Eq. [2.74a, b] centers on the determination of a value of ω that produces the best convergence. The quantity ω is called the **relaxation parameter** or the **acceleration parameter**.

The matrix form of the SOR method in terms of the matrices **D**, **V**, and **W** in Eq. [2.67a, b, c, d] is

[2.75a] $$\mathbf{x}^{(k)} = \mathbf{S}(\omega) \cdot \mathbf{x}^{(k-1)} + \omega(\mathbf{D} + \omega\mathbf{V})^{-1} \cdot \mathbf{f}$$

where $\mathbf{S}(\omega)$ is the SOR iteration matrix, which depends on ω and is given by

[2.75b] $$\mathbf{S}(\omega) = (\mathbf{D} + \omega\mathbf{V})^{-1} \cdot \{(1 - \omega)\mathbf{D} - \omega\mathbf{W}\}$$

When ω is equal to 1, $\mathbf{S}(\omega)$ is the Gauss–Seidel matrix **G** of Eq. [2.70b], and Eq. [2.75a] is equivalent to the Gauss–Seidel method of Eq. [2.70a].

The Jacobi iteration matrix **J**, the Gauss–Seidel iteration matrix **G**, and the SOR iteration matrix $\mathbf{S}(\omega)$ are related to each other in a particular way when the coefficient matrix **A** for the system $(\mathbf{A} \cdot \mathbf{x} = \mathbf{f})$ has certain properties. One of these properties is known as **Property A**, which means that **A** is **two cyclic**. A matrix is two cyclic if suitable row and column exchanges allow it to be expressed in block form

$$\begin{bmatrix} \mathbf{D_1} & \mathbf{F} \\ \mathbf{E} & \mathbf{D_2} \end{bmatrix}$$

in which **E** and **F** are rectangular submatrices, and $\mathbf{D_1}$ and $\mathbf{D_2}$ are diagonal submatrices. The matrix in Eq. [2.66a] may be shown to be two cyclic by exchanging columns 2 and 3, and then by exchanging rows 2 and 3. The resulting block form is

$$\begin{bmatrix} 15 & 0 & \vdots & -5 & -5 \\ 0 & 6 & \vdots & -2 & -2 \\ \cdots & \cdots & \cdots & \cdots & \cdots \\ -5 & -2 & \vdots & 12 & 0 \\ -5 & -2 & \vdots & 0 & 9 \end{bmatrix}$$

The second property is **consistent ordering.** We again refer to our brief discussion of eigenvalues in §2.2.7 and to the matrices $\mathbf{D}$, $\mathbf{V}$, and $\mathbf{W}$ of Eq. [2.67a,b,c,d], and define the coefficient matrix $\mathbf{A}$ to be consistently ordered if the eigenvalues of

$$\mathbf{C} = -(\alpha \mathbf{D}^{-1} \cdot \mathbf{V} + (1/\alpha)\mathbf{D}^{-1} \cdot \mathbf{W})$$

are independent of the (nonzero) scalar α. The matrix $\mathbf{C}$ for the coefficient matrix $\mathbf{A}$ in Eq. [2.66a] has zero-valued diagonal elements and is easily obtained. Its eigenvalues must satisfy the determinant relation

$$|\mathbf{C}| = \begin{vmatrix} -\lambda & 5/(15\alpha) & 0 & 5/(15\alpha) \\ 5\alpha/12 & -\lambda & 2/(12\alpha) & 0 \\ 0 & 2\alpha/6 & -\lambda & 2/(6\alpha) \\ 5\alpha/9 & 0 & 2\alpha/9 & -\lambda \end{vmatrix} = 0$$

It can be shown that the eigenvalues of $\mathbf{C}$ depend on α; thus the coefficient matrix for Eq. [2.66a] is **not** consistently ordered.

The relations that hold when the coefficient matrix $\mathbf{A}$ is two cyclic and is also consistently ordered are

[2.76a] $$\rho(\mathbf{G}) = \rho^2(\mathbf{J})$$

[2.76b] $$\rho^2(\mathbf{S}(\omega)) + [2(\omega - 1) - \omega^2 \rho(\mathbf{G})]\rho(\mathbf{S}(\omega)) + (\omega - 1)^2 = 0$$

Our previous results in Eq. [2.71] for the system of Eq. [2.66a] do not satisfy Eq. [2.76a] because the coefficient matrix for that system is not consistently ordered. Values of ω that produce convergence are found from Eq. [2.76b] to be in the range $(0 < \omega < 2)$, which includes underrelaxation. For strict overrelaxation, we have

[2.77] $$(1 < \omega < 2) \rightarrow \text{SOR convergence}$$

Another consequence of Eq. [2.76b] is the optimum value ω_{opt} of the relaxation parameter. The optimum value and the associated spectral radius are related by

[2.78] $$\omega_{\text{opt}} = 2 \left/ \left[1 + \sqrt{1 - \rho(\mathbf{G})}\right]\right. = 1 + \rho(\mathbf{S}(\omega_{\text{opt}}))$$

Although Eq. [2.76b] and [2.78] refer to $\rho(\mathbf{G})$, Eq. [2.76a] is useful because it is generally easier to determine the eigenvalues of the Jacobi matrix $\mathbf{J}$ directly than to obtain those of the more complex Gauss–Seidel matrix $\mathbf{G}$.

We now demonstrate the SOR method by considering the problem of torsion applied to one end of a shaft while the other end is clamped. The shaft has a rectangular cross section on which we overlay a square cartesian grid as illustrated

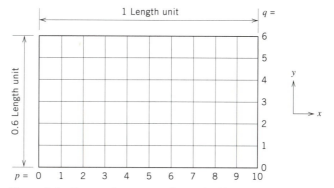

Figure 2–5 Rectangular cross section and uniform grid.

in Fig. 2–5. The nondimensional governing equation for this problem is the Poisson equation

$$\partial^2\phi/\partial x^2 + \partial^2\phi/\partial y^2 = -2$$

in which ϕ is a stress function whose value on the perimeter of the shaft is zero. When the value of the dependent variable is specified everywhere on the boundary of the domain, the boundary condition is known as a **Dirichlet** condition. Problems of this type are considered in more detail in Chapter 9; for now, we simply state that a formulation for computing ϕ at the grid intersections (p,q) defined in Fig. 2–5 is given by

[2.79] $\phi_{p,q} = 0.005 + (\phi_{p-1,q} + \phi_{p+1,q} + \phi_{p,q-1} + \phi_{p,q+1})/4$

The solution of the torsion problem requires Eq. [2.79] to be applied to all of the 45 interior points of the grid in Fig. 2–5. This means that the coefficient matrix **A** for the system of equations is a (45×45) matrix. The coefficient matrix is sparse since it contains only four nonzero elements per row, and it is also two cyclic and consistently ordered.

A practical way to establish Property A for a physical problem such as the torsion problem begins with assigning the colors black and white to the grid intersections or nodes in a checkerboard fashion. For our problem, we could assign black to nodes for which $(p + q)$ is even and white to nodes for which $(p + q)$ is odd. Property A is established if the solution variable at the black nodes can be found exclusively in terms of values at the white nodes and *vice versa*.

The properties of the coefficient matrix make it suitable for SOR. We take advantage of the sparseness by not even constructing the matrix. Instead, an iteration of the SOR method consists of applying Eq. [2.79] in a scalar manner that is consistent with Eq. [2.74a,b]. Note then that the only storage requirements are for the solution values $\phi_{p,q}$. Note also that the subscripts p and q do *not* indicate row and column positions of the coefficient matrix; rather, they refer to the physical location on the grid.

We begin the SOR solution by setting the initial estimate of the solution to zero at all grid nodes. Note that it is not generally necessary to initialize the solution to zero; however, this initialization includes the Dirichlet boundary conditions for the problem. The norm δ_k of the solution vector change from iteration $(k-1)$ to k is computed as the ∞-norm and monitored during the solution process. Iterations are ended when δ_k becomes smaller than 10^{-6}. A program for the SOR solution is given in Pseudocode 2–7.

Pseudocode 2–7   SOR solution of the torsion problem

Program:
\ Program for SOR solution of the torsion problem of Fig. 2–5 and Eq. [2.79]
Declare: ϕ \ two-dimensional solution array
For [$p = 0, 1, \ldots, 10$]   \ p = horizontal grid counter
   For [$q = 0, 1, \ldots, 6$]   \ q = vertical grid counter
      $\phi_{p,q} \leftarrow 0$ \ initialization
   End For   \ with counter q
End For   \ with counter p
Initialize: ω \ relaxation parameter
$k \leftarrow 0$ \ iteration counter
Repeat   \ start SOR iteration loop
   $k \leftarrow k + 1$ \ update iteration counter
   $\delta_k \leftarrow 0$ \ initialize norm for current iteration
   For [$p = 1, 2, \ldots, 9$]   \ sweep grid in p direction
      For [$q = 1, 2, \ldots, 5$] \ sweep grid in q direction
         temp $\leftarrow \phi_{p,q}$ \ save current ϕ value
         $\phi_{p,q} \leftarrow$ from Eq. [2.79]   \ Gauss–Seidel update
         $\phi_{p,q} \leftarrow \omega\phi_{p,q} + (1-\omega)$temp   \ SOR update
         $\delta_k \leftarrow \max(\delta_k, |\phi_{p,q} - \text{temp}|)$ \ update norm
      End For   \ with q counter
   End For   \ with p counter
   Write: k, δ_k \ monitor norm
Until [$\delta_k < 10^{-6}$]   \ end iteration loop
Write: ϕ \ output solution
End Program

The torsion problem that we are considering is a special case of a Dirichlet problem on a rectangular domain. The spectral radius for the Jacobi iteration matrix associated with Eq. [2.79] is known and is given by

[2.80] $$\rho(\mathbf{J}) = 0.5[\cos(\pi/M) + \cos(\pi/N)]$$

where M is the number of uniform horizontal grid spacings and N is the number of uniform vertical grid spacings. For our problem, M is equal to 10 and N is equal to 6; we may thus use Eq. [2.80], [2.76a], and [2.78] to compute

$$\rho(\mathbf{G}) = \rho^2(\mathbf{J}) = 0.825447; \quad \omega_{\text{opt}} = 1.41064$$

The convergence for our problem under the Jacobi, Gauss–Seidel, and optimum SOR schemes is illustrated in Fig. 2–6. Convergence to the level ($\delta_k < 10^{-6}$) requires 95 iterations for the Jacobi method, 52 iterations for the Gauss–Seidel method, and 19 iterations for the optimum SOR method.

The SOR method for the torsion problem requires one division (for Eq. [2.79]) and two multiplications (by ω and $[1 - \omega]$) for each of the 45 grid nodes at every

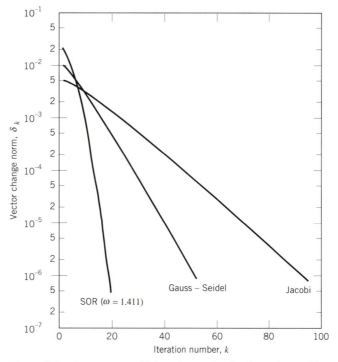

Figure 2–6 Convergence with iteration number for the torsion problem.

iteration. The cost of achieving a solution with SOR is 2565 operations over 19 iterations compared with 32385 operations by LU decomposition or Gauss elimination. The cost of computing the norm δ_k is relatively small, and the storage requirement is significantly less for SOR than for a direct method, which requires construction of the coefficient matrix.

In a general situation, we may not be able to compute $\rho(\mathbf{J})$ and thus ω_{opt} easily. An experimental approach to computing ω_{opt} is to start the SOR process as a Gauss–Seidel process with ω equal to 1. After a settling period of several iterations, the spectral radius $\rho(\mathbf{G})$ is approximated by (δ_k/δ_{k-1}) as stated in Eq. [2.73]. An illustration of this behavior is shown by the almost straight curves in the semilogarithmic plots of Fig. 2–6 after the first few iterations. Because the curves are not exactly straight, a good estimate of $\rho(\mathbf{G})$ is probably more likely to occur if we look at the vector change norm over several iterations; in other words, we may compute $\rho(\mathbf{G})$ from $(\rho^m(\mathbf{G}) = \delta_k/\delta_{k-m})$. The estimate of $\rho(\mathbf{G})$ may then be used in Eq. [2.78] to obtain an approximation for ω_{opt}, which is used for the remaining iterations.

A good estimate of ω_{opt} is especially beneficial in reducing the computational effort for very large systems of equations. It is better to overestimate ω_{opt} than to underestimate it. To understand this statement, let us look at the behavior of the spectral radius $\rho(\mathbf{S}(\omega))$ as ω varies. The SOR spectral radius for our torsion problem is obtained from Eq. [2.76b] with $\rho(\mathbf{G})$ equal to 0.825447 and with ω values ranging from 1 to 2.

The behavior of the spectral radius is shown in Fig. 2–7. The key features of the behavior shown in Fig. 2–7 are the infinite slope of the curve as ω_{opt} is approached from below, and the linear behavior for $\rho(\mathbf{S}(\omega))$ given by

[2.81] $$\rho(\mathbf{S}(\omega)) = \omega - 1; \qquad (\omega_{opt} \le \omega < 2)$$

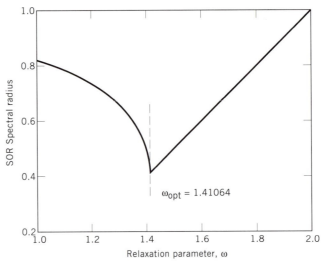

Figure 2–7   Behavior of the SOR spectral radius with ω.

A small underestimation of ω_{opt} will therefore cause a larger increase in the spectral radius than would a small overestimation of equal magnitude.

EXAMPLE   2.4

The torsion problem is started with the Gauss–Seidel method for 15 iterations. Norms of δ_{10} equal to 0.0283547 and δ_{15} equal to 0.0111593 are found.

Estimate the optimum relaxation parameter, and estimate the iteration number n at which δ_n first becomes smaller than 10^{-6} if SOR with the estimated value of ω_{opt} is used after iteration 15.

From Eq. [2.73]: $\delta_{15}/\delta_{10} = \rho^5(\mathbf{G}) \; \rightarrow \; \rho(\mathbf{G}) \simeq 0.829855$

From Eq. [2.78]: $\omega_{opt} \simeq 1.415943$

The true value of ω_{opt} is 1.41064; thus the approximate value overestimates the true value by 0.005303.

For $\omega \geq \omega_{opt}$, $\rho(\mathbf{S}(\omega))$ versus ω is linear; thus we have from Eq.[2.81]:

$$\rho(\mathbf{S}(1.415943)) = 0.415943$$

From Eq. [2.73]: $(0.415943)^{n-15} = \delta_n/\delta_{15} < 10^{-6}/0.0111593$

$\rightarrow n - 15 > \ln(10^{-6}/0.0111593)/ \ln(0.415943)$

$\rightarrow n - 15 > 10.62466$

Therefore: $n = 26$

Greater efficiencies can be achieved for the torsion problem whose solution is symmetric about the grid lines ($p = 5$) and ($q = 3$) of Fig. 2–5. We may use the symmetry to reduce the solution vector to 15 elements instead of the original 45. Computational effort is then reduced by more than a factor of 3 because the smaller problem generally has a faster rate of convergence.

The SOR method that we have described is a *point iterative* method because elements of the solution vector are updated in a point-by-point manner during each iteration. We may extend the method to block iterations in which a subset of the solution elements is updated simultaneously within an iteration. One popular form for the torsion problem of Fig. 2–5 is to update ϕ simultaneously at all grid intersections with a given q subscript. The method is then *line iterative*, and it uses the tridiagonal solver of §2.2.8.

2.3.4   The Conjugate Gradient Method

The *conjugate gradient* method (Ref. 6) was developed as long ago as 1952. It is rarely used as a primary method for solving linear systems; rather, its more common applications arise in solving differential equations and when other iterative methods converge very slowly.

The conjugate gradient method is a variational approach in which we seek to minimize a measure of the error in the solution of $(\mathbf{A} \cdot \mathbf{x} = \mathbf{f})$. Let $\mathbf{y}$ be a vector approximation of $\mathbf{x}$; then the function to be minimized is given in terms of the residual vector $\mathbf{r}$ and its transpose by

$$E(\mathbf{y}) - (1/2)(\mathbf{y} - \mathbf{x})^{\mathrm{T}} \cdot \mathbf{A} \cdot (\mathbf{y} - \mathbf{x}) = (1/2)\mathbf{r}^{\mathrm{T}} \cdot \mathbf{A}^{-1} \cdot \mathbf{r}; \qquad \mathbf{r} = \mathbf{f} - \mathbf{A} \cdot \mathbf{y}$$

$E(\mathbf{y})$ has a minimum value of zero when $\mathbf{y}$ is equal to $\mathbf{x}$.

The process is started by specifying an initial estimate $\mathbf{x}^{(0)}$ at iteration 0, and by computing the initial residual vector from

[2.82] $$\mathbf{r}^{(0)} = \mathbf{f} - \mathbf{A} \cdot \mathbf{x}^{(0)}$$

We then obtain improved estimates $\mathbf{x}^{(k)}$ from the iterative process

[2.83] $$\mathbf{x}^{(k)} = \mathbf{x}^{(k-1)} + \alpha_{k-1}\mathbf{p}^{(k-1)}$$

where $\mathbf{p}^{(k-1)}$ is a search direction expressed as a vector and α_k is chosen to minimize the value of $E(\mathbf{x}^{(k)})$.

In a related method called the **method of steepest descent**, $\mathbf{p}^{(k-1)}$ is chosen as the residual vector

$$\mathbf{p}^{(k-1)} = \mathbf{f} - \mathbf{A} \cdot \mathbf{x}^{(k-1)}$$

Such a choice tends to cause slow convergence after the first few iterations and tends to produce approximations $\mathbf{x}^{(k)}$ that oscillate about the solution. In the conjugate gradient method, we use $\mathbf{p}^{(0)}$ equal to $\mathbf{r}^{(0)}$ *only* at the beginning of the process. For all later iterations, we choose $\mathbf{p}^{(k-1)}$ to be *conjugate* to all previous direction vectors; that is,

$$(\mathbf{p}^{(k-1)})^{\mathrm{T}} \cdot \mathbf{A} \cdot \mathbf{p}^{(m)} = 0; \quad m = 0, 1, \ldots (k-2)$$

The computations of $\mathbf{p}^{(k-1)}$ and α_{k-1} for the conjugate gradient method are included in the following procedure.

■ **Procedure 2.1   The Conjugate Gradient Algorithm.**

1. Choose an initial estimate $\mathbf{x}^{(0)}$, compute $\mathbf{r}^{(0)}$ from Eq. [2.82], and set $\mathbf{p}^{(0)}$ equal to $\mathbf{r}^{(0)}$.

2. Compute $c = (\mathbf{r}^{(0)})^{\mathrm{T}} \cdot \mathbf{r}^{(0)}$

3. For iterations k equal to $(1, 2, \ldots)$ until convergence:

    a. Set $b = c$, compute $\mathbf{q} = \mathbf{A} \cdot \mathbf{p}^{(k-1)}$, and set $\alpha_{k-1} = b/[(\mathbf{p}^{(k-1)})^{\mathbf{T}} \cdot \mathbf{q}]$

    b. Compute $\mathbf{x}^{(k)}$ from Eq. [2.83]

    c. Compute $\mathbf{r}^{(k)} = \mathbf{r}^{(k-1)} - \alpha_{k-1}\mathbf{q}$

    d. Recompute $c = (\mathbf{r}^{(k)})^{\mathbf{T}} \cdot \mathbf{r}^{(k)}$

    e. Compute $\mathbf{p}^{(k)} = \mathbf{r}^{(k-1)} + (c/b)\mathbf{p}^{(k-1)}$ ■

Storage of the coefficient matrix $\mathbf{A}$ in the conjugate gradient procedure is not required if it is otherwise possible to construct the product $\mathbf{A} \cdot \mathbf{p}^{(k-1)}$ in Step 3a. We also note that the vectors $\mathbf{r}^{(k)}$ that are computed in Step 3c are ***not*** residual vectors, and vectors in a current iteration may overwrite those from the previous iteration.

The conjugate gradient method is applicable to symmetric, positive definite coefficient matrices. Convergence for a system of n equations occurs within n iterations if the procedure is performed with infinite precision. More than n iterations may be required in practical applications because of the introduction of round-off errors.

2.4 CLOSURE

The importance of systems of linear algebraic equations is reflected in the variety of methods that have been developed for their solution; indeed, we shall encounter such systems in many of the remaining parts of the book. The fundamental rules of linear algebra that help us to understand the methods of solution are presented in Section 2.1.

The direct Gauss elimination method and the iterative successive overrelaxation method are among the most popular because they are generally effective and are easily understood and implemented. The Gauss–Jordan variant of Gauss elimination is well suited to obtaining inverses of matrices at only a small extra cost. LU decomposition, another variant of Gauss elimination, is especially useful when the same coefficient matrix is used in repeated applications. The iterative residual correction procedure in Section 2.3 is based on some form of LU decomposition.

Special methods for systems with symmetric, positive definite matrices or with tridiagonal matrices yield dramatic reductions in computational effort. We may achieve even more efficiency if we are able to reduce sparse systems to easily solved block forms. Another effective approach for sparse systems with appropriate properties is to use iterative methods.

The successive overrelaxation method, or SOR, and its relation to the more fundamental Jacobi and Gauss–Seidel methods are discussed in Section 2.3. The success of SOR depends on making a good estimate of an optimum relaxation parameter. The need for estimating parameters is removed in the conjugate gradient method, which, although more complicated to code, can rival the SOR method in efficiency when dealing with large, sparse systems.

Ill-conditioned systems, discussed in Section 2.2, will produce unacceptable solutions if inadequate precision is used. The best way to deal with ill-conditioning is to avoid it by reformulating the problem.

2.5 EXERCISES

1. A cartesian coordinate system (x, y, z) is obtained from another such system (X, Y, Z) by a rotation α about the Z axis, a rotation β about the Y axis, and a rotation γ about the X axis. The coordinates in the two systems are related by

$$\begin{bmatrix} x \\ y \\ z \end{bmatrix} = \mathbf{A} \cdot \mathbf{B} \cdot \mathbf{C} \cdot \begin{bmatrix} X \\ Y \\ Z \end{bmatrix}$$

where

$$\mathbf{A} = \begin{bmatrix} 1 & 0 & 0 \\ 0 & \cos\alpha & \sin\alpha \\ 0 & -\sin\alpha & \cos\alpha \end{bmatrix}$$

$$\mathbf{B} = \begin{bmatrix} \cos\beta & 0 & -\sin\beta \\ 0 & 1 & 0 \\ \sin\beta & 0 & \cos\beta \end{bmatrix}$$

$$\mathbf{C} = \begin{bmatrix} \cos\gamma & \sin\gamma & 0 \\ -\sin\gamma & \cos\gamma & 0 \\ 0 & 0 & 1 \end{bmatrix}$$

   (a) Develop a general matrix for the product $\mathbf{A} \cdot \mathbf{B} \cdot \mathbf{C}$.

   (b) State an inverse of $\mathbf{A} \cdot \mathbf{B} \cdot \mathbf{C}$ by considering the fact that (X, Y, Z) is obtained from (x, y, z) by reversing the rotations (i.e., by using rotation angles $-\alpha, -\beta$, and $-\gamma$). Verify your statement by performing the matrix multiplication $(\mathbf{A} \cdot \mathbf{B} \cdot \mathbf{C}) \cdot (\mathbf{A} \cdot \mathbf{B} \cdot \mathbf{C})^{-1}$.

2. Use Cramer's rule to solve Eq. [2.53a, b] for the currents i_1 through i_4 with V_1 equal to 3 volts, V_2 equal to 9 volts, R_k equal to 5 ohms when k is odd, and R_k equal to 10 ohms when k is even.

3. Solve Problem 2 by Gauss elimination.

4. Solve Eq. [2.61a, b] by Gauss elimination when T is equal to 100 N and p is equal to 60 degrees.

5. Repeat Problem 4 with the Gauss–Jordan method, but with p equal to 75 degrees.

6. The performance parameters of a normally aspirated (neither supercharged nor turbocharged) car engine are the maximum torque T_m [lb·ft] and maximum power P_m [hp] along with the respective engine speeds k_t and k_p [thousands of rpm] at which they occur. An *ad hoc* procedure for computing the torque T and power P for a range of engine speeds k up to the redline speed k_m is now described.

T and P are related by

$$P = \pi k T / 16.5$$

At low and high engine speeds we assume the parabolic variations

$$
\begin{aligned}
T &= T_m + \alpha(k - k_t)^2; & \alpha &= -T_m/(k_t)^2; & 0 &\le k < k_t \\
P &= P_m + \beta(k - k_p)^2; & \beta &= -P_m/(k_p)^2; & k_p &< k \le k_m
\end{aligned}
$$

For engine speeds in the range $(k_t \le k \le k_p)$, we use a Taylor series expansion about k_p, noting that dP/dk is zero and d^2P/dk^2 is 2β at k_p, to obtain

$$P = P_m + \Delta^2\beta + (\Delta^3/6)P^{iii} + (\Delta^4/24)P^{iv} + (\Delta^5/120)P^v$$

in which Δ is equal to $(k - k_p)$ and the Roman numeral superscripts denote the derivatives of P with respect to k. Similar expansions for the first and second derivatives allow us to match the values from the expansions to those from the low-speed variation at k_t. The result is a system of equations

$$
\begin{bmatrix}
D^3/6 & D^4/24 & D^5/120 \\
D^2/2 & D^3/6 & D^4/24 \\
D & D^2/2 & D^3/6
\end{bmatrix}
\cdot
\begin{bmatrix}
P^{iii} \\
P^{iv} \\
P^v
\end{bmatrix}
=
\begin{bmatrix}
f_1 \\
f_2 \\
f_3
\end{bmatrix}
$$

in which

$$
\begin{aligned}
D &= k_t - k_p; \quad f_1 = \pi k_t T_m/16.5 - P_m - \beta D^2; \\
f_2 &= \pi T_m/16.5 - 2\beta D; \quad f_3 = 2\pi\alpha k_t/16.5 - 2\beta
\end{aligned}
$$

Use Gauss elimination to solve the system when T_m is 200 lb·ft, P_m is 195 hp, k_t is 3.5 thousand rpm, and k_p is 5.75 thousand rpm, and thus compute P for k from 1 to 6 in increments of 0.1.

7. Use the Gauss–Jordan method to solve the system in Problem 6.

8. Use the Gauss–Jordan method to obtain the inverse of the matrix **A** in Eq. [2.42b].

9. Use the Gauss–Jordan method to obtain the inverse of the coefficient matrix in Eq. [2.66a].

10. Find the Doolittle reduction factors **L** and **U** for the matrix in Eq. [2.61b].

11. Solve Problem 6 by the Doolittle reduction method.

12. Let **L** and **U** be the Doolittle reduction factors for a matrix **A**. Explain how to change **L** and **U** into the Crout reduction factors.

13. A polynomial passes through the points (3.0,2.0), (4.0,3.1), (5.0,4.4), and (6.0,6.0) in an (x, y) coordinate system. The polynomial gives y in terms of x by

$$y = a_1 + a_2 x + a_3 x^2 + a_4 x^3$$

Set up the system of equations and solve it for the coefficients a_1 through a_4 by Gauss elimination or by LU Decomposition. The matrix for this problem is a (4×4) *Vandermonde matrix*. Larger matrices of this type tend to be ill conditioned.

14. The supports at R and S for the truss in Fig. 2–1 are changed. S is now a pin support, which has two force components, and R is a wall anchor that provides a horizontal restraining force. Reformulate the problem and solve it for the internal and external forces by Gauss elimination or by LU decomposition when c is equal to 45 degrees, p is equal to 90 degrees, and the tension T in each of the cables supporting the sign is 300 N.

15. Consider Eq. [2.53a,b] with $R_1 = R_2 = R_3 = 5$ ohms, $R_5 = R_6 = R_7 = R_8 = 2$ ohms, $R_4 = 0$, and $V_1 = 3V_2 = 9$ volts. Solve for the currents i_1 through i_4 by Cholesky decomposition.

16. Repeat Problem 15 with the Thomas algorithm for tridiagonal matrices.

17. The nonzero elements of a pentadiagonal matrix are restricted to the diagonal and to two columns on either side of the diagonal element for a given row. Use the Thomas algorithm as a guide, and develop an algorithm for solving systems with pentadiagonal coefficient matrices.

18. Use a triangular block method to solve Eq. [2.61a,b,c] when T is equal to 80 lb and p is equal to 75 degrees.

19. The torsion problem described by Eq. [2.79] and Fig. 2–5 in §2.3.3 has a solution that is symmetrical about the grid lines ($p = 5$) and ($q = 3$). The boundary condition is that ϕ is zero at all points on the perimeter of the cross section in Fig. 2–5. Reformulate the problem so that only the grid intersections given by

$$p = 0, 1, 2, 3, 4, 5; \qquad q = 0, 1, 2, 3$$

are used in Jacobi, Gauss–Seidel, or successive overrelaxation iterations.

   (a) Solve the problem by either the Jacobi or Gauss–Seidel method. Monitor the 1-norm δ_k (Eq. [2.72b]) and stop iterating when δ_k becomes smaller than or equal to 10^{-5}.

   (b) Use the information for δ_k in Part (a) to estimate an optimum relaxation parameter ω_{opt}. Use this parameter to solve the problem again by SOR.

20. Points on the domain shown in Fig. 2–P20 are denoted by grid intersections (p, q). A variable u satisfies the Laplace equation

$$\partial^2 u / \partial x^2 + \partial^2 u / \partial y^2 = 0$$

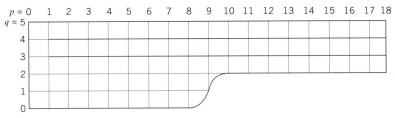

Figure 2–P20

and is approximated at all interior points of the domain by

$$4u_{p,q} - u_{p+1,q} - u_{p-1,q} - u_{p,q+1} - u_{p,q-1} = 0$$

Dirichlet boundary conditions are

$$
\begin{aligned}
u_{p,5} &= 30; & p &= 0, 1, \ldots, 18 \\
u_{p,0} &= 0 \ ; & p &= 0, 1, \ldots, 8 \\
u_{9,1} &= 0 \\
u_{p,2} &= 0 \ ; & p &= 10, 11, \ldots, 18
\end{aligned}
$$

and derivative conditions are

$$
\begin{aligned}
(\partial u/\partial x)_{0,q} &= 0; & q &= 1, 2, 3, 4 \\
(\partial u/\partial x)_{18,q} &= 0; & q &= 3, 4
\end{aligned}
$$

The derivative conditions are approximated by

$$
\begin{aligned}
4u_{0,q} - 2u_{1,q} - u_{0,q+1} - u_{0,q-1} &= 0; & q &= 1, 2, 3, 4 \\
4u_{18,q} - 2u_{17,q} - u_{18,q+1} - u_{18,q-1} &= 0; & q &= 3, 4
\end{aligned}
$$

This formulation may be used to model heat conduction in a plate with u representing the temperature. The derivative conditions correspond to insulated left and right boundaries. The formulation may also be used to model ideal fluid flow in a two-dimensional duct with u representing a stream function, which is constant on a streamline. The derivative conditions correspond to a flow with no y component of velocity at the inlet on the left and at the outlet on the right.

Solve the problem for 10 iterations by the Gauss–Seidel method and then to convergence by SOR. Monitor the norm δ_k and compute the Gauss–Seidel spectral radius after iteration 10 from

$$\rho(\mathbf{G}) = (\delta_{10}/\delta_6)^{0.25}$$

Use the value of $\rho(\mathbf{G})$ to estimate an optimum value of the SOR relaxation parameter, and use this estimate for the SOR iterations.

21. Use the conjugate gradient method to solve Eq. [2.53a,b] for the currents i_1 through i_4 with V_1 equal to 3 volts, V_2 equal to 9 volts, R_k equal to 2 ohms when k is odd, and R_k equal to 5 ohms when k is even.

3

Nonlinear Algebraic Equations

Many engineering phenomena are described by nonlinear algebraic equations; therefore, the determination of the roots of those equations is a fundamental problem in engineering analysis. Numerical methods are used to solve nonlinear algebraic equations when the equations prove intractable to ordinary mathematical techniques. These numerical methods are all iterative, and they may be used for equations that contain one or several variables. The procedures for all of the methods are broken down into major steps. Most of the procedures are supplemented by pseudocodes (see Appendix B).

Six methods for solving nonlinear equations in a single variable are presented in Section 3.1. These are the simple incremental-search and fixed-point iteration methods, the bisection and false-position bracketing methods, and the Newton–Raphson and secant methods which are based on calculus.

The calculus-based Newton–Raphson method is extended in Section 3.2 to deal with multivariable systems of nonlinear equations. The extended version is combined with synthetic division in Section 3.3 to develop Bairstow's method for obtaining real and complex roots of polynomial equations.

Several applied problems are used to illustrate the methods in this chapter. These deal with the ***deflection of a cantilevered beam***, determination of the ***critical Mach number*** for an airfoil, ***van der Waals equation of state***, the geometry of a ***catenary***, and ***natural frequencies of a spring–mass system***.

3.1 METHODS FOR EQUATIONS IN A SINGLE VARIABLE

The methods described in this section are aimed at solving equations that contain a single variable. We assume that the equation to be solved is written in the form

[3.1] $$f(x) = 0$$

A root of the general equation in Eq. [3.1] is a value of x that satisfies the equation; methods for solving the equation are thus called ***root-solving methods***.

3.1.1 The Incremental-Search Method

The ***incremental-search*** method is a numerical analog of finding a root of Eq. [3.1] by plotting $f(x)$ versus x to see where $f(x)$ crosses the x axis. The procedure is as follows.

■ **Procedure 3.1   The incremental-search method.**
   1. Set a counter i to zero, choose a starting value x_0, choose an increment h, and compute a reference value f_0 equal to $f(x_0)$.
   2. Increase i by 1, set x_i equal to $(x_0 + ih)$, and compute $f(x_i)$.
   3. If $\{f_0[f(x_i)]\} > 0$, return to Step 2; otherwise, go on to Step 4.
   4. Estimate the root x from: $x = x_i - h[f(x_i)]/[f(x_i) - f(x_i - h)]$ ■

The idea in Procedure 3.1 is to begin at a starting point x_0 and search along the x axis in increments h until we find two successive points $(x_i - h)$ and x_i whose line segment contains a root. This segment is indicated when the function first becomes zero or changes sign; that is, when the product $[f_0 f(x_i)]$ first becomes less than or equal to zero. The root is then estimated by linear interpolation in Step 4.

The magnitude of the increment h represents a crude error bound for the estimated solution because it is the length of the segment on which the exact solution lies. The sign of h dictates the search direction from the starting point x_0. If no solution exists in that direction, the method will obviously fail. The number of iterations should be limited to prevent infinite looping in such an instance. Failure and inifinite looping can also occur when h is large in magnitude and the equation has two closely spaced roots. If the line segment from $(x_i - h)$ to x_i contains both roots, the function values at the search points would have the same sign. The test of Step 3 would then fail to indicate the presence of the roots.

Another condition that may cause failure is a discontinuity. Consider, for example, the function

$$f(x) = \ln x - 1/(x - 1)$$

The function is discontinuous when x is equal to 1. It tends to $(-\infty)$ as x approaches 1 from above, and it tends to $(+\infty)$ as x approaches 1 from below. Application of the incremental search procedure to solving $f(x) = 0$ could cause the existence of a root to be falsely indicated when two search points lie on the opposite sides of $(x = 1)$.

To illustrate the incremental search method, we consider the deflection of a horizontal **cantilevered beam** when it is subjected to a uniform, vertical load. A beam extending from its clamped end $(x = 0)$ to its free end $(x = L)$ has a maximum deflection δ_{max} at $(x = L)$. The deflection δ at location $(x = \alpha L)$ is related to δ_{max} by

[3.2] $$f(\alpha) = \alpha^4 - 4\alpha^3 + 6\alpha^2 - 3\delta/\delta_{max} = 0$$

The determination of α for a given value of δ/δ_{max} in the range $(0 \leq \delta/\delta_{max} \leq 1)$ is shown in Example 3.1.

EXAMPLE 3.1

Use the incremental search method to solve Eq. [3.2] for the value of α at which δ/δ_{max} is equal to 0.75.

From the physical problem, we expect that a solution for α between 0 and 1 exists and is closer to 1 than it is to 0. We therefore choose a starting value α_0 equal to 1 and use a negative increment h. Too small a magnitude for h will cause excessive computations; on the other hand, too large a magnitude may result in a poor estimate of α from the final interpolation. A reasonable choice for h is (-0.05).

The results of the search with α_0 equal to 1, f_0 equal to 0.75 when computed from Eq. [3.1] with δ/δ_{max} equal to 0.75, and h equal to (-0.05) are tabulated as follows.

i	α_i	$f(\alpha_i)$	$f_0\, f(\alpha_i)$
1	0.95	0.550006	> 0
2	0.90	0.350100	> 0
3	0.85	0.150506	> 0
4	0.80	−0.048400	< 0

The interpolation of Step 4 in Procedure 3.1 now gives the estimate

$$\alpha \simeq 0.8 - (-0.05)(-0.0484)/(-0.0484 - 0.150506) \simeq 0.81217$$

The incremental-search algorithm is described in Pseudocode 3–1. It includes a limit on the number of search iterations and a simple check on the solution that is found.

The incremental-search method is simple to use; however, an excessive number of iterations may be required to perform a search with high resolution.

Pseudocode 3–1   The incremental-search method

Initialize: x_0, h, and N \ N = maximum number of increments

  $f_0 \leftarrow f(x_0)$ \ reference value of f
  $i \leftarrow 0$ \ initial counter

Repeat   \ search loop

  $i \leftarrow i + 1$ \ update counter
  $x \leftarrow x_0 + ih$ \ x at current search point
  $f_x \leftarrow f(x)$ \ function at x
  $p \leftarrow (f_0)(fx)$ \ product to be used in test to terminate loop

Until $[i = N$ **or not** $(p > 0)]$

If $[p > 0]$ **then**   \ because loop terminated with $i = N$
  Write: 'Solution not found' message

Else
  $x \leftarrow x - (h)(fx)/[fx - f(x - h)]$ \ interpolation
  $f \leftarrow f(x)$
  Write: The estimated solution x and the check value f

End If

3.1.2   Fixed-Point Iteration

The *fixed-point iteration method*, also known as the *successive approximation method*, requires us to rewrite the equation $[f(x) = 0]$ in the form

$$x = g(x)$$

The procedure starts from an initial estimate or guess of x, which is improved by iteration until convergence is achieved. For convergence to occur, the derivative (dg/dx) must be smaller than 1 in magnitude (at least for the x values that are encountered during the iterations). We shall establish convergence by requiring the change in x from one iteration to the next to be no greater in magnitude than some small quantity ϵ. The method is outlined in Procedure 3.2 and is applied to the beam deflection problem described earlier by Eq. [3.2] of §3.1.1

■ **Procedure 3.2   Algorithm for fixed-point iteration.**

  1. Guess a starting value x_0 and choose a convergence parameter ϵ.

  2. Compute an improved value x_{imp} from: $x_{imp} = g(x_0)$.

  3. If $|x_{imp} - x_0| > \epsilon$, set x_0 equal to x_{imp} and return to Step 2; otherwise, x_{imp} is the approximate root. ■

EXAMPLE 3.2

Use the fixed-point iteration method to solve Eq. [3.2] for the value of α at which δ/δ_{max} is equal to 0.75. Start with α_0 equal to 0.75, and use the criterion $|x_{imp} - x_0| \le 10^{-5}$ to indicate convergence.

Rewrite Eq. [3.2] in the form: $\alpha = g(\alpha) = \sqrt{(3\delta/\delta_{max} + 4\alpha^3 - \alpha^4)/6}$.

Then $\alpha_{imp} = g(\alpha_0)$ in accord with Step 2 of Procedure 3.2.

The sequence of α_{imp} values is tabulated for iteration numbers denoted by i.

i	α_0	α_{imp}	i	α_0	α_{imp}
1	0.750000	0.776863	9	0.811333	0.811682
2	0.776863	0.791745	10	0.811682	0.811889
3	0.791745	0.800240	11	0.811889	0.812011
4	0.800240	0.805166	12	0.812011	0.812084
5	0.805166	0.808048	13	0.812084	0.812127
6	0.808048	0.809743	14	0.812127	0.812152
7	0.809743	0.810742	15	0.812152	0.812167
8	0.810742	0.811333	16	0.812167	0.812176

The last computed value of α_{imp} is the estimated root: $\alpha \simeq 0.812176$

Important considerations for the fixed-point iteration method are the choice of the function $g(x)$ and the choice of the convergence parameter ϵ. The function $g(x)$ must yield convergence, and ϵ must be chosen to be compatible with the precision of the machine. A discussion on convergence criteria is deferred to §3.1.7 because the topic is also applicable to some of the other root-solving methods.

A code for the fixed-point iteration method is given in Pseudocode 3–2. A limit is placed on the number of iterations to prevent infinite looping in nonconvergent cases.

Pseudocode 3–2   The fixed-point iteration method

Initialize: starting value x_0
Initialize: ϵ and N \ N = limit on the number of iterations
    $i \leftarrow 0$ \ i = iteration counter
Repeat   \ iteration loop
        $i \leftarrow i + 1$ \ update counter
    $x_{imp} \leftarrow g(x_0)$ \ improved x
        $q \leftarrow |x_{imp} - x_0|$ \ used to test for convergence
        $x_0 \leftarrow x_{imp}$ \ reset x_0 to last value of x_{imp}
Until [$i = N$ **or not**($q > \epsilon$)]

If [$q > \epsilon$] **then**   \ because loop terminated with $i = N$
    Write: 'Nonconvergence to required limit' message
Else
    Write: Solution x_{imp} and check value $f(x_{imp})$
End If

In the event of nonconvergence, a history of x_{imp} values could be used to determine whether the method is convergent. If it is, but convergence is not attained, either the limit on the number of iterations or the criterion ϵ (or both) may be unrealistically small and should be modified.

The major advantage of the point-iterative method is ease of application when an appropriate function $g(x)$ can be found. Unfortunately, it is often impossible to find a function that produces convergence. Even in cases where convergence can be achieved, the convergence rate may be slow; it is therefore important to recognize that failure to converge to a specified limit may simply be the result of an iteration limit that is too small.

3.1.3   The Bisection Method

The **bisection** method is classified as a bracketing method. It is applicable to equations of the form $[f(x) = 0]$ when we can find two limiting values x_a and x_b such that the function $f(x)$ changes sign **once** for x values in the range ($x_a \le x \le x_b$). The limiting values therefore **bracket** the root.

The bisection method is also known as the **interval halving** method because the strategy is to bisect or halve the interval from x_a to x_b and then to retain the half interval whose ends still bracket the root. The fact that the function is required to change sign only once gives us a way to determine which half interval to retain; we keep the half on which $f(x)$ changes sign or becomes zero.

We can reduce the size of the interval that contains the root to an arbitrarily small value by iteratively applying the bisection process. The process ends when the interval becomes smaller than or equal to some prescribed value. The bisection method and an added interpolation step to obtain a refined estimate of the solution are described in Procedure 3.3, and the interval halving process is illustrated in Fig. 3–1.

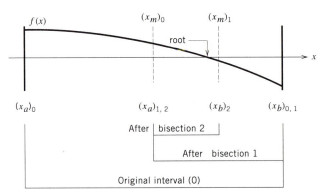

Figure 3–1   Illustration of the bisection method.

■ **Procedure 3.3 The bisection method.**

1. Choose limiting values x_a and x_b (with $x_b > x_a$).
2. Compute either $f_a = f(x_a)$ or $f_b = f(x_b)$.
3. Compute the interval midpoint $x_m = (x_a + x_b)/2$, and compute $f_m = f(x_m)$.
4. Use either (i) or (ii) depending on which of f_a or f_b is available from Step 2:

    (i) If $(f_a \, f_m) > 0$, reset x_a to x_m; otherwise, reset x_b to x_m
    (ii) If $(f_b \, f_m) > 0$, reset x_b to x_m; otherwise, reset x_a to x_m

5. If $(x_b - x_a)$ is sufficiently small, that is, less than or equal to some small prescribed quantity ϵ, proceed to Step 6; otherwise, return to Step 3.
6. Use linear interpolation to estimate the root x from either of

$$x = x_a - (x_b - x_a)f(x_a)/[f(x_b) - f(x_a)]$$

or

$$x = x_b - (x_b - x_a)f(x_b)/[f(x_b) - f(x_a)] \quad ■$$

The major step in Procedure 3.3 is Step 4 in which we determine which half of the previous interval we are to retain. Because there is only one root on the initial interval from x_a to x_b, f_m has the same sign as f_a if x_m and x_a are on the same side of the root, and f_m has the same sign as f_b if x_m and x_b are on the same side of the root. The product tests in Step 4 are to determine if f_m has the same sign as f_a or f_b rather than to find the actual numerical value of the product; it is therefore unnecessary to recompute $f(x_a)$ or $f(x_b)$ with each bisection. Note, however, that the actual values of $f(x_a)$ and $f(x_b)$ are required for the linear interpolation of Step 6.

The interval ϵ used to end the bisection process can be chosen directly; alternately, the number of iterations required to produce the condition

$$(x_b - x_a) \le \epsilon$$

can be computed. We note that the initial interval $(x_b - x_a)_{\text{init}}$ is reduced to a final interval $(x_b - x_a)_{\text{final}}$ given by

$$(x_b - x_a)_{\text{final}} = [(x_b - x_a)_{\text{init}}]/2^n$$

where n is the number of bisections. Therefore,

$$(x_b - x_a)_{\text{final}} \le \epsilon \rightarrow 2^n \ge [(x_b - x_a)_{\text{init}}]/\epsilon$$
$$\rightarrow n \ge \ln\{[(x_b - x_a)_{\text{init}}]/\epsilon\}/(\ln 2)$$

The final interval contains the root and its length is thus an upper bound on the error in estimating the solution.

We demonstrate the use of the bisection method by considering the problem of finding the **critical Mach number** for an aircraft. Mach number refers to the ratio of airspeed to the speed of sound. Subsonic aircraft (flying at Mach numbers less than 1) experience accelerated air flow over wing surfaces. The critical Mach number is the flight Mach number at which the flow at some point on the wing reaches the speed of sound. Flight Mach numbers of only about 2 to 5% above the critical produce significant increases in the drag.

The minimum pressure coefficient C_p on an airfoil is usually defined so that it is negative and corresponds to the maximum flow speed on the airfoil. At the critical Mach number M, the expression for C_p (in air) is

$$C_p = \{[(2 + 0.4M^2)/2.4]^{3.5} - 1\}/\{0.7M^2\}$$

Preliminary tests for an airfoil may be done at low speeds where compressibility effects are negligible. We assume that the minimum pressure coefficient C_{pi} is obtained for incompressible flow, and relate it to C_p through the Kármán–Tsien relation

$$C_p/C_{pi} = \left\{ \sqrt{1 - M^2} + (M^2 C_{pi}/2) \Big/ \left[1 + \sqrt{1 - M^2}\right] \right\}^{-1}$$

To determine M, we substitute the expression for C_p into the Kármán–Tsien relation and solve the resulting equation for M. The equation to be solved is

[3.3] $$f(M) = \{[(2 + 0.4M^2)/2.4]^{3.5} - 1\}/\{0.7M^2 C_{pi}\}$$

$$- \left\{ \sqrt{1 - M^2} + (M^2 C_{pi}/2) \Big/ \left[1 + \sqrt{1 - M^2}\right] \right\}^{-1} = 0$$

Let us assume that we are given a value of (-0.383) for the incompressible pressure coefficient C_{pi}, and that we are seeking the value of M, which we expect to be somewhere between 0.6 and 0.9. The behavior of the function $f(M)$ in Eq. [3.3] is important if we are to use the bisection method. The first term in $f(M)$ (corresponding to the left side of the Kármán–Tsien relation) has a value of $(+\infty)$ when M is zero and C_{pi} is negative, and decreases as M increases. The second term in $f(M)$ (corresponding to the right side of the Kármán–Tsien relation) increases with M from a finite value until it becomes $(+\infty)$ when M^2 is equal to $[4(1 - C_{pi})/(2 - C_{pi})^2]$. Thus for a value of C_{pi} equal to (-0.383), the function $f(M)$ decreases from $(+\infty)$ to $(-\infty)$ as M increases from 0 to 0.987. We cannot of course work with infinite values on a computer; therefore, we must place some restrictions on our initial bracketing values for the bisection method.

EXAMPLE 3.3

Use the bisection method to solve Eq. [3.3] when C_{pi} is equal to (-0.383). Use limit values $(M_a = 0.18)$ and $(M_b = 0.98)$, and stop bisecting when $(M_b - M_a)$ becomes smaller than or equal to 0.01.

At $M_b = 0.98$, $f(M_b)$ is negative (its value is -21.8382). We therefore reset M_b to M_m when f_m is negative, and reset M_a to M_m when f_m is positive or zero. The results of the computations are given in a table that shows the starting values M_a and M_b, the interval midpoint M_m, and the function value $f(M_m)$.

Bisection	Starting values of M_a	M_b	M_m	$f(M_m)$
1	0.18000	0.98000	0.58000	2.44757
2	0.58000	0.98000	0.78000	-0.51476
3	0.58000	0.78000	0.68000	0.79287
4	0.68000	0.78000	0.73000	0.12313
5	0.73000	0.78000	0.75500	-0.19607
6	0.73000	0.75500	0.74250	-0.03705
7	0.73000	0.74250	0.73625	0.04284

After bisection 7, M_a is 0.73625 and M_b is 0.74250; thus $(M_b - M_a) \leq 0.01$. The interpolation in Step 6 of Procedure 3.3 yields the estimated solution

$$M \simeq 0.73960, \text{ at which } f(M) = -4.3062 \times 10^{-5}$$

Pseudocode 3–3 gives one version of the bisection method. Other forms of logic are possible as indicated in Procedure 3.3.

Pseudocode 3–3   The bisection method

Initialize: x_a, x_b, ϵ \qquad\qquad $\backslash$ $x_a < x_b$
$\quad f_b \leftarrow f(x_b)$ \qquad\qquad $\backslash$ reference function value
While $[(x_b - x_a) > \epsilon]$
$\quad x_m(x_a + x_b)/2$ \qquad\qquad $\backslash$ interval midpoint
$\quad$ **If** $[(f_b \, f(x_m)) > 0]$ **then**
$\qquad x_b \leftarrow x_m$ \qquad\qquad $\backslash$ reset right limit
$\quad$ **Else**
$\qquad x_a \leftarrow x_m$ \qquad\qquad $\backslash$ reset left limit
$\quad$ **End If**
End While

$\quad x \leftarrow x_b - (x_b - x_a)f(x_b)/[f(x_b) - f(x_a)]$ \quad $\backslash$ interpolate
$\quad f \leftarrow f(x)$
Write: Estimated solution x and check value f

The bisection method allows a solution to be obtained to an arbitrary degree of accuracy (indicated by ϵ) within the limits of precision imposed by the computer. Its major disadvantages are relatively slow convergence and the requirement that there be only one solution on the line segment contained by the initial limits x_a and x_b. If the method is incorrectly applied to a segment on which there is no solution, x_a and x_b will approach one of the original limits. If more than one solution exists on the starting segment, one of them may be found or x_a and x_b may approach some common value that does not correspond to any approximate solution.

3.1.4   The False-Position Method

The *false-position* method may be thought of as an attempt to improve the convergence characteristics of the bisection method. We begin with limiting values x_a and x_b such that $f(x)$ changes sign only *once* on the interval from x_a to x_b. An approximate root is found by linear interpolation between x_a and x_b and serves as an intermediate value x_{int}. The new interval containing the root is now either from x_a to x_{int} or from x_{int} to x_b. The logic for determining which interval is retained is the same as in the bisection method. Because the intermediate point x_{int} is found by interpolation, the separate interpolation step used at the end of the bisection method is now unnecessary; instead, the last computed value of x_{int} is taken to be the approximate solution. The method is detailed in Procedure 3.4, and it is illustrated in Fig. 3–2.

■ **Procedure 3.4   The false-position method.**
1. Choose limiting values x_a and x_b (with $x_b > x_a$).
2. Compute either $f_a = f(x_a)$ or $f_b = f(x_b)$, and set a counter i to zero.
3. Increase i by 1, and compute the intermediate point x_{int} from either

$$x_{int} = x_a - (x_b - x_a)f(x_a)/[f(x_b) - f(x_a)]$$

or

$$x_{int} = x_b - (x_b - x_a)f(x_b)/[f(x_b) - f(x_a)]$$

4. Compute $f_{int} = f(x_{int})$
5. Use either (i) or (ii) depending on which of f_a or f_b is available from Step 2:
   (i)   If $(f_a\ f_{int}) > 0$, reset x_a to x_{int}; otherwise, reset x_b to x_{int}
   (ii)   If $(f_b\ f_{int}) > 0$, reset x_b to x_{int}; otherwise, reset x_a to x_{int}
6. If $|f(x_{int})|$ is sufficiently small, that is, less than or equal to some small prescribed quantity ϵ, or if i reaches an iteration limit N, take x_{int} as the approximate root; otherwise, return to Step 3. ■

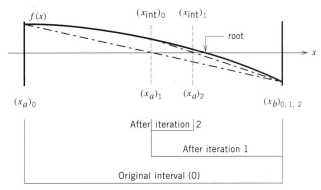

Figure 3–2 Illustration of the false-position method.

Although the interval from x_a to x_b is reduced with each iteration, it is not always reducible to an arbitrarily small value as in the bisection method. In many cases, only one limiting value, x_a or x_b, will approach the root while the other remains fixed at its original position. The proximity of at least one limiting value to the root still allows us to estimate the root accurately even though the interpolation interval may be large. Because we cannot rely on reduction of the interval size, we test for convergence by evaluating the magnitude of the function, and we impose an iteration limit N so that too stringent a convergence criterion does not cause infinite looping.

The false-position method is now applied in Example 3.4 to the same problem as in Example 3.3. Example 3.4 demonstrates a case in which only one of the limiting values changes.

EXAMPLE 3.4

Use the false-position method to solve Eq. [3.3] when C_{pi} is equal to (-0.383). Use limit values $(M_a = 0.18)$ and $(M_b = 0.98)$, and end iterations either when $|f(M_{int})|$ becomes smaller than or equal 10^{-2} or when 9 iterations have been completed.

At $M_b = 0.98$, $f(M_b)$ is negative (its value is -21.8382). The value of M_{int} is found by interpolation as in Step 3 of Procedure 3.4; the logic of Step 5 is then used to reset the interval for the next iteration. Results of the computations are shown in tabular form.

Iteration	Starting values of M_a	M_b	M_{int}	$f(M_{int})$
1	0.18000	0.98000	0.74306	−0.04414
2	0.18000	0.74306	0.74258	−0.03804
3	0.18000	0.74258	0.74217	−0.03278
4	0.18000	0.74217	0.74181	−0.02825
5	0.18000	0.74181	0.74151	−0.02435
6	0.18000	0.74151	0.74124	−0.02099
7	0.18000	0.74124	0.74101	−0.01809
8	0.18000	0.74101	0.74082	−0.01560
9	0.18000	0.74082	0.74065	−0.01345

The process is stopped by the iteration limit of 9. The estimated root is

$$M \simeq 0.74065, \qquad \text{at which} \quad f(M) = -0.01345$$

We see from Example 3.4 that although the false-position method initially converges more quickly than the bisection method in Example 3.3, it eventually fails in this case to give as good a solution as the bisection method. In addition, the computation of the intermediate point requires more effort than the midpoint computation for the bisection method, and the final interval of the false-position method may not be useful as a bound on the error. It does not seem, therefore, that the false-position method provides any major advantages over the bisection method; nevertheless, the concepts associated with it are conceptually useful in the development of other methods such as the secant method in Section 3.1.6.

In general, any advantage of the false-position method over the bisection method requires more knowledge about the bracketing values. If the root is closer to the bracketing value at which the magnitude of the function value is smaller, the false-position method will converge more quickly than the bisection method with the same bracketing values. Note, however, that knowledge of this condition may not be discernible *a priori*.

A modified form of the false-position method uses an interpolation from $(x_a, \omega f(x_a))$ to $(x_b, f(x_b))$ if x_b is reset at the end of a previous iteration, or an interpolation from $(x_a, f(x_a))$ to $(x_b, \omega f(x_b))$ if x_a is reset. This process can speed up convergence; however, an optimum parameter ω is difficult to estimate.

3.1.5 The Newton–Raphson Method

The *Newton–Raphson* method is calculus based and is one of the most popular root-solving methods for equations in the form of Eq. [3.1]. In this method, information about a function $f(x)$ and its first derivative $f'(x)$ are used to improve estimates of a root of $[f(x) = 0]$. Let x_0 be an initial estimate or guess of the root. We wish to find a change h such that $f(x_0 + h)$ is zero.

The Taylor series expansion for a function about a point x_0 is the infinite series

[3.4] $$f(x_0 + h) = f(x_0) + h \, df(x_0)/dx + (h^2/2!)d^2f(x_0)/dx^2$$
$$+ \cdots (h^n/n!)d^{nf}(x_0)/dx^n + \cdots$$

where $(n!)$ is the factorial product given by

$$n! = \begin{cases} (n)(n-1)(n-2)\ldots(2)(1); & n > 0 \\ 1; & n = 0 \end{cases}$$

If we assume that the function $f(x)$ can be approximated locally by a straight line (so that second and higher derivatives are taken to be zero), we can truncate the Taylor series in Eq. [3.4] immediately after the first derivative term. The local straight line approximation is then equivalent to the tangent line at x_0, and the value of h for which $f(x_0 + h)$ is zero can be approximated by

$$h = -f(x_0)/f'(x_0); \qquad f'(x_0) \equiv df(x_0)/dx$$

The improved estimate x_{imp} of the root is therefore given by

[3.5] $$x_{imp} = x_0 - f(x_0)/f'(x_0)$$

After x_{imp} is found from Eq. [3.5], it is replaced by x_0 and Eq. [3.5] is applied again. This iterative process continues until convergence is achieved as indicated by the condition $|f(x_0)| \leq \epsilon$, where ϵ is some small quantity. The Newton–Raphson method is described in Procedure 3.5 and is illustrated in Fig. 3–3.

■ **Procedure 3.5   The Newton–Raphson method.**

1.   Choose a starting value x_0.
2.   Compute $f(x_0)$
3.   If $|f(x_0)| \leq \epsilon$, x_0 is the estimated solution; otherwise, go on to Step 4.
4.   Compute x_{imp} from Eq. [3.5].
5.   Set x_0 equal to x_{imp} and return to Step 2.   ■

We demonstrate the application of the Newton–Raphson method by considering the ***van der Waals equation of state*** for gases. The ideal gas law is used to relate the pressure P (in Pascal [Pa]), temperature T (in degree Kelvin [° K]), and specific volume v (in cubic meter per kilogram [m³/kg]) of a gas by

$$Pv = RT$$

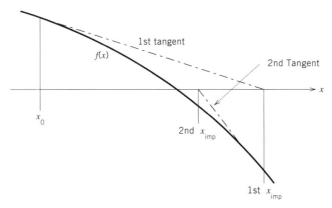

Figure 3–3   Illustration of the Newton–Raphson method.

where R is the gas constant (in Joules per kilogram per degree Kelvin [J/(kg $\cdot°$ K)]). The ideal gas law must be modified to account for the behavior of real gases. One modification is the van der Waals model, which accounts for intermolecular attractions and for the space occupied by molecules of finite size. The van der Waals equation is

[3.6]
$$(P + a/v^2)(v - b) = RT$$

in which a and b are the van der Waals constants for the gas.

To use the Newton–Raphson method to solve Eq. [3.6] for the specific volume v when P, T, and R are given, we must first recast it in the form of Eq. [3.1]; to do so, we rewrite Eq. [3.6] as

[3.7]
$$f(v) = (P + a/v^2)(v - b) - RT = 0$$

EXAMPLE   3.5

Use the Newton–Raphson method to solve Eq. [3.7] for the specific volume v of methane gas at a temperature T of 300 °K and a pressure P of $5(10^6)$ Pa. The gas constant R for methane is 518 J/(kg·°K), and the van der Waals constants have values of 887 Pa·m^6/kg^2 for a and 0.00267 m^3/kg for b. Use the ideal gas law to obtain a starting value v_0, and use a value of 0.1 for the convergence indicator ϵ.

Choose v_0 from the ideal gas law: $v_0 = RT/P = 0.03108$ m^3/kg
Obtain the derivative $f'(v)$: $f'(v) = P + a/v^2 - 2a(v - b)/v^3$
With v_0 and $f'(v)$, we iterate to convergence according to Procedure 3.5. Computations are tabulated as follows.

Iteration	v_0	$f(v_0)$	$f'(v_0)$	v_{imp}
1	0.0310800	1.27375(10^4)	4.23952(10^6)	0.0280755
2	0.0280755	2.16434(10^2)	4.08873(10^6)	0.0280226
3	0.0280226	8.03979(10^{-2})		

The required level of convergence is achieved after 2 improvements. The estimated solution is

$$v \simeq 0.0280226 \text{ m}^3/\text{kg}$$

Note that although 10^{-1} may appear to be a large value for ϵ, it is small relative to terms of magnitude RT in Eq. [3.7].

It is easy to see why the Newton–Raphson method is popular; although the determination of the derivative may be sometimes tedious for complicated equations, the implementation of the method is simple and convergence is fairly rapid. Nevertheless, the method may fail for a variety of reasons. One obvious difficulty occurs when $f'(x_0)$ in the denominator of Eq. [3.5] is zero. This situation may occur at the start of the procedure because of an unfortunate choice for the starting value x_0. The corrective action is to reset x_0 to a different value whenever $|f(x_0)|$ exceeds $A|f'(x_0)|$, where A is some large value based on the overflow characteristics of the computer.

Oscillations of x_{imp} values may indicate that there is no real root of an equation as shown in Fig. 3–4a. They may also occur when there is a root as shown in

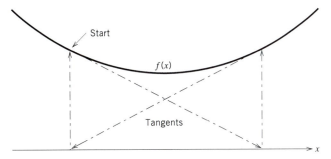

Figure 3–4a   Oscillations for a function with no real root.

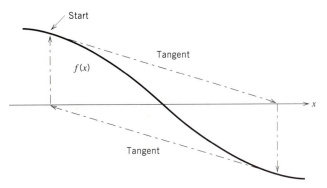

Figure 3–4b   Oscillations for a function with a real root.

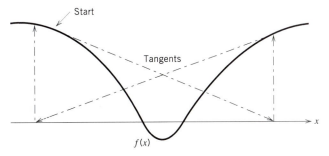

Figure 3–4c   Divergence for a function with two real roots.

Fig. 3–4b. For the case in Fig. 3–4b, round-off errors will break the cycle; but then divergence is just as likely to occur as convergence. Divergence is shown for a function with two real roots in Fig. 3–4c. For the functions in Fig. 3–4b and 3–4c, appropriate care in choosing the starting value x_0 is necessary to guarantee convergence.

We note also that the Newton–Raphson method will not necessarily converge to a root that is nearest to the starting point. For example, the equation

$$\sin x = 0; \qquad x \text{ in radians}$$

has an infinite number of roots $n\pi$, where n is an integer. If we start with x_0 equal to 1.5, the method converges to the root (-4π). It is therefore prudent to verify that the root found by the Newton–Raphson method is the desired one.

If a root of $f(x)$ occurs at the same x value at which df/dx is zero, the function has **multiple roots**. The Newton–Raphson method may still be used; however, the convergence to the solution will be slow. An effective approach in such cases is to use the **modified Newton–Raphson** method. In this approach, we solve for the root of a different function $h(x)$ that has the same, but not multiple, roots as $f(x)$. Such a function is given by

$$h(x) = f(x)/f'(x); \qquad f'(x) = df(x)/dx$$

The purpose of this approach is to improve the convergence rate for functions with multiple roots; we must still exercise great care in the computation of limiting values as we get very close to the root.

Pseudocode 3–4 for the Newton–Raphson method is now provided. A limit N on the number of iterations is imposed to prevent infinite looping.

Apart from possible nonconvergence or convergence to an unwanted root, a potential disadvantage of the Newton–Raphson method is tediousness or difficulty

Pseudocode 3–4   The Newton–Raphson method

Initialize: x_0, ϵ, N \ ϵ = convergence indicator, N = iteration limit
   $f_0 \leftarrow f(x_0)$ \ function value at x_0
   $i \leftarrow 0$ \ i = iteration counter
While [$i < N$ **and** $|f_0| > \epsilon$)]   \ iteration loop
    $i \leftarrow i + 1$ \ update iteration counter
    $x_0 \leftarrow x_0 - f_0/f'(x_0)$ \ improve estimate of root
    $f_0 \leftarrow f(x_0)$ \ function value at new estimate
End While

If [$|f_0| > \epsilon$] **then**   \ because loop terminated with $i = N$
   Write: 'Non-convergence to required limit' message
Else
   Write: Estimated solution x_0 and check value f_0
End If

in obtaining the first derivative analytically. Symbolic manipulator software is useful in performing this task; alternately, we can use a numerical approximation for the derivative. One second-order approximation of the derivative is

$$f'(x) \simeq [f(x + \Delta x) - f(x - \Delta x)]/(2\Delta x)$$

where Δx is small. Note, however, that too small a value of Δx can result in a poor approximation because of computer round-off error that occurs when a quantity is subtracted from another quantity of almost equal value. Another approximation for the derivative is embedded in the secant method that follows.

3.1.6 The Secant Method

The *secant* method is started from two distinct estimates x_a and x_b for the root of the equation $[f(x) = 0]$. An iterative procedure involving linear interpolation is used to update x_a and x_b and to converge to a root. The procedure is as follows.

■ **Procedure 3.6 The secant method.**

1. Choose x_a and x_b
2. Compute $f(x_b)$
3. If $|f(x_b)| \le \epsilon$ (a small, positive quantity), x_b is the estimated solution; otherwise, go on to Step 4.
4. Use linear interpolation to compute an intermediate point x_{int} from

$$x_{int} = x_b - (x_b - x_a)f(x_b)/[f(x_b) - f(x_a)]$$

5. Reset x_a to x_b, reset x_b to x_{int}, and return to Step 2. ■

Two iterations of the secant method are illustrated in Fig. 3–5, and the application of the procedure is demonstrated in Example 3.6.

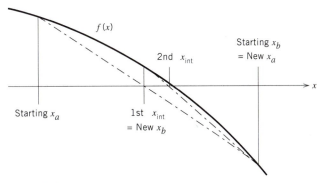

Figure 3–5 Illustration of the secant method.

EXAMPLE 3.6

Use the secant method to solve Eq. [3.7] for the specific volume v of methane gas at a temperature T of 300 °K and a pressure P of $5(10^6)$ Pa. The gas constant R for methane is 518 J/(kg·°K), and the van der Waals constants have values of 887 Pa·m^6/kg^2 for a and 0.00267 m^3/kg for b. Start with $(0.5RT/P)$ for v_a and $(1.5RT/P)$ for v_b, and use a value of 0.1 for ϵ.

The computations from Procedure 3.6 are tabulated below. The new v_b value is equal to the value of v_{int} from the previous iteration.

Iteration	v_a	v_b	$f(v_b)$	v_{int}
1	0.0155400	0.0466200	$8.22865(10^4)$	0.0263331
2	0.0466200	0.0263331	$-6.81592(10^3)$	0.0278850
3	0.0263331	0.0278850	$-5.61669(10^2)$	0.0280243
4	0.0278850	0.0280243	$7.18865(10^0)$	0.0280226
5	0.0280243	0.0280226	$-7.00575(10^{-3})$	

The estimated solution is the last value of v_b and is given by

$$v \simeq 0.0280226 \text{ m}^3/\text{kg}$$

The use of two starting values and the use of an interpolation step are similar to those in the false-position method in §3.1.4; however, the two starting values for the subsequent iterations are chosen in a different manner. The interpolation given in Step 4 of Procedure 3.6 can be rewritten as

$$x_{int} = x_b - f(x_b)/\{[f(x_b) - f(x_a)]/[x_b - x_a]\}$$

The divisor of $f(x_b)$ on the right-hand side is a first-order approximation of the derivative $f'(x_b)$; x_{int} is therefore obtained from x_b in the same way that x_{imp} is obtained from x_0 in the Newton–Raphson method. The secant method can then be properly thought of as a variation of the Newton–Raphson method. As such, we need to be concerned about the same issues regarding nonconvergence and choices of the starting values.

The secant method is a useful alternative to the Newton–Raphson method in cases where the first derivative is difficult to obtain. It can also be viewed as an improvement on the false-position method of §3.1.4. Indeed, if the starting points of the secant method bracket a single root as in the false-position method, the secant method will generally converge more quickly to a solution. Pseudocode 3–5 with an iteration limit N is now provided for the secant method.

3.1.7 Convergence Criteria

The methods presented in §3.1.1 through §3.1.6 are all iterative; the condition for terminating the iterative process is therefore important. The incremental-search and

Pseudocode 3–5 The secant method

Initialize: x_a, x_b, ϵ, N \qquad \ ϵ = convergence indicator, N = iteration limit

$\quad f_b \leftarrow f(x_b)$ \qquad \ function value at x_b
$\quad\quad i \leftarrow 0$ \qquad\qquad \ i = iteration counter

While $[i < N$ **and** $|f_b| > \epsilon)]$ \qquad\qquad \ iteration loop

$\quad\quad i \leftarrow i + 1$ \qquad\qquad\qquad\qquad\qquad \ update counter
$\quad x_{int} \leftarrow x_b - (x_b - x_a)f_b/[f_b - f(x_a)]$ \qquad \ intermediate value
$\quad\quad x_a \leftarrow x_b$ \qquad\qquad\qquad\qquad\qquad \ reset x_a
$\quad\quad x_b \leftarrow x_{int}$ \qquad\qquad\qquad\qquad\qquad \ reset x_b
$\quad\quad f_b \leftarrow f(x_b)$ \qquad\qquad\qquad\qquad\qquad \ function value at new x_b

End While

If $[|f_b| > \epsilon]$ **then** \qquad \ because loop terminated with $i = N$
$\quad$ **Write:** 'Nonconvergence to required limit' message
Else
$\quad$ **Write:** Estimated solution x_b and check value f_b
End If

---

bisection methods have simple termination conditions. We look for the function to become zero or change sign in the incremental-search method, provided that a solution exists in the direction of the search. For the bisection method that is properly applied with n decimal digits of precision, we look for a small bracketing interval whose length is no smaller than $10^{(1-n)}$ times the magnitude of the largest x value on the initial interval. Larger terminal intervals are often used in practice.

In all of the other methods (fixed-point iteration, false position, Newton–Raphson, and secant) we are looking for some indication that the solution has become sufficiently accurate. The two common ways of detecting convergence are to look for a sufficiently small change in the estimated solution from one iteration to the next and to look for a sufficiently small magnitude of the function $f(x)$.

To detect convergence based on the change in the estimated solution, we must be certain that the small change is a true indicator of convergence rather than an indicator of extremely slow convergence. Then, the size of the change must be compatible with the precision that is used and the magnitude of the solution.

In considering $|f(x)|$, we must be aware of how the function is computed. For example, the function in Eq. [3.7] has two terms of magnitude RT; one of these is subtracted from the other to give the function value. We cannot generally expect the magnitude of the error in the function value to be smaller than the magnitudes of the errors in the individual terms.

Strict procedures for establishing convergence require us to monitor both the change in the iterated solution and the magnitude of the function. As we noted earlier, a small change in the iterated solution may occur with slow convergence even if we are not close to the root. A small magnitude of the function value would not necessarily indicate convergence if the function approaches the root almost asymptotically.

Even if we choose our convergence criterion carefully, we must still be aware that a method may fail for other reasons. Imposition of an iteration limit prevents infinite looping under failure conditions. If a method is terminated because it reaches an iteration limit, a history of function values can give us an understanding of the function's behavior; we can then take appropriate steps to alleviate the difficulties that were encountered in solving the problem.

The number of iterations required for a method to converge depends on the method itself and on the starting point of the iterative process. We can often use our physical understanding of the problems we see in engineering to make very good estimates of the starting points. The starting points for the beam deflection problem in §3.1.1 and §3.1.2, the critical Mach number problem in §3.1.3 and §3.1.4, and the van der Waals problem in §3.1.5 and §3.1.6 are all chosen with the physical concepts of the problems in mind.

3.2 SYSTEMS OF NONLINEAR EQUATIONS

Several methods for solving simultaneous nonlinear equations may be derived from methods that are used for a single equation. The Newton–Raphson method of §3.1.5 is extended in this section to solve systems of nonlinear equations. These systems consist of n equations in n variables $(x_1, x_2, \ldots, x_n)$ and have the general form

$$f_k(x_1, x_2, \ldots, x_n) = 0; \qquad k = 1, 2, \ldots, n$$

Our development of the procedure for solving systems of nonlinear equations requires the concepts of matrices, vectors, norms, and solutions of linear systems that are discussed in Chapter 2. We begin by defining a column vector $\mathbf{x}$ in transpose form as

$$\mathbf{x}^T = [x_1 \; x_2 \; \ldots \; x_n]$$

and rewrite the system of equations more compactly as

[3.8]
$$f_k(\mathbf{x}) = 0; \qquad k = 1, 2, \ldots, n$$

As in the Newton–Raphson method, we start the solution process with an initial estimate $\mathbf{x_0}$ of the solution vector. We write the initial vector as

$$(\mathbf{x_0})^T = [(x_1)_0 \; (x_2)_0 \; \ldots \; (x_n)_0]$$

and seek an improved estimate

[3.9]
$$\mathbf{x_{imp}} = \mathbf{x_0} + \mathbf{h}$$

where

$$(\mathbf{x_{imp}})^T = [(x_1)_{imp} \ (x_2)_{imp} \ \cdots \ (x_n)_{imp}]$$

$$\mathbf{h}^T = [h_1 \ h_2 \ \ldots \ h_n]$$

$$(x_k)_{imp} = (x_k)_0 + h_k; \qquad k = 1, 2, \ldots, n$$

The procedure for determining **h** is based on the truncated Taylor series for a function of several variables (see Appendix C). If the truncation takes place immediately following the first derivative terms of the series, we have

$$f_k(\mathbf{x_0} + \mathbf{h}) \simeq f_k(\mathbf{x_0}) + \sum_{j=1}^{n} [h_j \ \partial f_k(\mathbf{x_0})/\partial x_j]; \qquad k = 1, 2, \ldots, n$$

We then obtain the vector **h** for which $f_k(\mathbf{x_0} + \mathbf{h})$ is zero from the *linear* system of equations given by

[3.10a]
$$\sum_{j=1}^{n} [h_j \ \partial f_k(\mathbf{x_0})/\partial x_j] = -f_k(\mathbf{x_0}); \qquad k = 1, 2, \ldots, n$$

The matrix form of Eq. [3.10a] is

[3.10b]
$$\mathbf{J} \cdot \mathbf{h} = -\mathbf{f}$$

in which **f** is the column vector given by

$$\mathbf{f}^T = [f_1(x_0) \ f_2(x_0) \ \ldots \ f_n(x_0)]$$

and **J** is the partial derivative matrix known as the *Jacobian* and is given by

[3.10c]
$$\mathbf{J} = \begin{bmatrix} \partial f_1(\mathbf{x_0})/\partial x_1 & \partial f_1(\mathbf{x_0})/\partial x_2 & \cdots & \partial f_1(\mathbf{x_0})/\partial x_n \\ \partial f_2(\mathbf{x_0})/\partial x_1 & \partial f_2(\mathbf{x_0})/\partial x_2 & \cdots & \partial f_2(\mathbf{x_0})/\partial x_n \\ \vdots & \vdots & & \vdots \\ \partial f_n(\mathbf{x_0})/\partial x_1 & \partial f_n(\mathbf{x_0})/\partial x_2 & \cdots & \partial f_n(\mathbf{x_0})/\partial x_n \end{bmatrix}$$

The vector **h** in Eq. [3.10b] may be easily found by any of several methods when n is small. For large systems, we recommend the use of either the formal matrix methods that are discussed in Chapter 2 or of mathematical subroutine packages that are available on most computer systems.

The procedure for the multivariable Newton–Raphson method is now given. It follows the same sequence of steps given in Procedure 3.5 of §3.1.5.

■   **Procedure 3.7   The Newton–Raphson method for systems of equations.**

    1. Choose a starting vector $\mathbf{x_0}$.

    2. Compute $f_k(\mathbf{x_0})$ for $k = 1, 2, \ldots, n$

    3. If $\|\mathbf{f}\| \leq \epsilon$, $\mathbf{x_0}$ is the estimated solution; otherwise, go on to Step 4.

    4. Obtain $\mathbf{h}$ from Eq. [3.10b] and compute $\mathbf{x_{imp}}$ from Eq. [3.9].

    5. Set $\mathbf{x_0}$ equal to $\mathbf{x_{imp}}$ and return to Step 2.   ■

The convergence criterion used in Step 3 of Procedure 3.7 is similar to the one used in Procedure 3.5, except that the absolute value of a single function is now replaced by the norm of the vector whose components are the n function values $f_k(\mathbf{x_0})$. The ∞-norm, which is the component with the largest magnitude, provides an upper bound on the magnitudes of all components of $\mathbf{f}$.

The discussions in §3.1.7 on convergence criteria, nonconvergence, and convergence to an unwanted solution are just as applicable to the multivariable Newton–Raphson method as they are to the single-variable version. The corrective actions in the event of failure are the same as for the single-variable version; however, the presence of several variables instead of only one may make corrective steps more difficult to accomplish.

We now illustrate the application of the multivariable Newton–Raphson method to a *catenary* problem. When a cable of uniform density is suspended between two points and supports only its own weight, the shape that it assumes is known as a catenary. The cable in Fig. 3–6 has a length L and is suspended between points A and B, which are separated by a horizontal distance d. The elevations of A and B differ by a nonnegative amount h, so that point B is never above point A.

A convenient coordinate system for describing the catenary has a horizontal x axis that is a distance c below the lowest point O on the cable, and it has a vertical y axis that passes through O. The equation of the catenary in this coordinate system

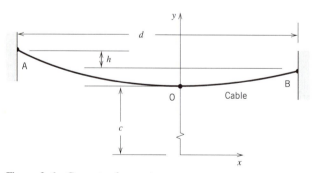

Figure 3–6   Geometry for a catenary.

and the arc length λ measured along the cable from point O and positive to the right are given by

[3.11a]
$$y = c \cosh(x/c)$$

[3.11b]
$$\lambda = c \sinh(x/c)$$

We continue our discussion of the catenary though a two-part example. The first part shows the formulation of a problem to produce a system of equations.

EXAMPLE 3.7a

The length L of the cable in Fig. 3–6, the horizontal distance d, and the vertical distance h are given. Let z be the x coordinate of point A. Formulate a system of equations to determine the coordinate system parameters c and z, and determine expressions for the partial derivative elements of the Jacobian **J**.

Denote the x coordinate of B by x_b; then $x_b = z + d$.
We have from Eq. [3.11a]:

$$h = c[\cosh(z/c) - \cosh((z + d)/c)]$$

From Eq. [3.11b]:

$$L = c[\sinh((z + d)/c) - \sinh(z/c)]$$

Our system of equations therefore consists of:

$$f_1(c, z) = c[\cosh(z/c) - \cosh((z + d)/c)] - h = 0$$
$$f_2(c, z) = c[\sinh((z + d)/c) - \sinh(z/c)] - L = 0$$

The required derivatives are:

$$(f_1)_c \equiv \partial f_1/\partial c, \quad (f_1)_z \equiv \partial f_1/\partial z, \quad (f_2)_c \equiv \partial f_2/\partial c, \quad (f_2)_z \equiv \partial f_2/\partial z$$

Define:

$$p = \cosh(z/c), \qquad q = \cosh((z + d)/c)$$
$$r = \sinh((z + d)/c), \qquad s = \sinh(z/c)$$

Then we have:

$$f_1 = c(p - q) - h, \qquad f_2 = c(r - s) - L$$
$$(f_1)_z = -(L + f_2)/c, \qquad (f_2)_z = -(h + f_1)/c$$
$$(f_1)_c = [rd - z(f_1)_z]/c - (f_2)_z, \qquad (f_2)_c = -[qd + z(f_2)_z]/c - (f_1)_z$$

These provide a simple way to evaluate the functions and their derivatives.

We continue with the second part of the example to show the mechanics of the solution process for the problem formulated in Example 3.7a.

EXAMPLE 3.7b

Determine c and z in Example 3.7a for L equal to 65 ft, d equal to 50 ft, and h equal to 10 ft. Use the ∞-norm and ϵ equal to 10^{-4} to establish convergence.

We choose starting values c_0 and z_0 by considering the symmetric case with h set to zero. The z_0 value for this case is equal to $(-d/2)$ or (-25) ft, and the expression for c_0 is

$$c_0 \sinh(d/(2c_0)) = L/2$$

Expand the sinh function to two terms (see Appendix C): $\sinh w = w + w^3/3! + \cdots$. We then obtain a crude but simple approximation

$$(c_0)^2 = d^3/[24(L-d)] \to c_0 \simeq 19 \text{ ft}$$

Start with $c_0 = 19$ and $z_0 = -25$; then compute the changes Δc and Δz to obtain improved estimates

$$c_{\text{imp}} = c_0 + \Delta c, \qquad z_{\text{imp}} = z_0 + \Delta z$$

The changes are obtained by solving the system corresponding to Eq. [3.10b]:

$$\begin{bmatrix} (f_1)_c & (f_1)_z \\ (f_2)_c & (f_2)_z \end{bmatrix} \cdot \begin{bmatrix} \Delta c \\ \Delta z \end{bmatrix} = \begin{bmatrix} -f_1 \\ -f_2 \end{bmatrix}$$

The sequence of computations is as follows:

Iteration	c_0	z_0	$f_1(c_0, z_0)$	$f_2(c_0, z_0)$
1	19.00000	−25.00000	−1.0000(10^{+1})	7.2917(10^{-1})
2	19.40546	−27.89065	−2.7796(10^{-1})	7.4787(10^{-1})
3	19.88129	−28.07877	−9.2309(10^{-3})	3.0681(10^{-2})
4	19.90228	−28.08639	−9.5203(10^{-6})	5.6955(10^{-5})

The c_0 and z_0 values for a given iteration in the computation table are the improved values c_{imp} and z_{imp} from the previous iteration. The final estimates for c and z are

$$c = 19.0228 \text{ ft}, \qquad z = -28.08639 \text{ ft}$$

3.3   ROOTS OF POLYNOMIALS

The final root-solving method that we shall discuss is **Bairstow's** method (Ref. 4) for finding all of the roots (real and complex) of a polynomial equation. **Synthetic division** and the multivariable Newton–Raphson method of Section 3.2 are combined in this method to obtain quadratic factors of a polynomial.

Let $P_n(x)$ be a polynomial of degree n defined in the equation

[3.12] $$P_n(x) \equiv a_0 \, x^n + a_1 \, x^{n-1} + \cdots + a_{n-1} \, x + a_n = 0; \qquad a_0 = 1$$

Division of $P_n(x)$ by a quadratic

[3.13] $$Q = (x^2 + ux + v)$$

yields at most a linear remainder R. For later convenience, we write R as

[3.14] $$R = (x + u) \, R_1(u, v) + R_2(u, v)$$

with the coefficients $R_1(u, v)$ and $R_2(u, v)$ depending on u and v. Q is a factor of $P_n(x)$ when R is zero; in that case, the two roots of $(Q = 0)$ are clearly roots of Eq. [3.12]. The major objective in Bairstow's method for solving Eq. [3.12] is to find quadratic factors by solving the equation $(R = 0)$; that is, to solve the two-equation system

[3.15] $$R_1(u, v) = 0; \qquad R_2(u, v) = 0$$

Let us use the definitions in Eq. [3.13] and [3.14] to express the polynomial $P_n(x)$ in Eq. [3.12] by

[3.16] $$P_n(x) = Q \, S_{n-2}(x) + (x + u) \, R_1(u, v) + R_2(u, v)$$

The quotient $S_{n-2}(x)$ of $P_n(x)/Q$ is a polynomial of degree $(n - 2)$ and is written as

[**3.17**] $$S_{n-2}(x) = b_0 \, x^{n-2} + b_1 \, x^{n-3} + \cdots + b_{n-3} \, x + b_{n-2}$$

Determination of the coefficients b_k and of $R_1(u, v)$ and $R_2(u, v)$ may be performed by the common technique of synthetic division (so called because the result of the division is obtained without actually performing the division). Rearrangement of Eq. [3.16] so that only the remainder terms are on the right-hand side gives us

$$P_n(x) - Q \, S_{n-2}(x) = (x + u) \, R_1(u, v) + R_2(u, v)$$

Then we expand the left-hand side of the rearranged equation by using the expressions for $P_n(x)$, Q, and $S_{n-2}(x)$ from Eqs. [3.12], [3.13], and [3.17], respectively. The result is

$$
\begin{aligned}
a_0\, x^n + \; a_1\, x^{n-1} + \;\; a_2\, x^{n-2} + \cdots + \;\; a_{n-2}\, x^2 + \; a_{n-1}\, x + \; a_n \\
- b_0\, x^n - \;\; b_1\, x^{n-1} - \;\; b_2\, x^{n-2} - \cdots - \;\; b_{n-2}\, x^2 \\
- u b_0\, x^{n-1} - u b_1\, x^{n-2} - \cdots - u b_{n-3}\, x^2 - u b_{n-2}\, x \\
- v b_0\, x^{n-2} - \cdots - v b_{n-4}\, x^2 - v b_{n-3}\, x - v b_{n-2} \\
= (x + u)\, R_1(u, v) + R_2(u, v)
\end{aligned}
$$

We next introduce

[3.18] $$b_{n-1} \equiv R_1(u, v); \qquad b_n \equiv R_2(u, v)$$

and equate the coefficients of each power of x to obtain the recursive system

[3.19] $$
\begin{cases}
b_0 = a_0 = 1 \\
b_1 = a_1 - u b_0 \\
b_k = a_k - u b_{k-1} - v b_{k-2}; \qquad k = 2, 3, \ldots, n
\end{cases}
$$

Recall that our objective is to find the values of u and v so that R_1 and R_2 (or b_{n-1} and b_n according to Eq. [3.18]) are zero. The values of R_1 and R_2 are computed from Eq. [3.19] for given values of u and v. To improve u and v by the Newton–Raphson method, we must also determine the partial derivatives of R_1 (or b_{n-1}) and R_2 (or b_n) with respect to u and v.

We denote the partial derivatives of b_k with respect to u and v by

[3.20a] $$c_k = \partial b_k / \partial u; \qquad k = 0, 1, \ldots, n$$

[3.20b] $$d_k = \partial b_{k+1} / \partial v; \qquad k = 0, 1, \ldots, n-1$$

On differentiation of the quantities in Eq. [3.19], the c_k values are found to be identical to the d_k values for k equal to $(0, 1, \ldots, n-1)$. The recursion for the c_k values is

[3.20c] $$
\begin{cases}
c_0 = 0 \\
c_1 = -b_0 \\
b_k = -b_{k-1} - u c_{k-1} - v c_{k-2}; \qquad k = 2, 3, \ldots, n
\end{cases}
$$

The required partial derivatives of R_1 and R_2 with respect to u and v are

[3.21] $$
\begin{cases}
\partial R_1 / \partial u = c_{n-1}; & \partial R_2 / \partial u = c_n \\
\partial R_1 / \partial v = c_{n-2}; & \partial R_2 / \partial v = c_{n-1}
\end{cases}
$$

We are now ready to summarize the steps of the multivariable Newton–Raphson method for solving Eq. [3.15] to find the coefficients u and v of a quadratic factor Q. These steps are described in Procedure 3.8.

■ **Procedure 3.8   The multivariable Newton–Raphson method for Eq. [3.15].**

1. Choose initial estimates u_0 and v_0 for u and v.
2. Compute b_0 through b_n from Eq. [3.19].
3. If $\|[b_{n-1}\ b_n]^T\| \le \epsilon$, where ϵ is some small quantity, u_0 and v_0 are the estimated coefficients, and b_0 through b_{n-2} are the coefficients of $S_{n-2}(x)$. Otherwise, go on to Step 4.
4. Compute the partial derivatives in Eq. [3.21] from Eq. [3.20c].
5. Obtain the changes Δu and Δv required to improve u_0 and v_0, respectively, by solving

$$\begin{bmatrix} c_{n-1} & c_{n-2} \\ c_n & c_{n-1} \end{bmatrix} \cdot \begin{bmatrix} \Delta u \\ \Delta v \end{bmatrix} = \begin{bmatrix} -b_{n-1} \\ -b_n \end{bmatrix}$$

The solutions for Δu and Δv from Cramer's rule in Chapter 2 are

$$\Delta u = (b_n\ c_{n-2} - b_{n-1}\ c_{n-1})/([c_{n-1}]^2 - c_n\ c_{n-2})$$
$$\Delta v = (b_{n-1}\ c_n - b_n\ c_{n-1})/([c_{n-1}]^2 - c_n\ c_{n-2})$$

6. Reset u_0 to $(u_0 + \Delta u)$ and v_0 to $(v_0 + \Delta v)$, and return to Step 2.   ■

The Newton–Raphson method does not necessarily converge for arbitrary starting values u_0 and v_0. The procedure is usually started with (u_0, v_0) equal to $(0,0)$. If convergence does not occur within a limited number of iterations, the procedure is restarted with new values of u_0 and v_0. Common practice in this case is to use the last three coefficients of $P_n(x)$ to set u_0 equal to (a_{n-1}/a_{n-2}) and to set v_0 equal to (a_n/a_{n-2}), provided that a_{n-2} is nonzero.

The overall method applied to Eq. [3.12] involves reducing $P_n(x)$ by quadratic factors until the quotient is either a quadratic expression (for even n) or a linear expression (for odd n). The general algorithm for Bairstow's method is given by Pseudocode 3–6.

In the Bairstow algorithm, the polynomial is reduced by degree 2 as long as n is greater than 2. The last factor is either the quadratic $S_2(x)$ with a final n value of 2 or the linear term $S_1(x)$ with a final n value of 1.

No pseudocode is given for Procedure 3.8 or for Procedure 3.7 from which it is derived; however, the pseudocode for the single-variable Newton–Raphson method of §3.1.5 may be used as a model. It should be recognized that Procedure 3.8 is only a part of the more general algorithm for Bairstow's method. It would therefore be convenient to provide a way to restart the module for Procedure 3.8 with new values of u_0 and v_0 if convergence does not occur within an imposed iteration limit.

Pseudocode 3–6   Bairstow's method

Declare: a, b, p \ coefficient arrays with subscripts $0, 1, \ldots, n$
Initialize: n, a_0 through a_n, ϵ \ $n > 0$, $a_0 = 1$, ϵ = convergence indicator
For $[k = 0, 1, \ldots, n]$
     $p_k \leftarrow a_k$ \ save original coefficients to evaluate original P_n later
End For
While $[n > 2]$ \ loop while reduced polynomial has degree 3 or more
     NewtonRaphson$(a, n, \epsilon, b, u, v)$ \ module to find u, v, and b
                                                  \ from Procedure 3.8

     $r1 \leftarrow$ 1st root of $Q = x^2 + ux + v = 0$ \ may be complex
     $r2 \leftarrow$ 2nd root of $Q = x^2 + ux + v = 0$ \ may be complex
     $p1 \leftarrow P_n(r1)$ \ with original polynomial
     $p2 \leftarrow P_n(r2)$ \ with original polynomial
     Write: Coefficients u and v; roots $r1$ and $r2$; check values $p1$ and $p2$
     $n \leftarrow n - 2$ \ reduce degree of polynomial
     For $[k = 0, 1, \ldots n]$
         $a_k \leftarrow b_k$ \ coefficients for the reduced polynomial
     End For
End While
If $[n = 2]$ **then**   \ for last quadratic term
     $r1 \leftarrow$ 1st root of $S_2(x) = x^2 + a_1 x + a_2 = 0$ \ may be complex
     $r2 \leftarrow$ 2nd root of $S_2(x) = x^2 + a_1 x + a_2 = 0$ \ may be complex
     $p1 \leftarrow P_n(r1)$ \ with original polynomial
     $p2 \leftarrow P_n(r2)$ \ with original polynomial
     Write: Coefficients a_1 and a_2; roots $r1$ and $r2$; check values $p1$ and $p2$
Else   \ for last linear term
     $r1 \leftarrow$ root of $S_1(x) = x + a_1 = 0$
     $p1 \leftarrow P_n(r1)$ \ with original polynomial
     Write: Coefficient a_1; root $r1$; check value $p1$
End If

A limit on the number of restarts should also be imposed to prevent infinite looping in the event that the convergence criterion is unrealistic or that the equation is truly pathological.

The root of the linear equation $[S_1(x) = 0]$ is obviously $(-a_1)$. The roots of a quadratic equation of the form $(x^2 + ux + v = 0)$ may be distinct or identical real roots, or they may be a complex conjugate pair. They are obtained from the following familiar formula or from the more sophisticated techniques of Appendix C.

$$r1, r2 = \left(-u \pm \sqrt{u^2 - 4v}\right)\Big/2$$

Polynomial methods such as Bairstow's should not be used simply because an equation is in polynomial form. For example, Eq. [3.7] for the van der Waals equation of state could easily be rewritten as a cubic polynomial; however, it is much more efficient to solve that equation by the Newton–Raphson method for the single real root in which we are interested.

More suitable types of problems for polynomial methods are those in which all or most of the roots are required. Examples of such types are eigenvalue problems, which are treated in Chapter 8. Here, we consider a *spring-mass system* shown in Fig. 3–7. The system consists of masses $2m$ and m, which are connected as shown to a fixed frame and to each other by linear springs of stiffnesses $2k$ and k. The system has four degrees of freedom because in-plane motions of each mass may be described in terms of two orthogonal directions. These coordinate directions are denoted by x_1, x_2, x_3, and x_4.

We consider small vibrations so that the rotation of the springs is negligible. Then for displacements x_i in the four coordinate directions, the spring forces are either tensions equal to k times the spring extension or compressions equal to k times the spring compression. If a_i denotes the acceleration (d^2x_i/dt^2) in each direction x_i, the equations of motion are obtained from Newton's Second Law to be

$$2ma_1 + 2kx_1 - 2k(x_3 - x_1) = 0$$
$$2ma_2 + 2kx_2 = 0$$
$$ma_3 + kx_3 + 2k(x_3 - x_1) = 0$$
$$ma_4 + kx_4 = 0$$

A natural frequency ω for the system in Fig. 3–7 is one for which each displacement x_i may be written as

$$x_i = A_i C \cos(\omega t + \phi)$$

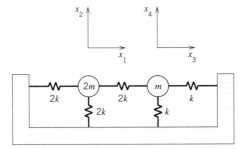

Figure 3–7 Schematic of a spring-mass system.

in which A_iC is an amplitude, t is time, and ϕ is a phase angle. We see then that the accelerations a_i are equal to $(-\omega^2 x_i)$, and we express the equations of motion in matrix form by

$$\begin{bmatrix} 4k - 2m\omega^2 & 0 & -2k & 0 \\ 0 & 2k - 2m\omega^2 & 0 & 0 \\ -2k & 0 & 3k - m\omega^2 & 0 \\ 0 & 0 & 0 & k - m\omega^2 \end{bmatrix} \cdot \begin{bmatrix} x_1 \\ x_2 \\ x_3 \\ x_4 \end{bmatrix} = \begin{bmatrix} 0 \\ 0 \\ 0 \\ 0 \end{bmatrix}$$

A nontrivial solution for the x_i components occurs when the coefficient matrix is singular; that is, when its determinant is zero. If we divide the equations by k, use λ to denote $(m\omega^2/k)$, and expand the determinant according to Eq.[2.17a] and [2.18], we obtain the polynomial equation

[3.22] $P_4(\lambda) = \lambda^4 - 7\lambda^3 + 15\lambda^2 - 13\lambda + 4 = 0$

The solution of Eq. [3.22] gives us the four natural frequencies of the system in Fig. 3–7.

EXAMPLE 3.8

A factor Q of Eq. [3.22] is found by an application of Bairstow's method to be $(\lambda^2 - 1.9997615\lambda + 0.9997615)$. Use synthetic division to obtain the other quadratic factor, and compute the four roots of Eq. [3.22].

The coefficients of $P_4(\lambda)$ are:

$$a_0 = 1, \quad a_1 = -7, \quad a_2 = 15, \quad a_3 = -13, \quad a_4 = 4$$

The other quadratic factor of $P_4(\lambda)$ is $[S_2(\lambda) = \lambda^2 + b_1\lambda + b_2]$. The coefficients b_k are obtained from Eq. [3.19] with $(u = -1.9997615)$ and $(v = 0.9997615)$:

$$b_0 = 1, \quad b_1 = -5.0002385, \quad b_2 = 4.0009541$$

Thus: $S_2 = \lambda^2 - 5.0002385\lambda + 4.0009541$

The roots of the quadratic Q are: $\lambda_1 = 0.9999925$, $\lambda_2 = 0.9997689$
The roots of the quadratic S_2 are: $\lambda_3 = 4.0000000$, $\lambda_4 = 1.0002385$

The exact factors and roots of Eq. [3.22] are

$$P_4(\lambda) = (\lambda^2 - 2\lambda + 1)(\lambda^2 - 5\lambda + 4)$$
$$\lambda = 1, 1, 4, \text{ or } 1$$

The four real roots are guaranteed because of the physical nature of the problem. Even in such cases, however, small round-off errors can create situations in which the numerically obtained roots are complex with a small imaginary part. A physical understanding of the problem would allow us to interpret the solution correctly in those cases.

3.4 CLOSURE

The choice of a root-solving method should be dictated by the equation to be solved. The simple incremental-search technique is generally inefficient, and fixed-point iteration depends on our ability to find a splitting of the equation that produces convergence. Bracketing methods such as the bisection and false position methods are useful if we are sure that a single root is contained between two initial starting points. The bisection method is particularly attractive in such cases because we can obtain a solution to an almost arbitrary level of accuracy.

The Newton–Raphson method is by far the most commonly used technique because of its simplicity and its generally faster convergence than any of the previous methods. The Newton–Raphson method is extended to systems of nonlinear equations in Section 3.2, and is used in Bairstow's method for polynomials in Section 3.3. We must remember, however, that the Newton–Raphson method may be sometimes unreliable. The probability of success with the Newton–Raphson method is enhanced greatly by good starting values.

Another disadvantage of the Newton–Raphson method occurs when the the function in the equation to be solved has a derivative that is difficult to obtain. This may occur if the function is complicated or if it is not expressed analytically. The secant method is a good alternative in those cases, or we may approximate the derivative numerically.

The root-solving methods of this chapter can be implemented successfully for a large majority of the equations that we are likely to encounter. Nevertheless, there are many pathological cases that can cause the methods to fail. Even worse than the usually obvious failure of a method is the situation in which a false, incorrect, or significantly inaccurate solution is obtained and accepted without question. Therefore, the effort should always be made to verify that solutions satisfy both the original equation to be solved and the constraints of the problem under consideration. It is also important to keep in mind the potential for computer round-off errors to accumulate with undesirable consequences.

Many failures that occur with root-solving methods are caused by poor choices of starting points. In most engineering applications, some forethought or simple preliminary analysis can provide a good idea of the general behavior of the functions involved in the equations. This knowledge is useful in choosing good starting points for a procedure, thereby improving the chances for a method to be successful. Good choices of starting points will also reduce computational effort by reducing the number of iterations or steps required to obtain a solution.

3.5   EXERCISES

1.   The time equation for elliptic orbits has the form

$$M = \phi - E \sin\phi; \qquad \phi \text{ in radians}$$

where M is known as the mean anomaly, ϕ is known as the eccentric anomaly, and E (between 0 and 1) is the eccentricity of the elliptic orbit. For a given value of E and for a given value of M in the range

$$n\pi \le M \le (n+1)\pi; \qquad n = \text{an integer}$$

the solution for ϕ behaves as follows:

$$n\pi \le \phi \le (n+1)\pi; \quad \phi \le M \text{ when } n \text{ is even}; \quad \phi \ge M \text{ when } n \text{ is odd}$$

Use a calculator or write a computer program to solve for ϕ by one or more of the methods listed. In each case, explain the choice of starting points and other necessary quantities (if any), and obtain a solution that satisfies

$$|M - \phi + E \sin\phi| < 0.001$$

   a.   The incremental-search method ($E = 0.12$, $M = 2.3$).
   b.   The fixed-point iteration method ($E = 0.12$, $M = 5.6$).
   c.   The bisection method ($E = 0.23$, $M = 5.6$).
   d.   The false-position method ($E = 0.23$, $M = 6.7$).
   e.   The Newton–Raphson method ($E = 0.23$, $M = 7.8$).
   f.   The secant method ($E = 0.34$, $M = 7.8$).

2.   The Colebrook formula for fully turbulent flows in smooth pipes relates a friction factor f to the diameter-based pipe Reynolds number R. This formula may be written as

$$\sqrt{f}\,\ln(R\,\sqrt{f}/2.51) - 1.1513 = 0$$

The formula is generally used with R values greater than about 3000 (where f is approximately equal to 0.0435). The value of f decreases as R increases from 3000.
   Use a calculator or write a computer program to determine f by any of the methods listed. In each case, explain the choice of starting points and other necessary quantities (if any) and obtain a solution such that the left-hand side of the given equation is less than 0.0005 in magnitude.

   a.   The incremental-search method ($R = 5000$).
   b.   The fixed-point iteration method ($R = 10,000$).
   c.   The bisection method ($R = 20,000$).
   d.   The false-position method ($R = 40,000$).

e. The Newton–Raphson method ($R = 80,000$).

f. The secant method ($R = 160,000$).

3. A square plate with sides of unit length has its center at the origin of a Cartesian (x, y) coordinate system. The sides are parallel to either the x or y axis. A circular hole of radius r is removed from the plate. The center of the hole is on the x axis at a distance r from the left side of the plate. The centroid of the remaining piece has an x coordinate equal to c, which is given by

$$c = \pi r^2 (0.5 - r)/(1 - \pi r^2)$$

The value of r that maximizes c satisfies the equation

$$\pi r^3 - 3r + 1 = 0$$

Use a calculator or write a computer program to determine r and thus c by one or more of the incremental-search, fixed-point iteration, bisection, false-position, Newton–Raphson, and secant methods. In each case, explain the choice of starting points and other necessary quantities (if any), and obtain a solution such that the left-hand side of the equation for r is less than 10^{-4} in magnitude.

4. The normal depth y for flow in a rectangular open channel of width w is related to the volume flow rate Q, the channel slope s, and the Manning friction coefficient n through the equations

$$y[wy/(w + 2y)]^{2/3} = c = nQ/(w\sqrt{s})$$

Use a calculator or write a computer program to determine y to four significant figures by either the fixed-point iteration or Newton–Raphson method for the set of data:

$$w = 15 \text{ m}; \quad Q = 20 \text{ m}^3/\text{s}; \quad n = 0.015; \quad s = 0.001$$

Assume that the solution for y will be much smaller than w and use the assumption to estimate a starting value of y.

5. An electrical circuit with a resistance R, a capacitance C, and an inductance L has an initial charge q_0 across the capacitor. When the circuit is closed, the charge is dissipated in time t to a value q given by

$$q/q_0 = e^{-Rt/(2L)} \cos\left[t\sqrt{(1/(LC) - [R/(2L)]^2}\right]$$

Determine the value of L required for (q/q_0) to attain a value of 0.1 in a time t of 0.03 s when R is 200 Ω and C is 10^{-4} F by one of the bracketing methods or by the secant method.

Hint: Use the requirements for a real square root operation to establish limits for $(1/L)$ and solve the equation for $(1/L)$.

6.   A variation of the bisection method is the trisection method in which the interval from x_a to x_b is divided into three equal subintervals. Develop an algorithm to solve $[f(x) = 0]$ by this method when there is only one solution on the original interval.

7.   Rearrange the van der Waals equation given by Eq. [3.6] and [3.7], and then use the fixed-point iteration method to find the specific volume v of water vapor at a pressure P of $5(10^6)$ Pa and a temperature T of 600 °K. R is 461 J/(kg·°K), a is 1703 Pa·m^6/kg^2, and b is 0.00169 m^3/kg for water vapor.

8.   The thickness t of an airfoil in an NACA airfoil family (NACA preceded the National Aeronautics and Space Administration) is described by

$$t/T = g(x) = 2.969 \sqrt{x} - 1.260x - 3.516x^2 + 2.843x^3 - 1.015x^4$$

The chord extends from the leading edge at $(x = 0)$ to the trailing edge at $(x = 1)$, the T is the **nominal** maximum thickness. The true maximum thickness t_{max} is slightly larger than T. An airfoil with T equal to 0.15 is shown in Fig. 3–P8.

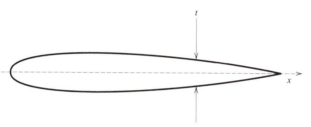

Figure 3–P8

a.   Use the bisection or false-position method to solve the equation $(dt/dx = 0)$ to find the value of x on $(0 < x < 1)$ at which t is maximum. Explain why these methods are not suitable for the almost equivalent problem of solving the equation $[g(x) = 1]$.

b.   Use the result of Part (a) and the bisection or false-position method to determine the two values of x on $(0 < x < 1)$ at which t is equal to $0.2T$. Explain why the Newton–Raphson method is not suitable for finding the smaller of these values when it is started from an arbitrary value between 0 and the value of x at which t is maximum.

9.   The engine speed of a car [revolutions per minute] and the rotational speed of the drive wheels are related through the gear ratios of the transmission and a fixed final drive ratio by

$$\text{Wheel rpm} = \text{Engine rpm}/(\text{Gear Ratio} \times \text{Final Drive Ratio})$$

The car's road speed V [mph] may be written in terms of the engine rpm Ω and a gear ratio G by

$$V = k\Omega/G$$

A conventionally geared car with a four-speed automatic transmission typically has a third gear ratio close to 1 and an overdrive fourth gear ratio of about 0.7. The fourth gear ratio G for a certain car is to be selected so that the car's top speed V corresponds to the engine rpm Ω at which maximum power is delivered. The drag D on the car [pounds force] is given by

$$D = 0.0215V^2 + 0.543V + 37.0; \qquad V \text{ in mph}$$

k is 0.0159, and maximum power output P of the engine is 195 horsepower when Ω is equal to 5750 rpm. The equation to be satisfied for the top speed V of the car is

$$DV = 375P; \quad D \text{ in lb}, \quad V \text{ in mph}, \quad P \text{ in hp}$$

Use the Newton–Raphson or the secant method to determine V and thus to obtain the gear ratio G.

10. A cable of length L equal to 17 m is suspended between two points A and B. The coordinates of A and B in a horizontal x and vertical y coordinate system are, respectively, (x_a, y_a) and (x_b, y_b) with

$$x_b - x_a = 13 \text{ m}; \qquad y_b - y_a = 2 \text{ m}$$

The shape assumed by the cable is a catenary whose lowest point has coordinates (x, y) equal to $(0, c)$. The description of the catenary is

$$y = c \cosh(x/c); \qquad \cosh t = (e^t + e^{-t})/2$$

so that we have

$$y_b - y_a = c \cosh(x_b/c - [x_b - 13]/c)$$

The length L is given by

$$L = c \sinh(x_b/c - [x_b - 13]/c); \qquad \sinh t = (e^t - e^{-t})/2$$

Use the multivariable Newton–Raphson method to determine x_b and c. Note that x_b must be greater than 6.5 m and that c must be large enough to prevent computer overflow during the solution process.

11. The displacement x [m] of a mass undergoing a damped oscillation varies with time t [s] according to the model

$$x = -0.1e^{\beta t}[\cos(\omega t) - (\beta/\omega)\sin(\omega t)]$$

in which β and ω have units of s^{-1}. Measurements give a displacement x_1 of 0.0162 m at a time t_1 of 0.41 s, and a displacement x_2 of (-0.0026) m at a time t_2 of 0.83 s. The values x_1 and x_2 are near the maximum and minimum displacements, respectively. Use these values in the model for x to determine β and ω. Initial estimates for β and ω may be found from the proximity of x_1 and x_2 to the extrema of the displacement. These estimates are

$$\beta = [\ln(-x_2/x_1)]/[t_2 - t_1]; \qquad \omega = \pi/[t_2 - t_1]$$

12. The average monthly temperatures T_m [°F] for a city (Chicago) correspond to the months denoted by m as shown:

Month, m:	1	2	3	4	5	6	7	8	9	10	11	12
T_m [°F]:	21	26	36	49	59	69	73	72	65	54	40	28

An engineer at a utility plant wishes to model the temperature T with the month m according to

$$T(m) = T_a + A\sin(m\pi/6 - \beta); \quad (m\pi/6 - \beta) \text{ in radians}$$

The constants T_a, A, and β for the model are to be determined by a nonlinear least-squares analysis (see Chapter 4) in which we seek to minimize

$$f(T_a, A, \beta) = \sum_{m=1}^{12} [T_m - T(m)]^2$$

In other words, we must solve the three simultaneous equations

$$\delta f/\delta T_a = 0; \quad \delta f/\delta A = 0; \quad \delta f/\delta \beta = 0$$

Use the multivariable Newton–Raphson method to obtain T_a, A, and β. The initial estimate of T_a may be taken as the average of T_m, the initial estimate of the amplitude A may be set to a half of the (positive) difference between the minimum and maximum T_m values, and the phase angle β (which has multiple solutions) may be initially estimated by assuming that the maximum T_m value occurs when the quantity $(m\pi/6 - \beta)$ is equal to $\pi/2$.

13. Find all real roots of the cubic equation for r in Problem 3 by Bairstow's method.

14. Recast the open channel flow equation in Problem 4 as the fifth-degree polynomial equation in y and find all real roots of the polynomial by Bairstow's method.

15. Replace the mass $2m$ in Fig. 3–7 by a mass $3m$ and solve for the natural frequencies of the resulting system.

16. The Birge–Vieta method for roots of a polynomial involves finding the linear factors $\{L = (x - w)\}$ of $P_n(x)$. The general relation for this method is

$$P_n(x) = LS_{n-1}(x) + R(w)$$

where $S_{n-1}(x)$ is a polynomial of degree $(n - 1)$ and $R(w)$ is the remainder.

Develop the synthetic division relations for finding R and the coefficients of $S_{n-1}(x)$.

PART II

DATA ANALYSIS

4

Statistics and Least-Squares Approximation

Engineers generate and collect numerical data as they attempt to understand the phenomena with which they are dealing. Data obtained from tests or experiments are usually in a "raw" form. The purpose of data analysis is to develop models from the raw data so that we can predict the behavior of the phenomenon that was tested. Among the tools of data analysis are statistics and methods for fitting curves to the measured data. The two topics of this chapter are elementary statistics and the least-squares method for fitting curves to data. Other curve-fitting methods will be treated in Chapter 5.

A ***product sampling*** to predict durability is used as a case study throughout the discussion of statistics. The treatment of least-squares approximations contains an example of a ***cooling*** problem.

4.1 ELEMENTARY STATISTICS

In elementary statistics, we seek a central tendency of a quantity that is subject to random variation, a measure of that variation, and a way to predict the probability that the quantity of interest will lie in a given range. Information on the quantity is sought for an entire set or ***population***; however, we usually obtain this information by testing only a part of the population. The part of the population on which measurements are

Table 4–1   Rebound Height x in Inches for a Sample of 50 Racquetballs

Ball	x	Ball	x	Ball	x	Ball	x	Ball	x
1	69.4	11	69.4	21	68.7	31	71.3	41	69.7
2	70.1	12	69.9	22	71.2	32	70.3	42	69.3
3	70.6	13	70.2	23	69.3	33	71.8	43	67.7
4	68.4	14	68.1	24	68.7	34	71.1	44	71.6
5	73.0	15	70.7	25	71.1	35	67.5	45	71.5
6	68.2	16	72.7	26	72.1	36	69.5	46	69.1
7	71.3	17	69.3	27	70.8	37	68.4	47	70.2
8	69.6	18	71.6	28	72.2	38	71.2	48	69.4
9	69.8	19	70.9	29	69.2	39	72.3	49	71.2
10	70.9	20	69.9	30	71.1	40	69.9	50	67.1

made is known as a *sample*, the number of measurements is the *sample size* n, and the values of the measurements are denoted by x_1 through x_n.

Consider, for example, a test for the durability of racquetballs. The quantity to be measured is the rebound height of balls that are dropped from 100 inches above the floor, after the balls have been subjected to a certain period of simulated playing time. We can measure the rebound height for all of the balls that are produced on a given day. Such testing is time-consuming, and it is costly because of the testing itself and because the manufacturer will be left with a day's production of racquetballs that cannot be sold.

We can instead test a sample from the day's production. The balls in the sample should be chosen randomly and not according to some prearranged pattern such as every 20th ball. The raw data for the test is typically collected in a table in the same order as they are obtained. Table 4–1 is an example of the results of a 50-ball sample with the rebound height denoted by x and given in inches. This sample will be used as a vehicle for discussing statistical analysis.

4.1.1   Statistical Quantities from Individual Measurements

The data in Table 4–1 represent individual measurements of the rebound height which is denoted by x_j for ball j. Although the data can be analyzed in the existing form, it is customary to rearrange the x values in ascending order. After the rearrangement, x_j no longer refers to a particular ball j; it refers to the value of x in numerical position j. The rearrangement can be done on a computer by a sorting procedure. One of these is the *bubble sort* algorithm, which is described in many programming texts. For convenience, Pseudocode 4–1 gives one implementation of the procedure with n values of x_1 through x_n.

In essence, we are seeking to sort the sequence from the minimum value at the top of the stack ($j = 1$) to the maximum at the bottom ($j = n$). With each sweep,

Pseudocode 4–1   The bubble sort algorithm

Initialize: n, and x_1 through x_n in the original order

$\quad K \leftarrow n - 1$ $\qquad$ \ K = number of x pairs to be compared in the **For** loop

$\quad L \leftarrow 0$ $\qquad\quad$ \ L is a switch; 0 = sort incomplete, 1 = sort complete

While $[L = 0]$

$\quad L \leftarrow 1$ $\qquad\qquad$ \ Assume that the values are sorted

$\quad$ **For** $[\,j = 1, 2, \ldots, K\,]$

$\qquad$ **If** $[x_j > x_{j+1}]$ **then** $\qquad$ \ exchange x values; set L switch to 0

$\qquad\quad$ xtemp $\leftarrow x_j$

$\qquad\quad x_j \leftarrow x_{j+1}$

$\qquad\quad x_{j+1} \leftarrow$ xtemp

$\qquad\quad L \leftarrow 0$

$\qquad$ **End If**

$\quad$ **End For**

$\quad K \leftarrow K - 1$ $\qquad$ \ Reduce K for the next sweep of the x values

End While

Write: Sorted values x_1 through x_n

---

the largest of the first $(K + 1)$ values is pushed down to position $(K + 1)$ while the remaining values "bubble" to the top. The procedure ends when exchanges of x values no longer occur; the switch L remains at the value of 1 that is set prior to each sweep.

Application of the sort algorithm to the data in Table 4–1 produces a sorted set of rebound heights arranged by rows in Table 4–2.

By sorting the data, it is easy to see the range of values for x. To find a central tendency, we may use one of several measures. The ***median*** value is the middle value of the sorted data if the sample size n is odd, or it is half the sum of the two middle values if n is even. The sample size is 50 in our case; therefore, the median value of 70.15 in. is half the sum of values 25 and 26.

The ***arithmetic mean*** is usually denoted by $\bar{x}$; however, we shall refer to it by the symbol A because the overbar is often mistaken for a minus sign in some

Table 4–2   Sorted Rebound Heights x in Inches from Table 4–1

67.1	67.5	67.7	68.1	68.2	68.4	68.4	68.7	68.7	69.1
69.2	69.3	69.3	69.3	69.4	69.4	69.4	69.5	69.6	69.7
69.8	69.9	69.9	69.9	70.1	70.2	70.2	70.3	70.6	70.7
70.8	70.9	70.9	71.1	71.1	71.1	71.2	71.2	71.2	71.3
71.3	71.5	71.6	71.6	71.8	72.1	72.2	72.3	72.7	73.0

expressions. This mean is the most commonly used central measure and is often referred to simply as the mean value. It is defined by

[4.1]
$$A \equiv [1/n] \sum_{j=1}^{n} (x_j) = (x_1 + x_2 + \cdots + x_n)/n$$

Other means are the harmonic mean H and the geometric mean G. H is defined by

[4.2]
$$H \equiv n \left/ \left[\sum_{j=1}^{n} (1/x_j) \right] \right. = n/[1/x_1 + 1/x_2 + \cdots + 1/x_n]$$

and its name is due to the harmonic series in the denominator. The definition of G and alternative expressions for it are

[4.3a]
$$G \equiv \left[\prod_{j=1}^{n} (x_j) \right]^{(1/n)} = [(x_1)(x_2)\ldots(x_n)]^{(1/n)}$$

[4.3b]
$$G = \left[\prod_{j=1}^{n} (x_j)^{(1/n)} \right] = \left[(x_1)^{(1/n)}(x_2)^{(1/n)} \ldots (x_n)^{(1/n)} \right]$$

[4.3c]
$$G = \exp(\ln G) = \exp \left[(1/n) \sum_{j=1}^{n} [\ln x_j] \right]$$

Direct use of Eq. [4.3a] to compute G may cause machine overflow with large x values or machine underflow with small x values. Use of Eq. [4.3b] or [4.3c] usually eliminates such difficulties.

The values of A, H, and G for the data in Table 4–2 are, respectively, 70.17, 70.14, and 70.16 in. (rounded to four significant figures). These values and the median value are similar since they are all central measures, although they are defined differently.

The measure of how the values are dispersed about the mean value is the *sample standard deviation* s. The square of the sample standard deviation is known as the *sample variance*, which is a weighted average of the squares of the deviations of x_j from the arithmetic mean A. The definition of the sample variance is

[4.4a]
$$s^2 \equiv [1/(n-1)] \left[\sum_{j=1}^{n} (x_j - A)^2 \right]$$

The form of Eq. [4.4a] requires us to compute A before we can perform the summation on the right-hand side. An alternate form is

$$[4.4b] \qquad s^2 = [1/(n-1)]\left[\left\{\sum_{j=1}^{n}(x_j)^2\right\} - nA^2\right]$$

The advantage of Eq. [4.4b] is that the sum of the x values required for A and the sum of the squares of x required for s may be computed in the same loop. This feature is especially convenient if data are entered one value at a time. In such a case, data entry can also be done within the summation loop.

Although Eq. [4.4a] is less convenient to use than Eq. [4.4b], it guarantees a zero or positive variance. The alternate form of Eq. [4.4b] is subject to round-off errors that can become severe enough to produce a very inaccurate or even negative variance when the difference term on the right of the expression is small in comparison to the quantities that produce the difference. We therefore look at another "crash-proof" procedure that retains the convenience of a single loop for data input and iterated computations.

Let A_j be the arithmetic mean of the first j sample values, and let s_j be the corresponding sample standard deviation. From Eq. [4.1], we obtain

$$A_{j-1} = (x_1 + x_2 + \cdots + x_{j-1})/(j-1); \qquad A_j = (x_1 + x_2 + \cdots + x_j)/j$$

which may be combined to give

$$[4.5] \qquad jA_j = (j-1)A_{j-1} + x_j$$

From Eq. [4.4b], we obtain the relations

$$(j-1-1)(s_{j-1})^2 = (x_1)^2 + (x_2)^2 + \cdots + (x_{j-1})^2 - (j-1)(A_{j-1})^2$$
$$(j-1)(s_j)^2 = (x_1)^2 + (x_2)^2 + \cdots + (x_j)^2 - j(A_j)^2$$

which allow us to write

$$(j-1)(s_j)^2 = (j-2)(s_{j-1})^2 + (j-1)(A_{j-1})^2 + (x_j)^2 - j(A_j)^2$$

We then use Eq. [4.5] to replace $(A_{j-1})^2$ in this last expression and manipulate the result to obtain the recursive relation

$$[4.6] \qquad (j-1)(s_j)^2 = (j-2)(s_{j-1})^2 + j(x_j - A_j)^2/(j-1)$$

To use Eq. [4.5] and [4.6], the sample size n must be at least 2. The terms on the right of Eq. [4.6] are then guaranteed to be nonnegative. We begin the computation

---

Pseudocode 4–2   Computation of the mean and standard deviation

---

Initialize: n and x_1 \ n = sample size, x_1 = first data point
　　　$A \leftarrow x_1$
　　　ssq $\leftarrow 0$ \ ssq is used to denote the square of s
For [j = 2, 3, ..., n]
　　　Initialize: x_j \ get new x value
　　　　$A \leftarrow$ solution for A_j from Eq. [4.5]   \ update A_j
　　　　ssq $\leftarrow$ solution for $(s_j)^2$ from Eq. [3.6]   \ update ssq
End For
Write: A and the square root of ssq   \ These are the values of A and s

---

of A and s by initializing j to 1, A_1 to the first value x_1, and $(s_1)^2$ to 0. We then update A_j and $(s_j)^2$ for j values from 2 through n. A code that incorporates data entry in the computation loop is given in Pseudocode 4–2.

The value of s for the sample in Table 4–2 is 1.388 in. (rounded to four digits). When the data is for a population instead of just a sample, the measure of dispersion is the *population standard deviation* σ whose square is the *population variance* defined by

$$[4.7] \qquad \sigma^2 \equiv (1/n) \left[\sum_{j=1}^{n} (x_j - A)^2 \right]$$

Clearly, s and σ tend to each other as n increases. For a population of 1, σ is zero; when the sample size is 1, s is (0/0) or indeterminate. Note too that the sum in Eq. [4.7] is for the squares of the deviations of x_j from A. Division by n gives the mean square deviation; thus σ is also the *root mean square* (or *rms*) deviation of the data from the mean. The values of s and σ are related (for n representing both the sample and population sizes) by

$$[4.8] \qquad s^2 = n\sigma^2/(n-1)$$

The arithmetic mean and sample standard deviation may now be used to make predictions about the entire population from which the sample was drawn. This process will be deferred until we have discussed grouped data.

4.1.2   Statistical Quantities from Grouped Data

Examination of the data in Table 4–2 reveals that some of the x values are repeated. Such data can be grouped with the number of occurrences of a given x value defined as the frequency $f(x)$. The grouping can also be done with less resolution than indicated by the original data. For example, the data for x may be accurate only to the nearest inch despite the manner in which the measurement is reported. We may then group the data according to *class*. The procedure for this kind of grouping begins with a suitable choice of an interval containing all of the sample values. The interval is then divided into equal subintervals called *class intervals*.

For the data in Table 4–1 or 4–2, we might choose a class interval of 1 and locate the intervals so that the *class midpoint* or *interval midpoint* is a whole number. Members of a particular class are those that satisfy

$$(\text{Midpoint} - \text{Subinterval}/2) \le x < (\text{Midpoint} + \text{Subinterval}/2)$$

Such a classification corresponds to rounding and places an x value that is exactly at the boundary between two classes in the higher class.

Classification of the data in Table 4–1 or 4–2 gives m class midpoints. We shall denote these by X_j (to avoid confusion with the x_j notation that is used for individual measurements). The appropriate class midpoints are $(67, 68, \ldots, 73)$ so that the group size m is 7 and X_j can be defined by

$$X_j = 66 + j; \qquad j = 1, 2, \ldots, m$$

The frequency $[f_j \equiv f(X_j)]$ is the number of data values satisfying

$$(X_j - 0.5) \le x < (X_j + 0.5)$$

Because numbers may not be represented exactly on a computer, some care must be used in computer algorithms for finding the frequencies. For example, a value x at the boundary of two class intervals may be missed altogether if it falls into a between-interval gap created by inexact representation of numbers. The algorithm presented in Pseudocode 4–3 is suggested for our example.

Pseudocode 4–3   Grouping of statistical data

Initialize: sample size n and group size m
Initialize: individual data values x_1 through x_n
For $[j = 1, 2, \ldots, m]$
    $X_j \leftarrow 66 + j$ \ establish class midpoints
    $f_j \leftarrow 0$ \ set frequencies to initial values of zero
End For
For $[k = 1, 2, \ldots, n]$ \ examine individual data values
    $j \leftarrow$ integer part of $(x_k + 0.5) - 66$ \ establish class
    $f_j \leftarrow f_j + 1$ \ update frequency
End For
Write: Table of X_j and f_j values

Table 4–3 Frequency Table for the Rebound Heights of Table 4–2

Interval j	Class Midpoint X_j [in.]	Frequency f_j	Cumulative Frequency F_j
1	67	1	1
2	68	6	7
3	69	10	17
4	70	11	28
5	71	13	41
6	72	7	48
7	73	2	50

Table 4–3 shows a frequency table for the grouped data. In addition to the interval number j, class midpoint X_j, and frequency f_j, the table contains a **cumulative frequency**, which is the sum of the frequencies from f_1 through f_j. This last quantity is denoted by F_j and is another tool for quick interpretation of the data.

There are several ways to display the data visually. Among them are the histogram shown in Fig. 4–1 and the frequency polygon of Fig. 4–2. A histogram is similar to a bar chart with the width of the "bars" indicating the class interval and with the bars immediately adjacent to each other. The frequency polygon is a plot of straight-line segments between each (X_j, f_j) pair and extends to one midpoint on either side of the data in Table 4–3 to show where the frequencies become zero.

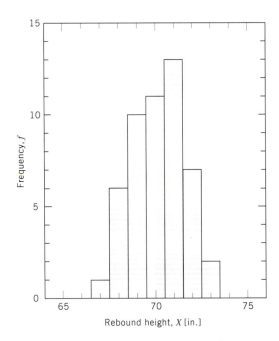

Figure 4–1 Histogram for data in Table 4–3.

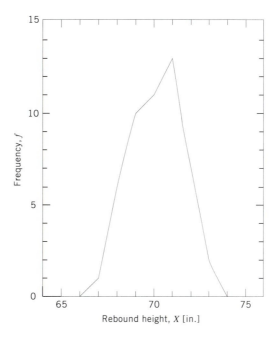

Figure 4–2   Frequency polygon for data in Table 4–3.

The median value of the grouped data is half the sum of measurements 25 and 26 as before. We see very quickly from the cumulative frequency in Table 4–3 that both of these are in interval 4, so that the median obtained from the grouped data is 70 in. (compared with 70.15 found earlier from the individual measurements). A new measure of central tendency that is appropriate for grouped data is the **mode** or most frequently occurring value. The mode for our example is 71 in.

The quantities found previously for individual measurements can be expressed easily in terms of the frequencies if it is recognized that summations within a class are now replaced with multiplications by f, and multiplications within a class are replaced with quantities raised to the power f. Conversely, the expressions for the grouped data can be used to recover those for individual measurements by treating each measurement as a group with frequency 1. The expression for the arithmetic mean A is

[4.9]
$$A = (1/n) \sum_{j=1}^{m} (f_j X_j); \quad n = \sum_{j=1}^{m} (f_j)$$

The expressions for the harmonic mean H and the geometric mean G (with sample size n as in Eq. [4.9]) are

[4.10]
$$H = n \left/ \left[\sum_{j=1}^{m} (f_j / X_j) \right] \right.$$

$$[4.11] \qquad G = \left[\prod_{j=1}^{m} (X_j)^{f_j} \right]^{(1/n)} = \prod_{j=1}^{m} (X_j)^{(f_j/n)}$$

$$= \exp\left[(1/n) \sum_{j=1}^{m} [f_j \ \ln(X_j)] \right]$$

The sample standard deviation s is defined by

$$[4.12] \qquad (n-1)s^2 = \sum_{j=1}^{m} [f_j(X_j - A)^2] = \left[\sum_{j=1}^{m} [f_j(X_j)^2] \right] - nA^2$$

The recursive relations for A and s^2 involve the cumulative frequencies F_j. The intermediate quantity A_j now denotes the mean of the first j **groups**, and s_j is the corresponding sample standard deviation. We first set F_1 equal to f_1, A_1 equal to X_1, and s_1 equal to 0 (where the group represented by j equal to 1 is the first group for which the frequency is not zero). Then the recursive forms for the cumulative frequency and for the group versions of Eq. [4.5] and [4.6] are

$$[4.13] \qquad\qquad\qquad F_j = F_{j-1} + f_j$$

$$[4.14] \qquad\qquad\qquad F_j A_j = (F_j - f_j)A_{j-1} + f_j X_j$$

$$[4.15] \qquad (F_j - 1)(s_j)^2 = (F_j - f_j - 1)(s_{j-1})^2 + F_j f_j (X_j - A_j)^2/(F_j - f_j)$$

There may be classes or intervals for which the frequency is zero; Eq. [4.13] through [4.15] are to be applied for j values from 2 through m only when f_j is not zero. If f_j is zero, the F_j, A_j, and $(s_j)^2$ are the same as for the previous group $(j-1)$. One particular instance in which the application of Eqs. [4.13] through [4.15] is disastrous for zero frequency occurs when f_1 is equal to 1 and f_2 is equal to zero. The solution of $(s_2)^2$ from Eq. [4.15] will then cause a division by zero to be attempted.

In dealing with the grouped data, we have introduced new notation and new indexing. It is important to recognize the difference between the value x and the class midpoint X, between the sample size n and the group size m, and between the frequency f and the cumulative frequency F. Values of A, H, G, and s for the grouped data of Table 4–3 are shown in Table 4–4 (rounded to three decimal places)

Table 4–4   Comparison of Statistical Quantities from Grouped and Individual Data

| | Values [inch] for | |
Quantity	Grouped Data	Individual Data
A	70.160	70.170
H	70.131	70.143
G	70.146	70.157
s	1.434	1.388

and are compared to values found previously from the individual measurements. Note that the standard deviation is considerably more sensitive than the mean values to the reduced resolution in the grouped data.

4.1.3 Prediction of Behavior from Statistical Quantities

We turn now to the use of the arithmetic mean A and the sample standard deviation s to make predictions about the entire population. We shall continue to use the racquetball sampling as a vehicle for discussion, and we shall use the values obtained for the grouped data of §4.1.2. The fundamental objective is to determine the probability that a measurement x for a randomly chosen member of the population will fall within a given range. To reach this objective, we employ a model of random behavior known as the **normal distribution** or **Gauss distribution model**.

The normal or Gauss distribution is characterized by a **density function** $g(x)$, which is defined by

[4.16] $$g(x) = e^{-[(x-A)/s]^2/2}/\left(s\sqrt{2\pi}\right)$$

The plot of $g(x)$ versus x is a bell-shaped curve, which is symmetric about the line $(x = A)$. The range of x is assumed to be from $-\infty$ to ∞ even if the phenomenon that is under investigation may not physically extend over such a range. The area between the x axis and the curve is 1. This fact allows us to relate $g(x)$ to the frequencies $f(X_j)$ for grouped data.

The area of a "bar" in a histogram for grouped data is $(f_j\Delta x)$ where f_j is the class frequency and Δx is the class interval. Therefore, the area of the entire histogram is found by summing the areas of the m "bars" to obtain (frequency sum $\times \Delta x$) or $(n\,\Delta x)$. The same result is obtained by summing areas of triangles and trapezoids to find the total area between the x axis and the frequency polygon. Division of the histogram or polygon area by $(n\,\Delta x)$ allows us to scale the area to a value of 1; consequently, the quantity $[f_j/(n\,\Delta x)]$ is a scaled, discrete representation of the continuous density function $g(x)$. In other words, $g(x)$ is a model for $[f_j/(n\,\Delta x)]$ for an infinite number of classes with infinitesimally small class intervals.

For the racquetball sampling that we discussed in §4.1.2, Δx is 1 and the sample size n is 50. The plot of $[f_j/(n\,\Delta x)]$ versus x is therefore equivalent to reducing the height of the frequency polygon in Fig. 4–2 by a factor of 50. The density function $g(x)$ and the scaled frequency polygon for the data of Table 4–3 are compared in Fig. 4–3. The supposition of the Gauss distribution model is that a larger sample size (or possibly one with smaller class intervals) will result in a scaled frequency polygon that more closely resembles the density function $g(x)$.

We now consider the problem of predicting probabilities from the normal distribution. The unit area between the density function and the x axis represents the probability of occurrence of all possible values of x. The probability that a randomly chosen value of x will be no greater than a given value x_a is then the ratio of the subarea bounded by the density function, the x axis, the line $x = -\infty$, and the line

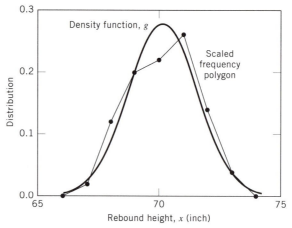

Figure 4–3   Density function and scaled frequency polygon for
the data of Table 4–3.

$x = x_a$ to the total unit area; that is, it is represented by the subarea. This probability is denoted by $P(x \leq x_a)$, or by $P(-\infty < x \leq x_a)$ in strict accord with the distribution model. Because the area under a curve is the physical representation of a definite integral, we may express the probabilities that x will occur in given ranges by

$$P(-\infty < x \leq x_a) = \int_{-\infty}^{x_a} g(x)\,dx; \quad P(-\infty < x \leq x_b) = \int_{-\infty}^{x_b} g(x)\,dx;$$

$$P(x_a < x \leq x_b) = \int_{x_a}^{x_b} g(x)\,dx$$

It is not possible to evaluate the integrals above by elementary methods, but we may represent them in terms of another integral $\Phi(z)$ known as the **cumulative distribution function** (which is usually shortened to distribution function and is not to be confused with the distribution itself). The distribution function is defined by

[4.17] $$\Phi(z) \equiv \int_{-\infty}^{z} g_s(u)\,du = \left(1/\sqrt{2\pi}\right)\int_{-\infty}^{z} e^{-u^2/2}\,du$$

The **standardized normal distribution** $g_s(u)$ in Eq. [4.17] is equivalent to $g(x)$ in Eq. [4.16] with the mean A equal to 0 and the standard deviation s equal to 1. The function $\Phi(z)$ is the probability $P(-\infty < u \leq z)$, which is equivalent to $P(-\infty < x \leq [sz + A])$.

The probabilities given earlier for the occurrence of x in given ranges may now be expressed in terms of the distribution function $\Phi(z)$ by

[4.18a] $$P(-\infty < x \leq x_a) = \Phi(z_a); \quad z_a = (x_a - A)/s$$

[4.18b] $$P(-\infty < x \le x_b) = \Phi(z_b); \quad z_b = (x_b - A)/s$$

[4.18c] $$P(x_a < x \le x_b) = \Phi(z_b) - \Phi(z_a); \quad z_b > z_a$$

The physical representation of the distribution function as it pertains to Eq. [4.18a, b,c] is shown in Fig. 4–4.

Although it is also not possible to evaluate the integral in Eq. [4.17] by elementary methods, the dependence on A and s for a specific distribution is removed. This fact allows us to use a single general algorithm for computing the probabilities. An infinite series expression for the integral in Eq. [4.17] is found by expanding the exponential function in series form and integrating the series term by term. The resulting expression for $\Phi(z)$ is

[4.19] $$\Phi(z) = 0.5 + \left(1/\sqrt{\pi}\right)\sum_{n=0}^{\infty} c_n;$$

$$c_0 = z/\sqrt{2}; \quad c_n = \{z^2(1 - 2n)/(2n[1 + 2n])\}c_{n-1} \text{ for } n > 0$$

The summation of the c_n terms in Eq. [4.19] is carried out until the magnitude of c_n becomes extremely small at some n. It is worthwhile to note that $\Phi(z)$ varies from a value of 0 when z is $(-\infty)$ to a value of 1 when z is ∞, and that

[4.20] $$\Phi(z) = 1 - \Phi(-z)$$

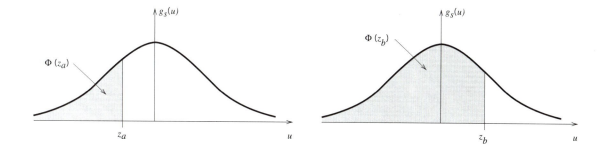

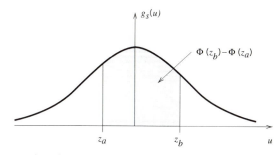

Figure 4–4 Area representation of the distribution function.

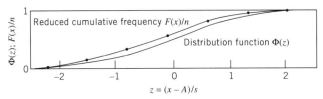

Figure 4–5 Comparison of $\Phi(z)$ and the reduced cumulative frequency from Table 4–3.

Thus, a function for computing $\Phi(z)$ may be restricted to nonnegative arguments and modified according to Eq. [4.20], if necessary, for negative arguments. The convergence of the series in Eq. [4.19] is rapid for z values up to 1, but it is slow for z values with large magnitudes. For example, the value of n at which c_n becomes smaller than 10^{-9} is 9 when z is 1, 16 when z is 2, 24 when z is 3, and 34 when z is 4. Even so, the cost of computing $\Phi(z)$ from Eq. [4.19] is not prohibitive, and the expression for $\Phi(z)$ is easy to code. For z values with large magnitudes, an alternative to the series computation is the asymptotic expansion approximation

[4.21] $$\Phi(z) \simeq 1 - e^{-z^2/2} \Big/ \left(z \sqrt{2\pi} \right); \quad z \gg 1$$

The values of $\Phi(z)$ from Eq. [4.19] and [4.21] agree to five significant digits when z is 4.

The discrete counterpart to the distribution function is the reduced cumulative frequency $F(x)/n$ from a grouped sample of size n, in which x is related to the variable z by

$$z = (x - A)/s$$

The reduced cumulative frequency from the racquetball sample of Table 4–3 is compared to the distribution function in Fig. 4–5. Values of $\Phi(z)$ are shown in Table 4–5 for several values of z.

Table 4–5 Values of the Distribution Function $\Phi(z)$
$[\Phi(0) = 0.5; \quad \Phi(-z) = 1 - \Phi(z)]$

z	$\Phi(z)$	z	$\Phi(z)$	z	$\Phi(z)$	z	$\Phi(z)$
0.1	0.53983	1.1	0.86433	2.1	0.98214	3.1	0.99903
0.2	0.57926	1.2	0.88493	2.2	0.98610	3.2	0.99931
0.3	0.61791	1.3	0.90320	2.3	0.98928	3.3	0.99952
0.4	0.65542	1.4	0.91924	2.4	0.99180	3.4	0.99966
0.5	0.69146	1.5	0.93319	2.5	0.99379	3.5	0.99977
0.6	0.72575	1.6	0.94520	2.6	0.99534	3.6	0.99984
0.7	0.75804	1.7	0.95543	2.7	0.99653	3.7	0.99989
0.8	0.78814	1.8	0.96407	2.8	0.99744	3.8	0.99993
0.9	0.81594	1.9	0.97128	2.9	0.99813	3.9	0.99995
1.0	0.84134	2.0	0.97725	3.0	0.99865	4.0	0.99997

To demonstrate the use of the distribution function, we return to the racquetball sample in the following example.

EXAMPLE 4.1

The rules of one racquetball body state that a ball dropped from 100 in. above the floor should rebound to a height of 68 to 72 in. Use the results $A = 70.16$ in. and $s = 1.4337$ in. for the grouped data to determine

a. The fraction of the racquetball population of §4.1.2 that is expected to meet the rules.

b. The fraction of the population that is expected to exceed the minimum rebound height of 68 in.

---

a. If we interpret the rules to mean that the rebound height x must satisfy the condition $(68 \leq x \leq 72)$, we are seeking the probability $P(68 \leq x \leq 72)$.

To get the form of Eq. [4.18c], which has a strict inequality, we write the probability as $P(68 - \epsilon < x \leq 72)$ where ϵ is an arbitrarily small positive quantity. Then we have

$$P(68 \leq x \leq 72) \simeq P(68 < x \leq 72)$$
$$= \Phi([72 - A]/s) - \Phi([68 - A]/s)$$
$$= \Phi(1.2834) - \Phi(-1.5066) = 0.900 - 0.066 = 0.834$$

b. Fraction $= P(x > 68) = P(68 < x \leq \infty) = \Phi(\infty) - \Phi(-1.5066)$
$$= 1 - 0.066 = 0.934$$

Some commonly used limits in statistics are the 1-, 2-, and 3-standard deviation limits. Probability values associated with these limits are shown in Table 4–6. The table indicates that 68.27% of a set of measurements is expected to fall within 1 standard deviation of the mean, 95.45% is expected to fall within 2 standard deviations of the mean, and 99.73% (or nearly all of the measurements) is expected to fall within 3 standard deviations of the mean.

When the probability is converted to a percentage, the percentage is known as the **confidence level**. Some commonly used confidence levels are 95%, 99%, and

Table 4–6 Probability $P(x_a < x \leq x_b)$ for Commonly Used Limits

x_a	x_b	P
$A - s$	$A + s$	0.6827
$A - 2s$	$A + 2s$	0.9545
$A - 3s$	$A + 3s$	0.9973

Table 4–7 Limits for Commonly Used Confidence Levels

Confidence Level	x_a	x_b
95%	$A - 1.96s$	$A + 1.96s$
99%	$A - 2.58s$	$A + 2.58s$
99.9%	$A - 3.29s$	$A + 3.29s$

99.9%. The limits x_a and x_b (equidistant from the mean) that give these levels are shown in Table 4–7.

The limits in Table 4–7 are special cases that are symmetric about the mean. There are more general situations that require the variable z to be determined when $\Phi(z)$ is given. The root-solving methods of Chapter 3 or interpolation may be used to obtain z in such cases. A simple two-point interpolation is demonstrated in the following example.

EXAMPLE 4.2

What is the rebound height h that 90% of the racquetball population in Example 4.1 is expected to exceed?

Let the height sought be h. Then we may write

$$P(x > h) = P(h < x \leq \infty) = \Phi(\infty) - \Phi(z_h) = 1 - \Phi(z_h) = 0.9$$

We may obtain h from

$$z_h = (h - A)/s$$

after finding z_h such that 0.1 is the value of $\Phi(z_h)$. A value of $\Phi(z)$ close to 0.1 is not shown in Table 4–5 because the corresponding z value is negative. However, we note from Eq. [4.20] that $\Phi(z)$ is equal to $[1 - \Phi(-z)]$. We therefore seek $(-z_h)$ such that 0.9 is the value of $\Phi(-z_h)$.

Table 4–5 shows bracketing values as follows:

$$\Phi(z) = 0.88493 \text{ when } z = 1.2; \qquad \Phi(z) = 0.90320 \text{ when } z = 1.3$$

Linear interpolation is used to find $(-z_h)$ from

$$(0.90320 - 0.90000)/(0.90320 - 0.88493) = (1.3 - [-z_h])/(1.3 - 1.2)$$

We find therefore that z_h is equal to -1.282485, and we compute h from

$$A = 70.16 \text{ in.}; \quad s = 1.4337; \quad h = A + s(z_h) = 68.3 \text{ in.}$$

4.1.4   The Chi-Squared Distribution

The Gauss distribution model is so widely used that we often fail to consider how appropriate it is for a given set of data. A test for goodness of fit is based on the *chi-squared distribution* (Ref. 7,8), which is described as follows.

Let us consider k independent variables u_1 through u_k, which are subject to random variations and which have a normal distribution with a zero mean and a unit standard deviation. The sum of the squares of these variables is denoted by

[4.22]
$$(\chi_k)^2 = \sum_{j=1}^{k} (u_j)^2$$

in which k is known as the *number of degrees of freedom*. The probability that $(\chi_k)^2$ belongs to the interval from χ^2 to $[\chi^2 + d(\chi^2)]$ is

[4.23]
$$p(\chi^2)d(\chi^2) = \left\{ (\chi^2)^{(k/2-1)} e^{-\chi^2/2} \Big/ \left[2^{k/2}\Gamma(k/2) \right] \right\} d(\chi^2)$$

where $p(\chi^2)$ is the density function for the chi-squared distribution and $\Gamma(\alpha)$ is the *gamma function* (Ref. 7,8,9) defined by

[4.24]
$$\Gamma(\alpha) \equiv \int_0^{\infty} e^{-t} t^{(\alpha-1)} dt$$

The gamma function may be thought of as a generalized factorial function; some properties that are useful for our discussion are

[4.25]
$$\Gamma(\alpha + 1) = \alpha\Gamma(\alpha); \quad \Gamma(1) = 1; \quad \Gamma(1/2) = \sqrt{\pi}$$

We find from Eq. [4.25] that $\Gamma(k/2)$ for integer k values greater than 2 may be obtained from

[4.26a]
$$\Gamma(k/2) = (k/2 - 1)!; \qquad k = 4, 6, \ldots$$

[4.26b]
$$\Gamma(k/2) = (k/2 - 1)(k/2 - 2)\ldots(1/2)\sqrt{\pi}; \qquad k = 3, 5, \ldots$$

The chi-squared density function $p(\chi^2)$ from Eq. [4.23] may be computed with the aid of Eq. [4.25] and [4.26a,b], is valid for χ^2 ranging from 0 to ∞, depends on a single parameter k, has a mean value equal to k, and has a variance equal to $2k$. The density function is shown in Fig. 4–6 for four values of k. The distribution curve increases with χ^2 from a value of zero to a maximum value, and then decreases again. For small values of k, the distribution is skewed strongly to the left of its mean value; as k increases, the distribution converges slowly to a normal distribution.

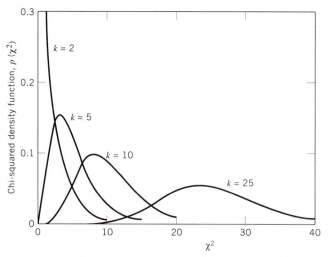

Figure 4–6   Chi-squared density function for 2, 5, 10, and 25 degrees of freedom.

The probability that χ^2 is in the the range $(0 \leq \chi^2 \leq c)$ is the chi-squared distribution function $P(\chi^2 \leq c)$. It is obtained by integrating the chi-squared density function to get

[4.27] $$P(\chi^2 \leq c) = \int_0^c p(\chi^2)d(\chi^2) = \gamma(k/2, c/2)/\Gamma(k/2)$$

The function $\gamma(\alpha, x)$ is one of several functions that are called **incomplete gamma functions** (Ref. 7,8,9), and it is defined by

$$\gamma(\alpha, x) \equiv \int_0^x e^{-t} t^{(\alpha-1)} dt$$

Note that the integrand is the same as in Eq. [4.24], but that the upper limit of integration is finite rather than infinite.

A series representation for $P(\chi^2 \leq c)$ is given by

[4.28] $$P(\chi^2 \leq c) = e^{[(k/2)\ln(c/2)-c/2]} \left[\sum_{n=0}^{\infty} [(c/2)^n/\Gamma(k/2+1+n)] \right]$$

Let us use T_n to denote the n-th term of the summation; that is, let us write

[4.29] $$T_n = (c/2)^n/\Gamma(k/2+1+n)$$

Then we have from Eq. [4.25]

[4.30] $T_0 = 2/[k\Gamma(k/2)]; \qquad T_n = cT_{n-1}/(k + 2n)$ for $n > 0$

Pseudocode 4–4 for computing $P(\chi^2 \leq c)$ with k degrees of freedom is shown as follows.

Pseudocode 4–4 Computation of the chi-squared distribution function

```
Initialize: k, c          \ k = positive integer; c ≥ 0
Initialize: ε             \ ε = a small positive value used to end summation
Gamma(k, G)               \ call module to obtain G = Γ(k/2)
T ← 2/(kG)                \ evaluate the term T₀
S ← T                     \ set the sum equal to the 0-th term
n ← 0                     \ initial summation index
While [T > ε]
    n ← n + 1             \ increment summation index
    T ← cT/(k + 2n)       \ compute new term Tₙ from Eq. [4.30]
    S ← S + T             \ update sum
End While
P ← from Eq. [4.28]       \ obtain the required result
Write: k, c, P

\ Program module to obtain Γ(k/2)
Module Gamma(k, G) :
If [k = even] then
    G ← 1                       \ value for k = 2
    If [k > 2] then
        For [j = 1, 2, ..., (k/2 − 1)]
            G ← jG              \ according to Eq. [4.26a]
        End For
    End If
Else          \when k is odd
    G ← square root of π        \ value for k = 1
    If [k > 1] then
        For [j = 1, 3, ..., (k − 2)]
            G ← jG/2            \ according to Eq. [4.26b]
        End For
    End If
End If
End Module Gamma
```

4.1.5   Testing for Goodness of Fit

We now consider the use of the chi-squared distribution function in testing for goodness of fit. Suppose that we have a measure $(\chi_k)^2$ with k degrees of freedom for the deviation of an actual distribution from a hypothetical one. We say that the hypothetical distribution is a good fit to the actual data if $(\chi_k)^2$ is not too large.

To determine if $(\chi_k)^2$ is too large, we first choose a percentage (typically 5%, 2.5%, or 1%), which is known as the *significance level* α. The significance level is the percent chance that χ^2 will exceed a certain value c. We then find the value c such that

[4.31]
$$P(\chi^2 \leq c) = 1 - \alpha/100$$

If the observed value $(\chi_k)^2$ is no greater than c, we accept the hypothesis that the hypothetical distribution is a good fit to the actual one; otherwise, we reject it.

We note that decreasing the significance level causes the limit c to increase. For example, a significance level of 0% produces an infinite value of c, which means that any hypothetical distribution will be considered a good fit. As we increase α, $(\chi_k)^2$ must meet more restrictive requirements. We should be wary of data for which $(\chi_k)^2$ meets extremely restrictive requirements. An observed value of $(\chi_k)^2$ that meets the requirement for a good fit at a significance level of 95% may be too good to be true.

The solution of Eq. [4.31] for c may be accomplished by a root-solving method from Chapter 3. Unless we are specifically interested in the limit c, there is no need to go to such lengths to establish goodness of fit. Instead, goodness of fit is more easily established by the condition

[4.32]
$$P(\chi^2 \leq [\chi_k]^2) \leq 1 - \alpha/100$$

Again, if the probability in Eq. [4.32] is found to be too small (less than about 0.05), we should be suspicious of the quality of the data in the actual distribution.

Our remaining task in establishing goodness of fit is to determine the value of $(\chi_k)^2$. Let the (infinite) range of the actual distribution be divided into ℓ intervals so that each interval contains at least five observations (so that the premises about the chi-squared distribution are valid). Let ν_j be the number of actual observations in the j-th interval, and let η_j be the expected number of observations from the hypothetical distribution. Then $(\chi_k)^2$ is computed from

[4.33]
$$(\chi_k)^2 = \sum_{j=1}^{\ell} [(\nu_j - \eta_j)^2/\eta_j]$$

The value of k for the number of degrees of freedom is ℓ minus the number of constraints that are imposed on the hypothetical distribution.

We illustrate the test for goodness of fit by considering the data in Table 4–3 and the Gauss distribution as the hypothetical distribution. The intervals for the test are the same as those in Table 4–3 except where modifications are necessary to meet the minimum of five observations in each interval. The test intervals are interval 1 from $(-\infty)$ to 68.5 with ν_1 equal to 7, interval 2 from 68.5 to 69.5 with ν_2 equal to 10, interval 3 from 69.5 to 70.5 with ν_3 equal to 11, interval 4 from 70.5 to 71.5 with ν_4 equal to 13, and interval 5 from 71.5 to ∞ with ν_5 equal to 9.

The expected number of observations η_j for each interval is computed by applying Eq. [4.18c] to obtain the probability that a measurement will fall within an interval. The expression for η_j for interval j from $[x_a]_j$ to $[x_b]_j$ is

$$\eta_j/n = \Phi(\{[x_b]_j - A\}/s) - \Phi(\{[x_a]_j - A\}/s)$$

in which n is the sample size (50), A is the mean (70.160 in.), and s is the sample standard deviation (1.434 in.).

The value of $(\chi_k)^2$ from Eq. [4.33] is found to be 0.77652 for this example. The number of intervals used in the test is 5, and the constraints consist of one for the sample size, one for the mean, and one for the standard deviation. The value of k for the number of degrees of freedom is thus equal to 2. If we use a significance level of 5% for our test, the limit c in Eq. [4.31] is found to be 5.9915, which is greater than our observed value for $(\chi_k)^2$. We therefore accept the hypothesis that the Gauss distribution is a good fit to the actual data.

Alternately, we could have applied Eq. [4.32] directly to find that

$$P(\chi^2 \leq \{[\chi_k]^2 = 0.77652\}) = 0.32176 < 1 - \alpha/100$$

The hypothesis is again validated by this result, and it would also have been accepted for significance levels almost as high as 68%.

4.2   THE LEAST-SQUARES APPROXIMATION

The concepts of elementary statistics that are discussed in Section 4.1 deal with random variations of a single entity. The *least-squares approximation* is a tool that allows us to model the behavior of a variable y in terms of several other variables x_1 through x_m when all of these variables are subject to random errors. Many texts treat the least-squares approximation as a part of curve fitting. We retain its treatment in this chapter because it is a tool for dealing with noisy data. As such, the basis of the least-squares approximation is more closely related to statistical concepts than it is to other forms of curve fitting.

Suppose that we have n data sets $(y, x_1, x_2, \ldots, x_m)_1$ through $(y, x_1, x_2, \ldots, x_m)_n$, and suppose that we wish to model y according to the function

$$y = f(\alpha_1, \alpha_2, \ldots, \alpha_k, x_1, x_2, \ldots, x_m); \qquad k < n$$

in which the α values are constant coefficients for the model. We may choose the values of the coefficients as those that minimize the function

$$S = \sum_{j=1}^{n} [y_j - f_j(\alpha_1, \alpha_2, \ldots, \alpha_k, x_1, x_2, \ldots, x_m)]^2$$

in which y_j is the value of y in the j-th data set, and f_j is evaluated with the x values from the j-th data set. The quantity S is the sum of the squares of the deviations or residuals of the model for y from the actual data values y_j. The least-squares approximation takes its name from the fact that we are seeking the set of α values that produces the least sum of the squares of the deviations.

In our treatment of the least-squares approximation, we shall cover straight-line models, other models that are formed by a linear combination of functions, and general nonlinear models for y as a function of an independent variable x.

4.2.1 Linear Regression

The process of determining the constants a and b for the line ($y = ax + b$) that best fits a set of n data pairs (x_1, y_1) through (x_n, y_n) in a least-squares sense is known as *linear regression*. The quantity to be minimized is

[4.34]
$$S = \sum_{j=1}^{n} [y_j - (ax_j + b)]^2$$

The values of a and b that minimize S must satisfy the partial derivative conditions

[4.35]
$$\partial S/\partial a = \partial S/\partial b = 0$$

The partial derivatives are given by

$$\partial S/\partial a = \sum\{-2x_j[y_j - (ax_j + b)]\}; \qquad \partial S/\partial b = \sum\{-2[y_j - (ax_j + b)]\}$$

where the summation is understood from now on to be performed for j values from 1 to n. Substitution of the expressions for the partial derivatives into Eq. [4.35] gives us the two linear algebraic equations

$$a\sum([x_j]^2) + b\sum(x_j) = \sum(x_jy_j); \qquad a\sum(x_j) + nb = \sum(y_j)$$

whose solutions are

[4.36a] $a = \{n\sum(x_jy_j) - [\sum(x_j)][\sum(y_j)]\}/\{n\sum([x_j]^2) - [\sum(x_j)]^2\}$

[4.36b] $$b = [\sum(y_j) - a \sum(x_j)]/n$$

The coefficients in Eq. [4.36a, b] are derived on the assumption that the values of the independent variable x are correct and that all errors are in the values of the dependent variable y. This situation is obviously not true; if the roles of x and y are reversed, we would generally obtain a slightly different line from the least-squares approximation.

A crude measure of the how well the data is explained by a straight line is given by the *linear correlation coefficient* r, which may be computed from

[4.37]
$$r = \{n \sum(x_j y_j) - [\sum(x_j)][\sum(y_j)]\}/\sqrt{t_x t_y};$$
$$t_x = n \sum([x_j]^2) - [\sum(x_j)]^2; \quad t_y = n \sum([y_j]^2) - [\sum(y_j)]^2$$

Values of r may range from (-1) to 1. If $|r|$ is exactly 1, the data is perfectly represented by the straight line.

The chi-squared test for goodness of fit is discussed in §4.1.4 and §4.1.5, and may also be used for the linear model. The value of χ^2 with $(n - 2)$ degrees of freedom is the value of S in Eq. [4.34] with the estimated coefficient values a and b. We may also relate χ^2 to the correlation coefficient through

$$\chi^2 = (1 - r^2)(n - 1)(s_y)^2$$

in which $(s_y)^2$ is the sample variance (defined in Eq. [4.4a]) of the y data.

The formulas for linear regression may be extended to models that can be recast in linear form. Let v be a variable that depends on another variable u, and let the model for v be described by

[4.38a] $$v = f(\alpha, \beta, u); \quad \alpha \text{ and } \beta \text{ are constant coefficients}$$

If we can recast Eq. [4.38a] in the form

[4.38b] $$h(v) = C_1(\alpha, \beta)g(u) + C_2(\alpha, \beta)$$

we may equate x, y, a, and b in our earlier discussion to the quantities $g(u)$, $h(v)$, $C_1(\alpha, \beta)$, and $C_2(\alpha, \beta)$, respectively. The constants α and β are then obtained from the estimates of a and b by solving the equations $[C_1(\alpha, \beta) = a]$ and $[C_2(\alpha, \beta) = b]$. Some simple models that can be recast in the linear form of Eq. [4.38b] are given in Table 4–8.

The choice of a model is usually based on the principles behind the phenomenon that produces the data to be analyzed. For example, we might use an exponential model (such as the last model in Table 4–8) if we were analyzing data for a

Table 4–8 Models that can Be Recast in Linear Form

Model	$g(u)$	$h(v)$	$C_1(\alpha, \beta)$	$C_2(\alpha, \beta)$
$v = \alpha/u + \beta$	$1/u$	v	α	β
$v = \alpha u^\beta$	$\ln(u)$	$\ln(v)$	β	$\ln(\alpha)$
$v = \alpha e^{\beta u}$	u	$\ln(v)$	β	$\ln(\alpha)$

phenomenon that exhibits exponential decay. Such a model is used for the cooling problem of Example 4.3.

EXAMPLE 4.3

A car is parked in the sun for a long time on a day when the ambient temperature T_a is 70 °F. As the car is driven away with minimal ventilation, the passenger compartment is slowly cooled. The compartment temperature T is measured to be 101 °F after 5 min of driving, 86 °F after 10 min of driving, and 77 °F after 15 min of driving.

A model for the cabin temperature T as a function of the driving time m is given by

$$T - T_a = \Delta T_0 e^{km}$$

where ΔT_0 is the value of $(T - T_a)$ when the car is first driven away, and k is a cooling rate constant. Use the linear least-squares approximation to estimate the value of k and ΔT_0.

The model can be recast in the linear form ($y = ax + b$) either by referring to Table 4–7 or by taking logarithms of both sides of the model. We have

$$\ln(T - T_a) = km + \ln(\Delta T_0)$$

The four sums for computing a and b from Eq. [4.36a,b] are:

$$\sum(x) = \sum(m) = 5 + 10 + 15 = 30$$
$$\sum(y) = \sum(\ln[T - T_a]) = \ln(31) + \ln(16) + \ln(7) = 8.1524861$$
$$\sum(x^2) = \sum(m^2) = 350$$
$$\sum(xy) = \sum(m \ln[T - T_a]) = 74.084475$$

The estimates from Eq. [4.36a,b] for 3 data pairs are:

$$a = k = [3(74.084475) - 30(8.1524861)]/[3(350) - 30^2]$$
$$\rightarrow k = -0.1488077 \text{ min}^{-1}.$$

$$b = \ln(\Delta T_0) = [8.1524861 - k(30)]/3 = 4.2055727$$
$$\rightarrow \Delta T_0 = 67.06 \text{ °F}$$

4.2.2 Linear Combinations of Functions

A more general form of linear least-squares approximation is obtained when the model for n pairs of (x, y) data is

[4.39] $$y = \alpha_1 f_1(x) + \alpha_2 f_2(x) + \cdots + \alpha_k f_k(x); \qquad k < n$$

The functions $f_p(x)$ in Eq. [4.39] are known as **basis** functions and must be completely specified in terms of x; in particular, they must not contain the constant coefficients α. The function to be minimized is now

[4.40] $$S = \sum \{y_j - [\alpha_1 f_1(x_j) + \cdots + \alpha_k f_k(x_j)]\}^2$$

in which the summation is again understood to be performed for j values from 1 to n.

The coefficients α_p must satisfy the conditions

[4.41] $$\partial S / \partial \alpha_p = 0; \qquad p = 1, 2, \ldots, k$$

These partial derivatives with respect to α_p are given by

$$\partial S / \partial \alpha_p = \sum (\{-2f_p(x_j)\}\{y_j - [\alpha_1 f_1(x_j) + \cdots + \alpha_k f_k(x_j)]\})$$

so that Eq. [4.41] yields a system of linear algebraic equations. The system is expressed in the matrix notation of Chapter 2 by

[4.42a] $$\mathbf{A} \cdot \boldsymbol{\alpha} = [a_{pq}] \cdot [\alpha_p] = \mathbf{h} = [h_p]$$

with the symmetric matrix $\mathbf{A}$ and the right-hand vector $\mathbf{h}$ given by

[4.42b] $$a_{pq} = \sum [f_p(x_j) f_q(x_j)]; \qquad h_p = \sum [y_j f_p(x_j)]$$

The matrix methods of Chapter 2 may be used to solve Eq. [4.42a].

A special application of the general linear least-squares approximation occurs when the model for y is the polynomial

[4.43] $$y = c_1 + c_2 x + c_3 x^2 + \cdots + c_k x^{k-1}$$

The basis functions $f_p(x)$ are equal to x^{p-1}, and the counterpart to Eq. [4.42b] is

[4.44] $$a_{pq} = \sum [(x_j)^{p+q-2}]; \quad h_p = \sum [y_j (x_j)^{p-1}]$$

The two equation system to be solved in the case of the straight-line model of §4.2.1 is a special case of Eq. [4.42a] and [4.44].

The degree of a polynomial used for least-squares fitting should not be too high. A high-degree polynomial can cause severe oscillations, especially when we have noisy data. Quadratics may be used for data that appear to produce a simple concave or convex curve; cubics may be used for data that seem to show an inflection point. Example 4.4 shows the details of the system of equations for least-squares fitting with a cubic.

EXAMPLE 4.4

Develop the system of linear algebraic equations to determine the coefficients c_j of the cubic polynomial

$$y = c_1 + c_2 x + c_3 x^2 + c_4 x^3$$

that best fits n pairs of (x, y) data in a least-squares sense.

It is understood that all summations are for an index j from 1 to n.
We seek the column vector

$$\mathbf{c} = [c_1 \quad c_2 \quad c_3 \quad c_4]^{\mathrm{T}}$$

from the solution of

$$\mathbf{A} \cdot \mathbf{c} = \mathbf{h}$$

The elements of $\mathbf{A}$ and the components of $\mathbf{h}$ are found directly from Eq. [4.44]. The resulting system is

$$
\begin{bmatrix}
n & \sum(x_j) & \sum[(x_j)^2] & \sum[(x_j)^3] \\
\sum(x_j) & \sum[(x_j)^2] & \sum[(x_j)^3] & \sum[(x_j)^4] \\
\sum[(x_j)^2] & \sum[(x_j)^3] & \sum[(x_j)^4] & \sum[(x_j)^5] \\
\sum[(x_j)^3] & \sum[(x_j)^4] & \sum[(x_j)^5] & \sum[(x_j)^6]
\end{bmatrix} \cdot \mathbf{c} =
\begin{bmatrix}
\sum(y_j) \\
\sum(x_j y_j) \\
\sum([x_j]^2 y_j) \\
\sum([x_j]^3 y_j)
\end{bmatrix}
$$

4.2.3 Nonlinear Models

The most general model for a dependent variable y in terms of an independent variable x and a set of constant coefficients α is expressed in the form

[4.45] $$y = f(x, \alpha_1, \alpha_2, \ldots, \alpha_k)$$

Assuming that we are unable to recast the model in the linear form of §4.2.2, we obtain the constants of the model for n given pairs of (x, y) data by minimizing the function

[4.46]
$$S = \sum [y_j - f_j]^2; \quad f_j = f(x_j, \alpha_1, \alpha_2, \ldots, \alpha_k)$$

To do so, we must solve the system of k nonlinear equations

[4.47]
$$\partial S / \partial \alpha_p = 0; \quad p = 1, 2, \ldots, k$$

The equations in Eq. [4.47] may be expressed by the alternate forms

[4.48]
$$\sum [f_j (\partial f_j / \partial \alpha_p)] - \sum [y_j (\partial f_j / \partial \alpha_p)] = 0$$

The solution of Eq. [4.48] for the coefficients α_p may be obtained by the methods of Chapter 3.

4.3 CLOSURE

The tools that are discussed in this chapter are used to analyze data that are subject to random errors. The methods of elementary statistics in Section 4.1 are used for a single entity. The least-squares approximation in Section 4.2 helps us to determine relations among two or more variables, which may have random errors. The statistical and least-squares techniques are intended to deal with noisy data. The noise may arise from sources such as uncertainties in measurements and poor control of conditions under which an experiment is performed.

The statistical methods focus on the normal or Gauss distribution and the use of the chi-squared distribution to test for goodness of fit. Some functions such as the Gauss distribution function, the gamma function, and one of the incomplete gamma functions are introduced in our coverage. These functions, along with others such as the error function, belong to a class of special functions that are treated in intermediate or advanced texts on mathematics (Ref. 7,9). The series expansions that are used in this chapter are relatively simple to understand and to implement. Other more efficient expansions based on continued fractions do exist and are used in subroutine libraries.

The least-squares approximation is a special form of curve-fitting to discrete data. It provides a best fit for a prescribed model only in the sense of minimizing the squares of the deviations of the model from the actual data. As with our statistical models, we should be aware that the resulting models may not always fit the data well. The methods for finding the coefficients of the models in §4.2.2 and §4.2.3

involve systems of equations, whose solutions are discussed in Chapters 2 and 3. These systems may sometimes be ill conditioned.

The topics that are presented in this chapter represent only a small part of a large field of study. The concepts they embody are based in the theory of continuous and discrete probability. Because our treatment is necessarily limited, great care should be used in interpreting the results that are obtained.

4.4 EXERCISES

1. Write a program to sort the first 30 values of x in Table 4–1 in ascending order.

2. Write a program to produce a frequency distribution for the last 30 values of x in Table 4–1. Use interval sizes of 1 and interval midpoints equal to whole numbers.

3. The stopping distances d_j [ft] from 60 mph to zero is measured for a sample of 100 cars. The frequency table is as follows.

j :	1	2	3	4	5	6	7	8	9	10	11	12	13	14
d_j:	120	125	130	135	140	145	150	155	160	165	170	175	180	185
f_j :	1	3	3	8	19	18	13	14	11	6	1	1	1	1

   a. Write a program to compute the mean stopping distance and sample standard deviation for the distribution.

   b. Use the results of part (a) to predict the percentage of cars in the population that require more than 175 ft to stop from an initial speed of 60 mph.

4. Determine the geometric and harmonic means of the distribution in Problem 3.

5. What are the median and mode for the distribution in Problem 3?

6. *Chauvenet's criterion* for discarding a datum point x in a sample of size n is based on finding a value Δ such that

$$P(A - \Delta < x \le A + \Delta) = 1 - 1/(2n)$$

   Points with x values that deviate more than Δ are then discarded. Determine Δ for the racquetball sample of Table 4–3. Use values of 70.160 in. for A and 1.434 in. for s.

7. Consider a normalized triangular density function

$$g(z) = \max\{0, 1 - |z|/3\}; \qquad z = (x - A)/s$$

   in which x is a value in the distribution, A is the mean, and s is the sample standard deviation. Develop a probability distribution function $\phi(z)$ for this model.

8. Test the model in Problem 7 for goodness of fit with the data in Table 4–3. Use values of 70.160 in. for A and 1.434 in. for s.

9. Test the Gauss distribution model for goodness of fit with the data in Problem 3.

10. The stopping distances of a car are d_{60} [ft] from 60 mph and d_{80} [ft] from 80 mph. Data for 6 cars are as follows:

$$d_{60}: 128 \ 142 \ 150 \ 162 \ 167 \ 179$$
$$d_{80}: 220 \ 250 \ 261 \ 278 \ 294 \ 320$$

It is expected that these distances are linearly related by

$$d_{60} = \alpha d_{80} + \beta$$

Use a linear least-squares analysis to determine α and β.

11. The lift coefficient C_L for a wing–body combination varies with the angle of attack α [degrees] as shown.

$$\alpha: \ 0.00 \ 2.50 \ 5.00 \ 7.50 \ 10.00$$
$$C_L: \ 0.03 \ 0.17 \ 0.31 \ 0.44 \ \ 0.56$$

For this range of α values, the lift coefficient behaves nearly linearly with α and may be expressed by the model

$$C_L = m\alpha + C_{L_0}$$

Use a linear least-squares analysis to estimate m and C_{L_0}.

12. The viscosity μ [centipoise] of a certain crude oil varies with temperature T [°C] as shown.

$$T: \ -12.0 \ 10.0 \ 38.0$$
$$\mu: \ \ \ 50.1 \ 10.0 \ \ 4.9$$

The model for the viscosity is

$$\mu = \mu_0 T^k$$

Use a least-squares analysis to estimate μ_0 and k from the data.

13. The number of cycles N required to cause failure of a part under a load w [lb] is given by the model

$$N = K/e^{cw}$$

Test data are

$$w: \ \ 100 \ \ \ 125 \ 150 \ 175$$
$$N: \ 9238 \ 1724 \ 323 \ \ 63$$

Estimate K and c by a least-squares analysis of the data.

14.   The speed v [ft/s] of a car from a coast-down test to determine its drag characteristics is sampled at intervals of 1 s. Eighteen data pairs for the variation of v with time t [s] are as follows.

t[s]	v[ft/s]	t[s]	v[ft/s]
1	65.658	10	62.912
2	65.346	11	62.576
3	65.041	12	62.317
4	64.835	13	61.993
5	64.385	14	61.698
6	64.103	15	61.417
7	63.809	16	61.096
8	63.485	17	60.804
9	63.210	18	60.519

The model for deceleration is

$$-dv/dt = c_1 + c_2 v + c_3 v^2$$

Estimate dv/dt at times t equal to $(2, 3, \ldots , 17)$ seconds from

$$dv(t)/dt = [v(t + 1) - v(t - 1)]/2$$

and use these values in a least-squares analysis to estimate the coefficients c_1, c_2, and c_3.

15.   The current i in a semiconductor diode varies with voltage V according to the model

$$i = I_s[\exp(\lambda V) - 1]$$

where I_s [ampere] is known as the reverse saturation current, and λ [volt^{-1}] is a characteristic value for the diode. Tests for a diode yield the following data.

$$V: \quad 0.03 \quad 0.06 \quad 0.09 \quad 0.12$$
$$i: \quad 0.041 \quad 0.133 \quad 0.342 \quad 0.819$$

a.   Use *only* the data at $V = 0.03$ and at $V = 0.12$ to form two equations and thus solve for I_s and λ.

b.   Use the results from part (a) as starting values, and estimate I_s and λ by a least-squares analysis of the full set of data.

16.   The density function $g(x)$ in Eq. [4.16] is a model for $(f_j/50)$ in the distribution of Table 4–3. The values X_j in the table are used as x values in Eq. [4.16]. Use the data of Table 4–3 to estimate the quantities A and s in the model by a least-squares approximation. Use the median of the distribution as a starting value for A, and use one-sixth of the range of x values as a starting value for s.

5

Curve Fitting

The fundamental curve-fitting problem in two dimensions is to predict from a discrete set of (x, y) data pairs (x_1, y_1) through (x_n, y_n) the value of y when the value of x is specified. The least-squares approximation of Chapter 4 is one form of curve fitting in which we seek the parameters of a given model that best fit the data. In this chapter, we look at other methods which do not account for random errors in the x and y values. We instead treat the data as if they were accurate and determine a curve that passes through the data points exactly.

Classical methods for fitting curves to given data involve polynomial models. Some of these are considered in Section 5.1. Alternatives in the form of cubic splines and of the trigonometric series of the discrete Fourier transform are treated in Sections 5.2 and 5.3. Problems that are used in the discussion of the methods deal with **performance data for a car** and the representation of a **time-varying signal.**

5.1 POLYNOMIAL INTERPOLATION

If we are given a set of n data values (x_1, y_1) through (x_n, y_n) to represent y as a single-valued function of x, we can find a unique polynomial of degree $(n - 1)$ that passes through the data points. For example, we can find a unique straight line that passes through two data points, and we can find a unique quadratic that passes through three data points.

We may model the polynomial equation for y by

$$y = a_0 + a_1 x + a_2 x^2 + \cdots + a_n x^{n-1}$$

and use the n data pairs to write n equations for the coefficients a_i. These equations are

$$y_j = a_0 + a_1 x_j + a_2 (x_j)^2 + \cdots + a_{n-1}(x_j)^{n-1}; \qquad j = 1, 2, \ldots, n$$

and form a system of linear algebraic equations. The solution of the system may be determined by the methods of Chapter 2. However, the coefficient matrix, known as a **Vandermonde matrix,** is prone to ill conditioning. Furthermore, solving the system of linear equations is an inefficient way to obtain a representation of y.

Other polynomial models give us more efficient ways of predicting y for a given value of x. The appearance of these models may be quite different from the basic model above; nevertheless, they produce the same unique curve through the n data points. We shall consider the Lagrange formula, Newton's general formula, and the solution produced by Neville's algorithm.

5.1.1   Lagrange Interpolation

The **Lagrange formula** for interpolating among n data pairs (x_1, y_1) through (x_n, y_n) is perhaps the best known of the classical formulas. It is used to represent y as a function of x in the form

[5.1a]
$$y(x) = \sum_{j=1}^{n} (L_j y_j)$$

in which the multiplying polynomials L_j are given by

[5.1b]
$$L_j = \prod_{\substack{i=1 \\ i \neq j}}^{n} [(x - x_i)/(x_j - x_i)]$$

Note that when x is equal to one of the given values x_k, each of the L_j polynomials in Eq. [5.1b] has a value of zero except for L_k, which has a value of 1.

The Lagrange formula is popular because it is well known and is easy to code. In addition, the data are not required to be specified with x in ascending or descending order. Although the computation of y is simple, the method is still not particularly efficient for large values of n. When n is large and the data for x is ordered, some improvement in efficiency can be obtained by considering only the data pairs in the vicinity of the x value for which y is sought. The price of this improved efficiency is the possibility of a poorer approximation to y.

5.1.2 Newton's General Interpolating Formula

Newton's general interpolating formula for data pairs (x_1, y_1) through (x_n, y_n) has the form

[5.2]
$$y(x) = a_0 + a_1(x - x_1) + a_2(x - x_1)(x - x_2) + \cdots$$
$$+ a_{n-1}(x - x_1)(x - x_2) \cdots (x - x_{n-1})$$

The coefficients a_j are generally obtained by computing a set of quantities known as *divided differences*.

The notation for divided differences is given by

[5.3a] $f[x_k] = y_k;$ $f[x_k, x_{k+1}] = (f[x_{k+1}] - f[x_k])/(x_{k+1} - x_k)$

[5.3b] $f[x_k, x_{k+1}, x_{k+2}] = (f[x_{k+1}, x_{k+2}] - f[x_k, x_{k+1}])/(x_{k+2} - x_x)$

and thereafter by

[5.3c] $f[x_k, x_{k+1}, \ldots, x_{k+j}]$
$$= (f[x_{k+1}, x_{k+2}, \ldots, x_{k+j}] - f[x_k, x_{k+1}, \ldots, x_{k+j-1}])/(x_{k+j} - x_k)$$

Let us denote the divided differences by

[5.4] $f_{k,j} = f[x_k, x_{k+1}, \ldots, x_{k+j}]$

The coefficients a_j of Eq. [5.2] are related to the divided differences by

[5.5] $a_j = f_{1,j}$

The recursion of Eq. [5.3a, b, c] for computing $f_{k,j}$ is illustrated in Fig. 5–1 for four pairs of (x, y) values.

The computation of the coefficients may be performed by treating the values of $f_{k,j}$ as elements F_k of an array at the j-th sweep of the computation process. At each

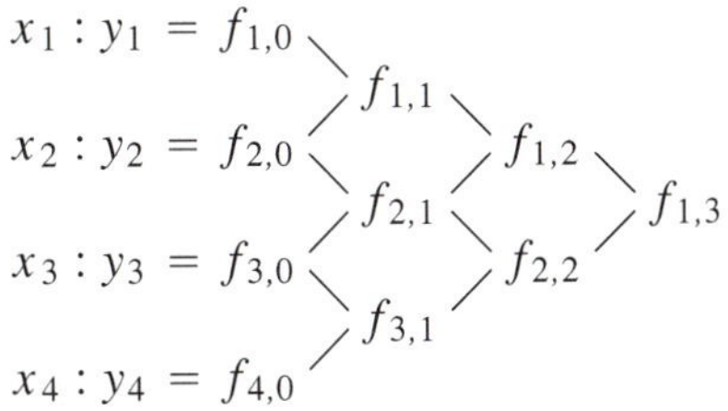

Figure 5–1   Recursion for divided differences with four data pairs.

Pseudocode 5–1   Computations of divided differences

Declare: x, y, F \ arrays with elements 1 through n
Declare: a \ array with elements 0 through $(n-1)$
Initialize: x, y, n \ given data values

For $[k = 1, 2, \ldots, n]$
     $F_k \leftarrow y_k$ \ values at the 0-th sweep
End For
$a_0 \leftarrow F_1$

For $[j = 1, 2, \ldots, (n-1)]$ \ j = sweep counter
     For $[k = 1, 2, \ldots, (n-j)]$ \ F elements to be updated
         $F_k \leftarrow (F_{k+1} - F_k)/(x_{k+j} - x_k)$
     End For   \ with k counter
     $a_j \leftarrow F_1$
End For   \ with j counter

Write: array elements of a and of F

---

sweep j, we update only elements F_1 through F_{n-j}. The process for computing a_j and F_k is shown in Pseudocode 5–1.

The coefficients a_j can now be used in Eq. [5.2]. The final elements F_k are the divided differences $f_{k,n-k}$, which correspond to those shown at the bottom of Fig. 5–1. To compute y for a given value of x from Eq. [5.2], we may use Pseudocode 5–2.

Once the coefficients a_j have been determined, it is somewhat easier to compute y from the Newton general form of Eq. [5.2] than it is to compute it from the Lagrange form of Eq. [5.1a, b]. If another (x, y) data pair is introduced, computation of the additional coefficient proceeds directly from the last computed values of F_k.

Pseudocode 5–2   Implementation of Eq. [5.2]

Initialize: xvalue   \ given x value
yvalue $\leftarrow a_0$ \ initialize y to constant term of Eq. [5.2]
$p \leftarrow 1$ \ initialize polynomial term

For $[j = 1, 2, \ldots, (n-1)]$ \ j = term counter
     $p \leftarrow p(\text{xvalue} - x_j)$ \ update polynomial for term j
     yvalue $\leftarrow$ yvalue $+ a_j p$ \ increase y by term j
End For

Write: xvalue, yvalue

EXAMPLE 5.1

The elapsed time t [s] for a car to accelerate to a speed v [mph] from an initial speed of 30 mph is described by the following data.

$$v \text{ [mph]:} \quad 30 \quad 40 \quad 55 \quad 70$$
$$t \text{ [s]:} \quad \quad 0.0 \quad 1.6 \quad 4.8 \quad 8.9$$

Evaluate the coefficients a_0 through a_3 for the Newton form of Eq. [5.2] with t as the dependent variable and v as the independent variable.

Following the pseudocode or Fig. 5–1, we have:

Sweep 0: $\quad F_1 = 0.0 = a_0; \quad F_2 = 1.6; \quad F_3 = 4.8; \quad F_4 = 8.9.$

Sweep 1: $\quad F_1 = (1.6 - 0.0)/10 = 1.6/10 = a_1;$
$\quad\quad\quad\quad F_2 = (4.8 - 1.6)/15 = 3.2/15;$
$\quad\quad\quad\quad F_3 = (8.9 - 4.8)/15 = 4.1/15.$

Sweep 2: $\quad F_1 = (3.2/15 - 1.6/10)/(55 - 30) = 1.6/750 = a_2;$
$\quad\quad\quad\quad F_2 = (4.1/15 - 3.2/15)/(70 - 40) = 0.9/450.$

Sweep 3: $\quad F_1 = (0.9/450 - 1.6/750)/(70 - 30) = -0.3/90000 = a_3.$

The fitted curve for this example is shown in Fig. 5–2.

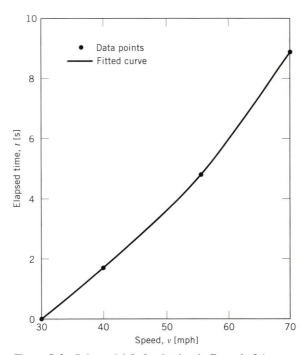

Figure 5–2 Polynomial fit for the data in Example 5.1.

5.1.3   Neville's Algorithm

The Newton interpolation of §5.1.2 involves two stages. The first of these is the computation of the coefficients; the second is the evaluation of the polynomial. *Neville's algorithm* (Ref. 8) is a method that requires about the same effort as the first stage of the Newton interpolation. It is used with n pairs of data (x_1, y_1) through (x_n, y_n) to obtain $y(x)$ directly for a given value of x.

The aim of the method is to evaluate a polynomial of degree $(n-1)$ by a process that is similar to the one for computing the coefficients of the Newton polynomial. Indeed, we may depict the process by Fig. 5–1; however, the quantities $f_{k,j}$ in Fig. 5–1 are now

[5.6a] $$f_{k,0} = y_k; \qquad k = 1, 2, \ldots, n$$

[5.6b] $$f_{k,j} = [(x - x_{k+j})f_{k,j-1} - (x - x_k)f_{k+1,j-1}]/[x_k - x_{k+j}];$$
$$j = 1, 2, \ldots, n - 1, \ k = 1, 2, \ldots, n - j$$

The interpolated value $y(x)$ for a given x is the last computed quantity

[5.7] $$y(x) = f_{1,n-1}$$

The pseudocode for computing $f_{k,j}$ is similar to the one in §5.1.2 for the coefficients of the Newton polynomial. We use F_k at sweep j to replace $f_{k,j}$ and use Eq. [5.6b] to update the values of F_k.

EXAMPLE   5.2

The elapsed time t [s] for a car to accelerate to a speed v [mph] from an initial speed of 30 mph is described by the following data.

v [mph]:	30	40	55	70
t [s]:	0.0	1.6	4.8	8.9

Estimate t when v is 50 mph by using Neville's algorithm.

We apply Eq. [5.6a, b] to obtain:

$j = 0:$ $f_{1,0} = 0.0;$ $f_{2,0} = 1.6;$ $f_{3,0} = 4.8;$ $f_{4,0} = 8.9.$

$j = 1:$ $f_{1,1} = [(50 - 40)f_{1,0} - (50 - 30)f_{2,0}]/[30 - 40] = 32/10$
         $f_{2,1} = [(50 - 55)f_{2,0} - (50 - 40)f_{3,0}]/[40 - 55] = 56/15$
         $f_{3,1} = [(50 - 70)f_{3,0} - (50 - 55)f_{4,0}]/[55 - 70] = 103/30$

$j = 2:$ $f_{1,2} = [(50 - 55)f_{1,1} - (50 - 30)f_{2,1}]/[30 - 55] = 272/75$
         $f_{2,2} = [(50 - 70)f_{2,1} - (50 - 40)f_{3,1}]/[40 - 70] = 109/30$

$j = 3:$ $f_{1,3} = [(50 - 70)f_{1,2} - (50 - 30)f_{2,2}]/[30 - 70] = 3.63$

From Eq. [5.7], elapsed time from 30 mph to 50 mph = 3.63 s.

5.2   CUBIC SPLINES

The curves produced by the polynomials that are discussed in Section 5.1 pass through the n specified data points (x_1, y_1) through (x_n, y_n); however, the behavior of the curve between the data points may exhibit strong oscillations when polynomials of high degree are used. For example, suppose that the data points are approximations to a straight line. By forcing a high-degree polynomial to pass through several points, the curve that is produced may deviate significantly from the line. To alleviate such undesirable conditions, we may use local approximations and patch them together to produce a curve fit. One type of fitting process uses what is known as a ***cubic spline***.

The cubic spline approximation is applied to n ***ordered*** pairs of data. We seek $(n - 1)$ curves that connect points 1 and 2, points 2 and 3, . . ., and points $(n - 1)$ and n. In addition, the two curves connecting points $(k - 1)$ and k and points k and $(k + 1)$ are required to have the same slope at point k. In this way, the resulting curve fit is both continuous and smooth.

A cubic polynomial is the polynomial of least degree that generally satisfies the conditions for the curve fit. Although the cubic can be expressed in standard form, we shall look at the expression for the second derivative y'' between two successive points k and $(k + 1)$. This derivative must vary linearly between the two points because we have restricted the curve fit to a cubic polynomial. We may therefore write y'' at x values from point k to point $(k + 1)$ as

[5.8]
$$y'' = A(y'')_k + B(y'')_{k+1}$$

in which A and B are

[5.9a]
$$A = (x_{k+1} - x)/(x_{k+1} - x_k)$$

[5.9b]
$$B = (x - x_k)/(x_{k+1} - x_k) = 1 - A$$

Integration of Eq. [5.8] with the expressions for A and B from Eq. [5.9a, b] yields the first derivative y' as

[5.10] $$y' = [-A(x_{k+1} - x)(y'')_k + B(x - x_k)(y'')_{k+1}]/2 + C_1$$

in which C_1 is the constant of integration. A second integration allows us to write an expression for y in the form

[5.11] $$y = [A(x_{k+1} - x)^2(y'')_k + B(x - x_k)^2(y'')_{k+1}]/6 + C_1 x + C_2$$

in which C_2 is a second constant of integration.

To determine C_1 and C_2, we write Eq. [5.11] at x equal to x_k and at x equal to x_{k+1}. We also note from Eq. [5.9a, b] that A is 1 and B is 0 when x is equal to

x_k, and that A is 0 and B is 1 when x is equal to x_{k+1}. The equations from Eq. [5.11] are therefore

$$y_k = (x_{k+1} - x_k)^2 (y'')_k / 6 + C_1 x_k + C_2$$
$$y_{k+1} = (x_{k+1} - x_k)^2 (y'')_{k+1} / 6 + C_1 x_{k+1} + C_2$$

The results for C_1 and C_2 are

[5.12a]
$$C_1 = [y_{k+1} - y_k] / [x_{k+1} - x_k]$$
$$+ [(y'')_{k+1} - (y'')_k][x_{k+1} - x_k] / 6$$

[5.12b]
$$C_2 = [x_{k+1}y_k - x_k y_{k+1}] / [x_{k+1} - x_k]$$
$$- [x_{k+1}(y'')_k + x_k(y'')_{k+1}][x_{k+1} - x_k] / 6$$

When we substitute the values of C_1 and C_2 from Eq. [5.12a,b] into Eq. [5.11], we obtain an expression for y as

[5.13]
$$y = Ay_k + By_{k+1}$$
$$+ [(A^3 - A)(y'')_k + (B^3 - B)(y'')_{k+1}][x_{k+1} - x_k]^2 / 6$$

By choosing arbitrary values of y'' at the n data points, we can use Eq. [5.9a, b] and Eq. [5.13] to construct curves between successive points; however, we will not obtain a smooth fit to the data. For smoothness, we must consider the first derivative y'. When the value of C_1 from Eq. [5.12a] is substituted into Eq. [5.10], we obtain

[5.14]
$$y' = [y_{k+1} - y_k] / [x_{k+1} - x_k]$$
$$+ [(3B^2 - 1)(y'')_{k+1} - (3A^2 - 1)(y'')_k][x_{k+1} - x_k] / 6$$

We then use Eq. [5.14] to equate $(y')_k$ for the curve between points k and $(k + 1)$ to $(y')_k$ for the curve between points $(k - 1)$ and k. The result is

[5.15]
$$[(x_k - x_{k-1})/6](y'')_{k-1} + [(x_{k+1} - x_{k-1})/3](y'')_k$$
$$+ [(x_{k+1} - x_k)/6](y'')_{k+1}$$
$$= (y_{k+1} - y_k)/(x_{k+1} - x_k) - (y_k - y_{k-1})/(x_k - x_{k-1});$$
$$k = 2, 3, \ldots, (n - 1)$$

The relations in Eq. [5.15] provide $(n - 2)$ equations for the n second derivatives. Two additional equations must be provided to obtain a complete system. One possibility is given by

[5.16]
$$(y'')_1 = (y'')_n = 0$$

The spline produced in this case is known as a ***natural spline***. Another possibility is to extrapolate linearly the values of $(y'')_1$ and $(y'')_n$.

The system of linear algebraic equations formed by Eq. [5.15] and by Eq. [5.16] or its alternatives is tridiagonal. The Thomas algorithm of Chapter 2 may be used to obtain solutions of $(y'')_k$ for k equal to $(1, 2, \ldots, n)$. Once we have found these derivatives, we may use Eq. [5.9a, b] and Eq. [5.13] to estimate a value of y for a given value of x in the interval $(x_k < x < x_{k+1})$.

The value k such that x_k and x_{k+1} contain x may be found by a sequential search. A more efficient method is a binary search, which is similar to the bisection method of Chapter 3. The process uses integer arithmetic and is described in Pseudocode 5–3.

Pseudocode 5–3   A binary search algorithm

```
\ Pseudocode segment to determine the index k such that
\ xvalue is between x_k and x_{k+1} of an n-element array x.
\ It is assumed that x is in ascending order and that x
\ and xvalue have been initialized.
\ It is also assumed that a check has been made to verify
\ that xvalue is within the limiting values of the array.
    k ← 1          \ lowest possible value
    u ← n          \ upper index limit
    While [(u − k) > 1]                          \ k not yet found
        m ← integer part of (k + u)/2            \ "midpoint"
        If [xvalue > x_m]
            k ← m                                \ increase estimate of k
        Else
            u ← m                                \ reduce upper limit
        End If
    End While
```

EXAMPLE   5.3

Use a natural spline to estimate the elapsed time t [s] for a car to accelerate from 30 mph to 50 mph from the data below.

v [mph]:	30	40	55	70
t [s]:	0.0	1.6	4.8	8.9

Let t'' denote d^2t/dv^2. For a natural spline, $(t'')_1 = (t'')_4 = 0$.

From Eq. [5.15], we have:

$$(25/3)(t'')_2 + (15/6)(t'')_3 = 3.2/15 - 1.6/10$$
$$(15/6)(t'')_2 + (30/3)(t'')_3 = 4.1/15 - 3.2/15$$

whose solutions are:

$$(t'')_2 = 0.00497297; \qquad (t'')_3 = 0.00475676$$

For $v = 50$ between v_3 and v_4, we use Eq. [5.9a, b] to obtain

$$A = 1/3; \qquad B = 2/3$$

Then from Eq. [5.13], we have

$$t = 1.6/3 + 2(4.8)/3 + [(-8/27)(t'')_2 + (-10/27)(t'')_3][15^2/6]$$
$$\rightarrow t = 3.612 \text{ s}$$

Compare this result with the one in Example 5.2.

5.3   THE DISCRETE FOURIER TRANSFORM

The *discrete Fourier transform* (Ref. 10) is often used in fields such as signal processing to determine the waveform components of a signal. It is also an alternative to polynomials and splines for fitting curves to data. We shall present the use of the transform in the usual complex number notation, which can be coded directly in FORTRAN. For other languages that do not contain intrinsic features for complex arithmetic, we shall give real number equivalents.

Let us consider a time-varying signal $f(t)$ over the time interval from t_a to t_b. To use the discrete Fourier transform, we must interpret the signal as a periodic function. Consider a signal as shown in Fig. 5–3a. One interpretation of periodicity is to assume that the period is equal to $(t_b - t_a)$. This interpretation forces the signal to return to the value $f(t_a)$ when t is equal to t_b; it is illustrated in Fig. 5–3b.

Another approach is to treat the signal as if it had a period as shown in Fig. 5–3c. In this case, the second half of the signal is a mirror image of the first, and the signal value is $f(t_a)$ when t is equal to $(t_a + 2(t_b - t_a))$.

In general, we use the period $(t_b - t_a)$ when $f(t_b)$ is equal to $f(t_a)$, and we use the period $2(t_b - t_a)$ when $f(t_b)$ is different from $f(t_a)$. We also scale and shift the time axis as needed so that we can represent the signal by $F(\tau)$ over a period from 0 to 1. The period T and the relation of τ to t are given by

[5.17a] $$T = c(t_b - t_a); \qquad c = \begin{cases} 1; f(t_b) = f(t_a) \\ 2; f(t_b) \neq f(t_a) \end{cases}$$

[5.17b] $$\tau = (t - t_a)/T$$

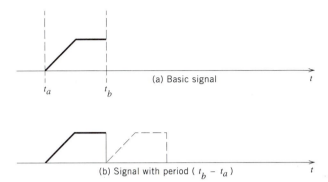

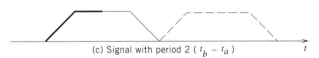

Figure 5–3   A signal and interpretations of its periodicity.

The discrete Fourier transform of $F(\tau)$ can be accomplished in several ways. The version that is presented here begins by selecting an odd number n of equally spaced values τ_j such that

[5.18] $$\tau_j = j/n; \qquad j = 0, 1, \ldots, (n-1)$$

Note that this set of τ_j values excludes the function when τ is equal to 1 because $F(0)$ is equal to $F(1)$ under our assumption of periodicity. We also choose a set of integers

[5.19] $$S = \{(-m), (-m+1), \ldots, m\}; \qquad m = (n-1)/2$$

With Eq. [5.18] and [5.19], we express the discrete Fourier transform as

[5.20] $$G(\tau) = \sum_{k \in S} [\alpha_k \exp(2\pi i k \tau)]$$

in which i is the unit imaginary number and the α_k coefficients are given by

[5.21] $$\alpha_k = (1/n) \sum_{j=0}^{n-1} [F(\tau_j) \exp(-2\pi i k \tau_j)]$$

The coefficient α_k contains the amplitude and phase of the waveform associated with k. The function $G(\tau)$ approximates the given function $F(\tau)$ on the interval $(0 \le \tau \le 1)$. We can therefore use Eq. [5.20] and [5.21] to fit a curve to data for a function $f(t)$ whose values are given at an even number $(n + 1)$ of equally spaced t values over one period.

The real counterparts of the expressions in Eq. [5.20] and [5.21] may be derived from the complex value relations

$$e^{ix} = \cos(x) + i\ \sin(x); \qquad (a + ib)(c + id) = ac - bd + i(ad + bc)$$

The real and imaginary parts of Eq. [5.21] are expressed by

[5.22a] $$a_k = \text{Re}(\alpha_k) = (1/n) \sum_{j=0}^{n-1} [F(\tau_j)\cos(2\pi k\tau_j)]$$

[5.22b] $$b_k = \text{Im}(\alpha_k) = (1/n) \sum_{j=0}^{n-1} [F(\tau_j)\sin(-2\pi k\tau_j)]$$

The function $G(\tau)$ is necessarily real when $F(\tau)$ is real; we may thus rewrite Eq. [5.20] as

[5.23] $$G(\tau) = \text{Re}(G(\tau)) = \sum_{k \in S} [a_k \cos(2\pi k\tau) - b_k \sin(2\pi k\tau)]$$

To illustrate the discrete Fourier transform, we apply it to a signal of the type in Fig. 5–3a. We represent the signal by

$$f(t) = \begin{cases} t; & 0 \le t \le 1 \\ 1; & 1 < t \le 2 \end{cases}$$

We take the period of the signal to be equal to 4 and express $F(\tau)$ by

$$F(\tau) = \begin{cases} 4\tau; & 0.00 \le \tau \le 0.25 \\ 1; & 0.25 < \tau \le 0.75 \\ 4 - 4\tau; & 0.75 < \tau \le 1.00 \end{cases}$$

The results of applying Eq. [5.20] and [5.21] or Eq. [5.22a,b] and [5.23] with n equal to 5 and with n equal to 11 are shown in Fig. 5–4. We see that the approximation with n equal to 11 is somewhat better than the approximation with n equal to 5. Larger values of n will yield better approximations.

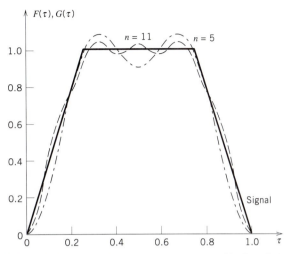

Figure 5–4 A signal and approximations with 5 and 11 waveforms.

The signal shown in Fig. 5–4 is for one period. If we repeat the signal, we see that it is an even function; that is, $F(\tau)$ is equal to $F(-\tau)$. In such cases, α_k is equal to α_{-k}, and the imaginary part b_k vanishes. If the signal is represented as an odd function so that $F(\tau)$ is equal to $[-F(-\tau)]$, α_k is equal to $[-\alpha_{-k}]$, and the real part a_k vanishes.

Although we have used an odd value of n for symmetry of the k values, an even value of n is not prohibited. For even n, we can replace the set S of k values in Eq. [5.19] by

$$S = \{-n/2, (-n/2 + 1), \ldots, (n/2 - 1)\}$$

Furthermore, use of a subset of the n values of k in Eq. [5.20] and [5.21] or in Eq. [5.22a, b] and [5.23] produces an approximation to $F(\tau)$ that is a best fit in a least-squares sense.

5.4   CLOSURE

The Lagrange and Newton polynomials and the Neville algorithm of Section 5.1 yield the same interpolation for a set of n (x, y) data pairs. The pairs are not required to be ordered, nor is the independent variable required to be equally spaced. The dependent variable is approximated as a single-valued function. The Newton polynomial is generally more efficient than the Lagrange form, and it can be adjusted easily for additional data. The Neville algorithm is best suited for interpolation at a few specific values of the independent variable.

Cubic spline approximations are patches among ordered data that maintain continuity and smoothness. They are more useful than polynomials for a large number of data points because high-degree polynomials may exhibit strong oscillatory behavior.

Finally, we looked at the discrete Fourier transform in Section 5.3. This is a very powerful tool for waveform analysis of signals, but can also be used for curve fitting. Fast Fourier transform techniques for improved efficiency are available.

5.5 EXERCISES

1. The distribution function $\Phi(z)$ that is used in statistics is given in Table 4–5 for several values of z. We wish to determine the value of z at which $\Phi(z)$ is equal to 0.95. Use the data for z equal to (1.5, 1.6, 1.7, 1.8) to obtain the desired value of z with the Lagrange interpolation polynomial.

2. Repeat problem 1 with the Newton interpolating polynomial.

3. The lift coefficient C_L for a wing–body combination is given for angles of attack α [degrees] near maximum lift as follows.

$$\alpha: \quad 15.00 \quad 17.50 \quad 20.00 \quad 22.50$$
$$C_L: \quad 0.783 \quad 0.843 \quad 0.887 \quad 0.891$$

   a. Use all of the data to find the coefficients of the Newton interpolating polynomial.
   b. Use the results of part (a) to find where $dC_L/d\alpha$ is zero, and thus obtain the maximum value of C_L.

4. Use the Neville algorithm with the data in Problem 3 to estimate the value of C_L when α is equal 18°.

5. A shaft of length L [in.] is fixed at one end and is subjected to a torque T [lb·in.] applied to the other end. The shaft has a rectangular cross section with sides of dimensions a and b [in.], with $b \le a$. The maximum shear stress S_{max} [psi] and the maximum angle of twist α_{max} [degrees] are given by

$$S_{max} = c_1 T/(ab^2); \quad \alpha_{max} = 180TL/(\pi c_2 Gab^3)$$

where G is the shear modulus [psi], and c_1 and c_2 are coefficients that depend on (a/b) as follows.

a/b	c_1	c_2
1.0	4.81	0.141
1.5	4.33	0.196
2.0	4.06	0.229
3.0	3.74	0.263
5.0	3.44	0.291

Write a program to estimate S_{max} and α_{max} for steel shafts ($G = 12 \times 10^6$ psi) when a and b are in the range ($1.0 \le a/b \le 5.0$). The coefficients c_1 and c_2 are to be approximated by an interpolation with the Neville algorithm. Execute the program for the following input data sets.

Set 1: $T = 7500$ lb·in, $L = 12.0$ in, $a = 1.00$ in, $b = 0.80$ in
Set 2: $T = 8000$ lb·in, $L = 10.0$ in, $a = 1.00$ in, $b = 0.60$ in
Set 3: $T = 9000$ lb·in, $L = 15.0$ in, $a = 1.50$ in, $b = 0.40$ in

6. Write a program to output the values of C_L for α values from $15°$ to $22°$ in increments of $0.2°$, based on a natural spline fit to the data in Problem 3.

7. Write a program to output values of c_1 and c_2 in Problem 5 for values of (a/b) from 1 to 5 in increments of 0.1, based on a natural spline fit to the data.

8. Splines are often used in graphics software to draw a smooth curve through a set of given data points. Points for a certain curve are given in (x, y) coordinates in the following sequence.

i:	1	2	3	4	5	6	7	8
x_i:	0.1	0.2	0.4	0.5	0.5	0.3	0.3	0.4
y_i:	0.6	0.8	1.0	1.0	0.8	0.6	0.4	0.3

Neither x nor y is a single-valued function of the other; nevertheless, we can generate a smooth curve by letting x and y depend on a common parameter p whose values for the data set are p_i equal to i. Use natural splines to fit x and y to p for values of p from 1 to 8 in increments of 0.2, and thus generate a table of (x, y) values. Plot the tabulated data.

9. Twelve data points (x_i, y_i) are given by

$$x_i = \cos(\pi i/6), \, y_i = \sin(\pi i/6); \quad i = 0, 1 \ldots, 11$$

These are obviously points on a unit circle with center at (0,0). To close the circle, we must get to the 13th point whose coordinates are identical to those of the first point. Use natural splines to fit x and y to a parameter p for p values from 0 to 12 in increments of 0.2, and thus generate a sequence of (x, y) points for plotting the circle. You must make sure that the slopes at $(p = 0)$ and $(p = 12)$ are the same. Use your generated data to draw the circle.

10. A time signal is described by

$$f(t) = \begin{cases} 1; & 0 \leq t \leq 1.5 \\ t - 0.5; & 1.5 < t < 2.5 \\ 2; & 2.5 \leq t \leq 3.0 \end{cases}$$

Determine the coefficients a_k and b_k (or the complex coefficients α_k) of the discrete Fourier transform for k equal to $(-30, -29, \ldots, 30)$ by assuming that the period of the signal is equal to 6. That is, assume that $f(3 + \Delta t)$ is equal to $f(3 - \Delta t)$ for Δt values ranging from 0 to 3. Compute and plot $G(\tau)$ from Eq. [5.20] or [5.23] for τ values from 0 to 1 in increments of 0.02.

11. Repeat Problem 10. Now, however, treat the signal as one with a period equal to 3. This means that $f(t)$ "jumps" from a value of 2 to a value of 1 when t is equal to 3.

12. Forty-five values of a smooth signal $F(\tau)$ are shown in sequence (by rows) for τ values $(0/45, 1/45, \ldots, 44/45)$. The signal has value 0 when τ is equal to 1. Determine the coefficients a_k and b_k (or the complex coefficients α_k) of the discrete Fourier transform for k equal to $(-22, -21, \ldots, 22)$, and determine $G(\tau)$ from Eq. [5.20] or [5.23] for τ values from 0 to 1 in increments of 0.02. Plot the variation of $G(\tau)$.

0.0000	0.0006	0.0038	0.0079	0.0069	−0.0045	−0.0250	−0.0440	−0.0457
−0.0193	0.0309	0.0837	0.1093	0.0854	0.0126	−0.0816	−0.1529	−0.1622
−0.0971	0.0189	0.1346	0.1960	0.1724	0.0731	−0.0570	−0.1588	−0.1880
−0.1359	−0.0313	0.0751	0.1369	0.1333	0.0759	−0.0010	−0.0603	−0.0805
−0.0626	−0.0252	0.0088	0.0252	0.0234	0.0123	0.0023	−0.0015	−0.0008

13. Add the value $F(1)$ equal to 0 to the data in Problem 12. Fit a natural spline to the data, and thus generate a curve to approximate $F(\tau)$ for τ values from 0 to 1 in increments of 0.02. Plot the approximate curve.

14. The curve in Problem 12 crosses the τ axis nine times other than at τ values of 0 and 1. Use four consecutive values — two positive and two negative — that straddle each crossing and use one of the polynomial interpolation methods to determine the nine τ values at which $F(\tau)$ is zero. Is there any pattern in the sequence of the τ values?

PART III

NUMERICAL CALCULUS

6

Differentiation and Integration

Differentiation and integration are the two basic processes of calculus. One or both of these processes will generally be encountered in applications where models are described in terms of rates.

In principle, it is always possible to determine an analytic form of a derivative for a given function. In some cases, however, the analytic form is very complicated, and a numerical approximation of the derivative may be sufficient for our purposes. For functions which are described only in terms of tabulated data, numerical differentiation is the only means of computing a derivative. Numerical differentiation is discussed in Section 6.1.

Numerical integration is used to integrate tabulated functions or to integrate functions whose integrals are either impossible or very difficult to obtain analytically. Integration methods are discussed in Section 6.2.

Problems that are used in the discussion of the methods include those on **particle dynamics**, the **normal distribution function** from statistics, **large oscillations of a simple pendulum**, and **area moments of inertia** of a nonconvex domain.

6.1 NUMERICAL DIFFERENTIATION

One approach to numerical differentiation is to fit a curve with a simple form to the function, and then to differentiate the curve-fit function. For example, the polynomial or spline methods of Chapter 5 can be used to fit a curve to tabulated data for the function, and the resulting polynomial or spline can then be differentiated.

199

Another approach is to obtain an approximation of a derivative directly in terms of the function values. The resulting expression for the approximation is known as a *difference formula*. Such formulas are especially useful for solving differential equations and may be derived either from curve fitting or from the Taylor series. We shall use the Taylor series approach in §6.1.1 as follows.

6.1.1   Difference Formulas

The Taylor series expansion for a function is

[6.1] $f(x + h) = f(x) + [(h/1!)(df/dx) + (h^2/2!)(d^2f/dx^2) + \cdots]|_x$

We can also write the Taylor series with a remainder term R_{n+1} as

[6.2] $f(x + h) = f(x) + [(h/1!)(df/dx) + (h^2/2!)(d^2f/dx^2)$
$$+ \cdots + (h^n/n!)(d^nf/dx^n)]|_x + R_{n+1}$$

in which the remainder term is given by

[6.3] $R_{n+1} = (h^{n+1}/[n + 1]!)(d^{n+1}f(\xi)/dx^{n+1}); \qquad x < \xi < x + h$

The error that occurs when the Taylor series is truncated immediately after the term containing the n-th derivative is R_{n+1} and is said to be *of order* h^{n+1} or $O(h^{n+1})$.

Simple difference formulas may be obtained by truncating the Taylor series after the first derivative term. That is, by writing

$$f(x \pm h) = f(x) \pm hf^i(x) + O(h^2); \qquad f^i(x) = df(x)/dx$$

we obtain the approximations

Forward Difference:

[6.4a] $f^i(x) = [f(x + h) - f(x)]/h + O(h)$

Backward Difference:

[6.4b] $f^i(x) = [f(x) - f(x - h)]/h + O(h)$

The formulas in Eq. [6.4a, b] have truncation error of order h and are expressed in terms of two function values. Higher-order approximations may be derived by using additional function values. For example, we use Roman superscripts to denote the derivatives of the function at x and write expansions for two function values as

$$f(x + h) = f(x) + hf^i + (h^2/2)f^{ii} + O(h^3)$$
$$f(x + 2h) = f(x) + 2hf^i + (4h^2/2)f^{ii} + O(h^3)$$

We then eliminate the second derivative term and solve for f^i to obtain

[6.5a] $$f^i(x) = [-3f(x) + 4f(x + h) - f(x + 2h)]/[2h] + O(h^2)$$

A similar second-order formula in terms of $f(x)$, $f(x - h)$, and $f(x - 2h)$ is

[6.5b] $$f^i(x) = [3f(x) - 4f(x - h) + f(x - 2h)]/[2h] + O(h^2)$$

Another way to obtain higher-order formulas is to recognize that the signs in the terms of the Taylor series expansions for $f(x + h)$ and $f(x - h)$ can produce cancellations. Thus we write

$$f(x + h) = f(x) + hf^i + (h^2/2)f^{ii} + O(h^3)$$
$$f(x - h) = f(x) - hf^i + (h^2/2)f^{ii} + O(h^3)$$

Subtraction gives the first-derivative formula

Central Difference:

[6.6] $$f^i(x) = [f(x + h) - f(x - h)]/(2h) + O(h^2)$$

Addition of the expansions for $f(x + h)$ and $f(x - h)$ through the third-derivative terms gives a central-difference formula for the second derivative as

[6.7] $$f^{ii}(x) = [f(x - h) - 2f(x) + f(x + h)]/h^2 + O(h^2)$$

The formulas involving three function values have been derived so far with constant spacing h between the x values. The Taylor series approach can also be used with unequal spacing as illustrated in the following example.

EXAMPLE 6.1

A particle crosses three sensors which are spaced at intervals of 0.3 m. Measurements of the time to travel from sensor 1 to sensor 2 and from sensor 2 to sensor 3 are 0.0550 s and 0.0554 s, respectively. Estimate the speed and acceleration of the particle as it crosses the second sensor.

Let x denote position, let t denote time, and let subscripts denote the sensors. Then

$$\text{Speed at sensor } 2 = dx/dt \text{ at time } t_2;$$
$$\text{Acceleration at sensor } 2 = d^2x/dt^2 \text{ at time } t_2$$

From Eq. [6.2] and [6.3], we have expansions about t_2 :

$$x_1 = x_2 - 0.0550(dx/dt) + (0.0550^2/2)(d^2x/dt^2) + O(0.0550^2)$$
$$x_3 = x_2 + 0.0554(dx/dt) + (0.0554^2/2)(d^2x/dt^2) + O(0.0554^2)$$

We neglect the remainder terms and solve the resulting linear equations by Cramer's rule to obtain the estimates:

$$v = dx/dt = [(x_1 - x_2)(0.0554^2/2) - (x_3 - x_2)(0.0550^2/2)]/D;$$
$$a = d^2x/dt^2 = [(x_3 - x_2)(-0.0550) - (x_1 - x_2)(0.0554)]/D$$

where

$$D = -(0.0550)(0.0554)(0.0554 + 0.0550)/2;$$
$$x_1 - x_2 = -0.3 \text{ m}; \qquad x_3 - x_2 = 0.3 \text{ m}$$

The results are

$$\text{speed } v = 5.434925 \text{ m/s}; \qquad \text{acceleration } a = -0.713460 \text{ m/s}^2$$

The estimates for the speed and acceleration in Example 6.1 could have been found (to a slightly different level of accuracy) by differentiation with respect to x. The procedures are shown in Example 6.2.

EXAMPLE 6.2

Repeat example 6.1 by using derivatives with respect to position.

$v = dx/dt = 1/(dt/dx)$. From the central-difference formula of Eq. [6.6],

$$v_2 = 1/[(t_3 - t_1)/(x_3 - x_1)] = 5.434783 \text{ m/s}$$
$$a = dv/dt = (dv/dx)(dx/dt) = v(dv/dx). \text{ Thus}$$
$$a = v_2[(v_3 - v_1)/(x_3 - x_1)]$$

We compute v_1 and v_3 from Eq. [6.5a, b] as follows:

$$v_1 = 1/[(-3t_1 + 4t_2 - t_3)/(x_3 - x_1)] = 5.4744526 \text{ m/s}$$
$$v_3 = 1/[(3t_3 - 4t_2 + t_1)/(x_3 - x_1)] = 5.3956835 \text{ m/s}$$

Then $a_2 = v_2(5.3956835 - 5.4744526)/(0.6) = -0.713489 \text{ m/s}^2$

The computation of partial derivatives is the same as for ordinary derivatives when differentiation is with respect to only one variable. For cross derivatives such as $[\partial^2 f/(\partial x \partial y)]$, we may derive formulas from the Taylor series for several variables (see Appendix C), or we may build up the derivative from one-variable forms. For example, we may use the central-difference formula of Eq. [6.6] to obtain

$$
\begin{aligned}
\partial^2 f(x,y)/(\partial x \partial y) &= \partial [\partial f(x,y)/\partial x]/\partial y \\
&= \{[f(x+\Delta x, y+\Delta y) - f(x-\Delta x, y+\Delta y)]/(2\Delta x) \\
&\quad - [f(x+\Delta x, y-\Delta y) - f(x-\Delta x, y-\Delta y)]/(2\Delta x)\}/(2\Delta y)
\end{aligned}
$$

6.1.2 Truncation and Round-Off Errors

Consider the forward-difference formula derived from Eq. [6.2] and [6.3]:

$$
f^i(x) = [f(x+h) - f(x)]/h - (h/2)f^{ii}(\xi)
$$

The approximation to the derivative df/dx is $[f(x+h) - f(x)]/h$, and the error in the approximation is $(h/2)f^{ii}(\xi)$. The error is known as the ***truncation error*** because it is caused by truncation of the Taylor series.

Finite machine precision causes another error, which is known as ***round-off error***. This error appears in the computation of the term $[f(x+h) - f(x)]$ in the numerator of the approximation. If we denote the error in the difference by ϵ, the round-off error in computing the approximation to df/dx is ϵ/h.

The truncation error $O(h)$ clearly decreases with h; at the same time, the round-off error increases. In general, errors in the approximation to df/dx are dominated by truncation error for large values of h and by round-off error for small values of h. An example of this behavior is shown in Fig. 6–1 for the error in the derivative of $(\sin x)$ at x equal to $(\pi/3)$ radian. The derivative is computed from the forward-difference formula of Eq. [6.4a] with IEEE single-precision arithmetic. All errors are negative for the range of h values that is used.

The variation of the absolute value of the error with h is shown on a log–log plot in Fig. 6–1. For larger h values where truncation error is dominant and is roughly proportional to h, the slope of the variation is approximately equal to 1. The error increases smoothly with h in this region, because the term $[f^{ii}(\xi)/2]$ that multiplies h is essentially constant. The slope at smaller h values where round-off error is dominant and increases with $(1/h)$ has an average value that is approximately equal to (-1). The scatter in the data for small h values is due to larger variations in ϵ caused by loss of precision as the terms in the difference $[f(x+h) - f(x)]$ approach each other in value.

The limitations imposed by the growth of round-off error as the step size h decreases may be bypassed by the ***Richardson extrapolation*** technique. Let us look

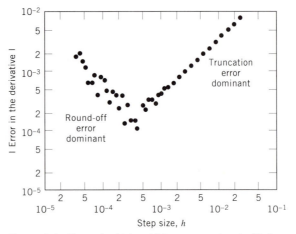

Figure 6–1 Errors in $d(\sin x)/dx$ at x equal to $(\pi/3)$ from Eq. [6.4a].

again at the first-order, forward-difference approximation to the first derivative. The expansions of Eq. [6.2] and [6.3] yield

[6.8a] $\qquad f^i(x) = [f(x + h) - f(x)]/h - (h/2)f^{ii}(x) - (h^2/6)f^{iii}(\xi)$
$\qquad\qquad = (f^i)_h - (h/2)f^{ii}(x) - (h^2/6)f^{iii}(\xi)$

in which $(f^i)_h$ is the approximate derivative with step size h. If we use a different step size (ch), the corresponding result is

[6.8b] $\quad f^i(x) = [f(x + ch) - f(x)]/(ch) - (ch/2)f^{ii}(x) - (c^2h^2/6)f^{iii}(\xi)$
$\qquad\qquad = (f^i)_{ch} - (ch/2)f^{ii}(x) - (c^2h^2/6)f^{iii}(\xi)$

By multiplying Eq. [6.8a] by c and subtracting the result from Eq. [6.8b], we eliminate the term in the second derivative and obtain

[6.8c] $\qquad (1 - c)f^i(x) = (f^i)_{ch} - c(f^i)_h - (c^2 - c)(h^2/6)f^{iii}(\xi)$

We may then rewrite Eq. [6.8c] as the second-order approximation

[6.8d] $\qquad\qquad f^i(x) = [(f^i)_{ch} - c(f^i)_h]/[1 - c] + O(h^2)$

The Richardson extrapolation may be continued by using other step sizes. Use of a third step size allows us to eliminate the third-derivative term, and so on. Rather than continuing in this fashion, we should recognize that the Richardson extrapolation is equivalent to finding higher-order approximations by using additional function values. For example, Eq. [6.8d] is equivalent to the formula in Eq. [6.5a] when c is equal to 2.

Central-difference formulas with equally spaced control points are generally of higher order than forward-difference or backward-difference formulas. As a result, central differencing is usually the method of choice, and forward and backward differencing are reserved for points that are near boundaries. On occasion, forward or backward differencing is deliberately used instead of central differencing. Such an occasion occurs in the modeling of certain types of partial differential equations where the error in the difference formula is used to damp the growth of errors.

We also note that higher order does not necessarily mean higher accuracy. This is because the difference formulas give the derivatives of curves that are ***fitted*** to the data. As we discussed in Chapter 5, the use of high-degree polynomials can produce strong oscillations in the fitted curves; the derivatives of such curves can then be highly inaccurate. In most cases with simple curvature, second-order formulas are adequate. For other cases that contain inflection points, third-order formulas may be more appropriate.

6.2 NUMERICAL INTEGRATION

Integrals arise when we wish to determine the change in a quantity y whose rate of change with respect to another variable x is described as a function $f(x)$. If we have a relation

[6.9a] $$dy/dx = f(x)$$

the change in y from y_a at x equal to x_a to y_b at x equal to x_b is

[6.9b] $$y_b - y_a = \int_{x_a}^{x_b} f(x)\ dx$$

The relation in Eq. [6.9a] is a simple form of an ordinary differential equation. Its solution is described by the definite integral in Eq. [6.9b]. The integration process is a summation process. We may think of the quantity $[f(x)dx]$ as the area of a rectangle with height $f(x)$ (which may be negative or zero) and with an infinitesimally small width dx. The definite integral in Eq. [6.9b] is the sum of all such areas for x values from x_a to x_b.

Numerical integration methods are used when $f(x)$ is difficult or impossible to integrate analytically, or when $f(x)$ is given as a set of tabulated values. The integration methods that we shall consider in §6.2.1 through §6.2.4 are the classic trapezoidal rule and Simpson's rule methods, Romberg integration, and Gaussian quadratures. Techniques for improper and multidimensional integrals are discussed in §6.2.5 and §6.2.6.

6.2.1   The Trapezoidal Rule

Consider a function $f(x)$ as x varies from x_i to x_{i+1}. If we use Lagrange interpolation from Chapter 5 to approximate the function from $[x_i, f(x_i)]$ to $[x_{i+1}, f(x_{i+1})]$ by the straight line

$$f(x) \simeq [(x - x_{i+1})/(x_i - x_{i+1})]f(x_i) + [(x - x_i)/(x_{i+1} - x_i)]f(x_{i+1})$$

and we integrate the approximate function analytically, we obtain

[6.10]
$$\int_{x_i}^{x_{i+1}} f(x)dx \simeq [x_{i+1} - x_i][f(x_i) + f(x_{i+1})]/2$$

The approximation given in Eq. [6.10] is known as the **_trapezoidal rule_** because the right side is the area of a trapezoid with base equal to $(x_{i+1} - x_i)$ and parallel sides equal to $f(x_i)$ and $f(x_{i+1})$ as shown in Fig. 6–2.

A simplistic use of the trapezoidal rule to evaluate integrals of the form in Eq. [6.9b] is as follows. Divide the interval from x_a to x_b into n subintervals by choosing values x_0 through x_n that satisfy

$$x_a = x_0 < x_1 < \cdots < x_{n-1} < x_n = x_b$$

The points x_i should be chosen so that the part of the curve $f(x)$ between x_i and x_{i+1} can be approximated well by a straight line. Then we have

[6.11]
$$\int_{x_a}^{x_b} f(x)dx = \sum_{i=0}^{n-1}\left[\int_{x_i}^{x_{i+1}} f(x)dx\right]$$

$$\simeq \sum_{i=0}^{n-1}\{[x_{i+1} - x_i][f(x_i) + f(x_{i+1})]/2\}$$

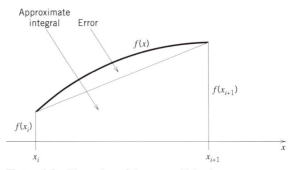

Figure 6–2   Illustration of the trapezoidal rule.

The result given by Eq. [6.11] does not contain any mechanism for controlling the accuracy of the approximation. It should therefore be used only for quick, informal estimates, or when experience allows the user to choose the control points x_i with good spacings, or when $f(x)$ is described by data values at specified control points.

Application of the trapezoidal rule to obtain estimates that are controlled by a criterion depend on equal spacings Δx between control points x_i and x_{i+1}. If we have equal spacings, the approximation in Eq. [6.11] may be rewritten as

[6.12]
$$I = \int_{x_a}^{x_b} f(x)dx \simeq (\Delta x/2)[f(x_0) + f(x_n)$$
$$+ \Delta x[f(x_1) + f(x_2) + \cdots + f(x_{n-1})]$$

The first term on the right-hand side is the trapezoidal rule approximation with only the end points as the control points. We begin with this term as the initial estimate of the integral I. Then we repeatedly refine our estimate by adding more control points at the midpoints of existing intervals until successive estimates show no significant change. An illustration of the process through two refinements is shown in Fig. 6–3.

For any refinement r (greater than zero), the spacing Δx_r is equal to $(\Delta x_{r-1}/2)$, the number k of *new* control points is equal to 2^{r-1}, and the new control points are located at

[6.13a]
$$x_{2m-1} = (x_a + [2m - 1]\Delta x_r); \qquad m = 1, 2, \ldots, k$$

Initial estimate (refinement 0):

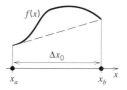

Refinement 1:

Refinement 2:

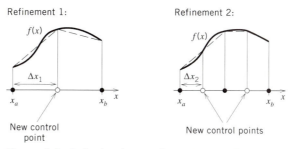

Figure 6–3 Refined estimates of an integral with the trapezoid rule.

Pseudocode 6–1   The trapezoidal rule

Module Trapezoid$(x_a, x_b, I, \epsilon, r_{max})$:
$\qquad r \leftarrow 0; \qquad \Delta x \leftarrow x_b - x_a; \qquad I \leftarrow \Delta x[f(x_a) + f(x_b)]/2$

$\quad$ **Repeat**
$\qquad\qquad r \leftarrow r + 1; \qquad \Delta x \leftarrow \Delta x/2; \qquad k \leftarrow 2^{r-1} \quad$ \ for next refinement
$\qquad\qquad s \leftarrow 0$
$\qquad\quad$ **For** $[m = 1, 2, \ldots, k]$
$\qquad\qquad\qquad x \leftarrow x_a + (2m - 1)\Delta x \qquad$ \ new control points
$\qquad\qquad\qquad s \leftarrow s + f(x) \qquad\qquad\qquad$ \ accumulate sum of function values
$\qquad\quad$ **End For**
$\qquad\qquad \Delta I \leftarrow s(\Delta x) - I/2 \qquad\qquad$ \ change in the estimate of I
$\qquad\qquad I \leftarrow I + \Delta I \qquad\qquad\qquad$ \ new estimate
$\quad$ **Until** $[r = r_{max}$ **or** $|\Delta I| < \epsilon]$
End Module Trapezoid

We then use Eq. [6.12] to compute a refined estimate I_r in terms of the old estimate I_{r-1} as

$$[6.13b] \qquad I_r = I_{r-1}/2 + \Delta x_r[f(x_1) + f(x_3) + \cdots + f(x_{2k-1})]$$

Pseudocode 6–1 shows the refinement process.

The quantity ϵ in the trapezoidal rule pseudocode is used to end the refinement process when the change in the estimate of I becomes sufficiently small. The quantity r_{max} is used to limit the number of refinements in case ϵ is chosen to be unrealistically small.

In general, the application that we have described is best used for x intervals on which the values of the integrand $f(x)$ are either all nonnegative or all nonpositive. If we observe this restriction, we can change the termination criterion to one in which the magnitude of ΔI becomes smaller than a small fraction of the magnitude of I. This fraction can be based easily on the precision level of the machine and is not as arbitrary as the choice of the quantity ϵ.

If we do not observe the restriction on the function values, we should specify a minimum number of refinements; otherwise, the refinement loop may be terminated too quickly and may produce an incorrect result. For example, consider the integrand

$$f(x) = e^{-x} \sin x$$

and integration limits x_a equal to 0 and x_b equal to 2π. Because $f(x)$ is zero at x_a and at x_b and also at $(x_a + x_b)/2$, the estimates I_0 and I_1 will both be zero and will cause premature termination of the process. To overcome that problem, we should either specify a minimum number of refinements or split the integral into two integrals from 0 to π and from π to 2π.

6.2.2   Simpson's Rule

An obvious way to improve the approximation of an integral is to use a better model for the integrand $f(x)$. One attempt at a better model is to use a quadratic instead of a linear approximation. A Lagrange interpolating polynomial through the points $[x_i, f(x_i)]$, $[x_{i+1}, f(x_{i+1})]$, and $[x_{i+2}, f(x_{i+2})]$ provides a quadratic approximation. When the spacings $(x_{i+1} - x_i)$ and $(x_{i+2} - x_{i+1})$ are chosen to be equal, analytical integration of the approximating quadratic yields **Simpson's rule:**

[6.14]
$$\int_{x_i}^{x_{i+2}} f(x)dx \simeq [x_{i+2} - x_i][f(x_i) + 4f(x_{i+1}) + f(x_{i+2})]/6$$

As with the trapezoidal rule, we may apply Simpson's rules in a simplistic way to integrals of the form in Eq. [6.9b] by dividing the interval from x_a to x_b into an **even** number of subintervals. We do so by choosing control points x_0 through x_{2j} that satisfy

$$x_a = x_0 < x_1 < \cdots < x_{2j-1} < x_{2j} = x_b;$$
$$x_{i+2} - x_{i+1} = x_{i+1} - x_i \text{ for even values of } i$$

The Simpson's rule approximation of the integral is then

[6.15]
$$\int_{x_a}^{x_b} f(x)dx = \sum_{p=0}^{j-1}\left[\int_{x_{2p}}^{x_{2p+2}} f(x)dx\right]$$
$$\simeq \sum_{p=0}^{j-1}\{[x_{2p+2} - x_{2p}][f(x_{2p}) + 4f(x_{2p+1}) + f(x_{2p+2})]/6\}$$

The relation between Simpson's rule and the trapezoidal rule may be seen by writing Eq. [6.14] in the form

$$\int_{x_i}^{x_{i+2}} f(x)dx \simeq [x_{i+2} - x_i][f(x_i) + 4f(x_{i+1}) + f(x_{i+2})]/6$$
$$= (4/3)\{[x_{i+1} - x_i][f(x_i) + f(x_{i+1})]/2$$
$$+ [x_{i+2} - x_{i+1}][f(x_{i+1}) + f(x_{i+2})]/2\}$$
$$- (1/3)[x_{i+2} - x_i][f(x_i) + f(x_{i+2})]/2$$

The term multiplied by $(4/3)$ is the trapezoidal rule approximation to the integral with two intervals of equal size $(x_{i+2} - x_{i+1})$ or $(x_{i+1} - x_i)$, and the term multiplied by $(1/3)$ is the trapezoidal rule approximation with one interval of size $(x_{i+2} - x_i)$.

In general, let $(I_s)_{2j}$ be the Simpson's rule approximation to the integral with $2j$ intervals of size Δx, let $(I_t)_{2j}$ be the trapezoidal rule approximation with the same intervals, and let $(I_t)_j$ be the trapezoidal rule approximation with j intervals of size $2\Delta x$. Then

[6.16] $$(I_s)_{2j} = (4/3)(I_t)_{2j} - (1/3)(I_t)_j$$

A Simpson's rule evaluation of an integral may therefore be performed by modifying the trapezoidal rule pseudocode in §6.2.1 to include the line

$$I \leftarrow I + \Delta I/3$$

immediately after the refinement loop.

The formulas for the trapezoidal rule and for Simpson's rule are special cases of the *closed Newton–Cotes* integration formulas (see Appendix C). These formulas are obtained by fitting N–th degree polynomials to the integrand function at $(N + 1)$ control points. Closed formulas include the limits of the integral in the set of control points. For example, the end values x_i and x_{i+2} are two of the three control points for Simpson's rule. Open formulas exclude the end values from the set of control points.

The Simpson's rule application given by Eq. [6.15] is useful for tabulated data or for informal estimates. The form given by Eq. [6.16] is in fact a special case of Romberg integration, which is discussed in the following section.

6.2.3   Romberg Integration

Extrapolation of the estimates of an integral found by the trapezoidal rule in the manner of the Richardson extrapolation of §6.1.2 is known as *Romberg integration*. The technique is based on the fact that the error in a trapezoidal rule estimate with n intervals may be expressed as

$$\text{error} = O(1/n^2) + O(1/n^4) + \cdots$$

Let $I_{r,0}$ be the trapezoidal rule estimate with 2^r intervals for r equal to $(0, 1, \ldots)$. If we interpret $O(1/n^2)$ as $\alpha(1/n^2)$ with α taken to be a constant multiplier, a true integral I may be expressed to the leading error term as

$$I = I_{r,0} + \alpha(1/2^{2r}) = I_{r+1,\,0} + \alpha\,(1/[4(2^{2r})])$$

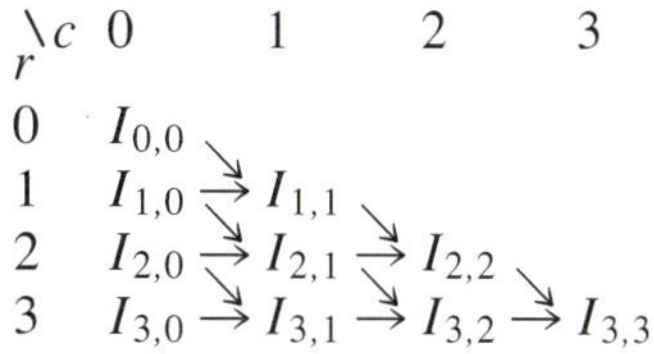

$$\begin{array}{lllll}
\backslash c & 0 & 1 & 2 & 3 \\
r & & & & \\
\end{array}$$

$$\begin{array}{llll}
0 & I_{0,0} & & & \\
1 & I_{1,0} \rightarrow I_{1,1} & & \\
2 & I_{2,0} \rightarrow I_{2,1} \rightarrow I_{2,2} & \\
3 & I_{3,0} \rightarrow I_{3,1} \rightarrow I_{3,2} \rightarrow I_{3,3}
\end{array}$$

Figure 6–4   Sequence of Romberg integration estimates up to level 3.

A first-level improvement in the estimate is obtained by eliminating the leading error term according to

$$I_{\text{improved}} = [4I_{r+1,0} - I_{r,0}]/3$$

This improvement is the same as the Simpson's rule given in Eq. [6.16] and has a leading error term $O(1/[2^{4r+4}])$. Further extrapolation is used to eliminate the new leading term as many times as desired. Each elimination represents a *level* of improvement c. The estimate at level c with 2^r intervals is denoted by $I_{r,c}$ and is generally expressed as

[6.17] $$I_{r,c} = I_{r,c-1} + (I_{r,c-1} - I_{r-1,c-1})/(4^c - 1)$$

The Romberg integration estimate for a level c is built up from the level 0 trapezoidal rule estimates as illustrated in Fig. 6–4. The arrows indicate the derivation of $I_{r,c}$ from the estimates $I_{r,c-1}$ and $I_{r-1,c-1}$ according to Eq. [6.17]. Improved accuracy is achieved either by increasing r (and thus the number of intervals) or by increasing the level c. The advantage of the latter approach is that fewer evaluations of the integrand function are required.

The computation of an integral by the Romberg technique proceeds from $I_{0,0}$ until the absolute difference between a current estimate $I_{r,r}$ and a previous estimate $I_{r-1,r-1}$ is less than a prescribed value ϵ. As we approach convergence, there is an apparent danger of loss of accuracy due to subtractive cancellation when we compute the right-hand side of Eq. [6.17]. Fortunately, this danger is never realized with sensible values of ϵ, because the nature of the convergence and the size of the term $(4^c - 1)$ make errors due to subtractive cancellation insignificant.

Romberg integration is shown in Pseudocode 6–2; it builds on the pseudocode given earlier for the trapezoidal rule. The required result is denoted by Int, and the terms $I_{r,c}$ of the Romberg integration are stored in a two-dimensional array for convenience.

Pseudocode 6–2   Romberg integration

Module Romberg(x_a, x_b, Int, ϵ, r_{max}):
Declare: I \ with each of the two subscripts varying from 0 to r_{max}

$\quad$ $r \leftarrow 0$; $\Delta x \leftarrow x_b - x_a$; $I_{0,0} \leftarrow \Delta x[f(x_a) + f(x_b)]/2$

$\quad$ **Repeat**

$\qquad$ $r \leftarrow r + 1$; $\Delta x \leftarrow \Delta x/2$; $k \leftarrow 2^{r-1}$ \ for next refinement
$\qquad$ $s \leftarrow 0$
$\qquad$ **For** $[m = 1, 2, \ldots, k]$
$\qquad\qquad$ $x \leftarrow x_a + (2m - 1)\Delta x$ \ new control points
$\qquad\qquad$ $s \leftarrow s + f(x)$ \ accumulate sum of $f(x)$
$\qquad$ **End For**
$\qquad$ $\Delta I \leftarrow \Delta x(s) - I_{r-1,0}/2$ \ change in trapezoidal rule value
$\qquad$ $I_{r,0} \leftarrow I_{r-1,0} + \Delta I$ \ new trapezoidal rule estimate
$\qquad$ **For** $[c = 1, 2, \ldots, r]$ \ loop for Romberg estimates
$\qquad\qquad$ $I_{r,c} \leftarrow I_{r,c-1} + (I_{r,c-1} - I_{r-1,c-1})/(4^c - 1)$
$\qquad$ **End For**

$\qquad$ Int $\leftarrow I_{r,r}$ \ best Romberg estimate so far

$\quad$ **Until** $[r = r_{max}$ **or** $|$Int$- I_{r-1,r-1}| < \epsilon]$
End Module Romberg

To illustrate the effectiveness of Romberg integration, we show in Table 6–1 the $I_{r,c}$ values for the integral

$$[6.18] \qquad I = \int_0^{\pi/2} (e^{-x} \sin x)\, dx = \left[1 - e^{-\pi/2}\right]/2$$

The table is terminated at r equal to 3 when $|I_{3,3} - I_{2,2}|$ first becomes smaller than 10^{-4}. The number of evaluations of the integrand is $(2^3 + 1)$ or 9. To achieve the same level of accuracy with the trapezoidal rule requires an r value of 11 or 2049 evaluations of the integrand. The efficiency of the Romberg method is especially significant when it takes considerably more effort to evaluate an integrand than it takes to perform the computations required by Eq. [6.17].

Table 6–1   Values of $I_{r,c}$ for Romberg integration of Eq. [6.18]

r	0	1	2	3
			c	
0	0.1632682375			
1	0.3348440848	0.3920360339		
2	0.3805906044	0.3958394442	0.3960930049	
3	0.3921828620	0.3960469479	0.3960607815	0.3960602700

6.2.4 Gaussian Quadrature

The integration formulas of Eq. [6.10] for the trapezoidal rule and Eq. [6.14] for Simpson's rule are called *interpolatory quadratures* because the integrand is approximated by an interpolating polynomial at control points x_0 through x_n. The control points are equally spaced for n values greater than 1. Extended forms of these rules such as those in Eq. [6.11] and [6.15] and in Romberg integration are known as *compound quadratures*. In all cases, the form of the quadrature is

$$[6.19] \qquad \text{Integral} = \sum_{i=0}^{n} \alpha_i f(x_i)$$

in which the multipliers α_i are chosen after the control points x_i have been selected.

The major concept in *Gaussian quadrature* is the *simultaneous* selection of the multipliers and control points to satisfy a given condition. The general integral to be approximated by Gaussian quadrature is usually represented as

$$\int_{-1}^{1} \omega(x) f(x) dx$$

in which $\omega(x)$ is a weight function. In the special case where $\omega(x)$ is equal to 1, the resulting Gaussian quadratures are more formally called *Gauss–Legendre quadratures*.

The Gaussian quadrature problem is to determine α_i and x_i so that the quadrature represented by

$$[6.20] \qquad \int_{-1}^{1} f(x) dx = \sum_{i=0}^{n} \alpha_i f(x_i)$$

is exact when the integrand is any polynomial up to degree $(2n + 1)$. The polynomial with the highest degree has $(2n + 2)$ coefficients; this value matches the number of parameters x_0 through x_n and α_0 through α_n that are to be selected.

An example of the selection process is shown with n equal to 1. All polynomials up to degree 3 must be exactly represented by Eq. [6.20]. We replace the integrand function $f(x)$ in Eq. [6.20] by x^k to form the four equations

$$\int_{-1}^{1} x^k dx = \alpha_0(x_0)^k + \alpha_1(x_1)^k = \begin{cases} 0; & k = \text{odd} \\ 2/(k+1); & k = \text{even} \end{cases} ; \quad k = 0, 1, 2, 3$$

The solutions of the four equations are

$$\alpha_0 = \alpha_1 = 1; \qquad -x_0 = x_1 = 1/\sqrt{3}$$

In general, the control points x_0 through x_n for Gauss–Legendre quadrature are the roots of the **Legendre polynomial** $P_{n+1}(x)$, which is defined by

[6.21a] $$P_0(x) = 1; \qquad P_1(x) = x;$$
$$P_{n+1}(x) = [(2n + 1)xP_n(x) - nP_{n-1}(x)]/(n + 1); \qquad n = 2, 3, \ldots$$

The multipliers α_0 through α_n are then obtained from

[6.21b] $$\alpha_i = 2[1 - (x_i)^2]/[(n + 1)P_n(x_i)]^2$$

Roots and multipliers are tabulated in various mathematical handbooks and textbooks (Ref. 11) for several values of n. Values of x_i and α_i for n from 0 to 4 are as follows.

One-Point Quadrature $(n = 0)$:

[6.22a] $$x_0 = 0; \qquad \alpha_0 = 2$$

Two-Point Quadrature $(n = 1)$:

[6.22b] $$-x_0 = x_1 = 1/\sqrt{3}; \qquad \alpha_0 = \alpha_1 = 1$$

Three-Point Quadrature $(n = 2)$:

[6.22c] $$-x_0 = x_2 = \sqrt{0.6}; \quad x_1 = 0; \quad \alpha_0 = \alpha_2 = 5/9; \quad \alpha_1 = 8/9$$

Four-Point Quadrature $(n = 3)$:

[6.22d] $$p = \left[30 + \sqrt{480}\right]/70; \qquad q = \left[30 - \sqrt{480}\right]/70;$$
$$-x_0 = x_3 = \sqrt{p} = 0.8611363; \qquad -x_1 = x_2 = \sqrt{q} = 0.3399810;$$
$$\alpha_0 = \alpha_3 = 1 - \alpha_1 = 1 - \alpha_2 = (1 - 3q)/(3[p - q]) = 0.3478548$$

Five-Point Quadrature $(n = 4)$:

[6.22e] $$p = \left[70 + \sqrt{1120}\right]/126; \qquad q = \left[70 - \sqrt{1120}\right]/126;$$
$$-x_0 = x_4 = \sqrt{p} = 0.9061798; \qquad -x_1 = x_3 = \sqrt{q} = 0.5384693;$$
$$x_2 = 0;$$
$$\alpha_0 = \alpha_4 = (3 - 5q)/(15p[p - q]) = 0.2369269;$$
$$\alpha_1 = \alpha_3 = (5p - 3)/(15q[p - q]) = 0.4786287;$$
$$\alpha_2 = 128/225;$$

Values in Eq. [6.22a] through [6.22e] are given in such a way that they can be evaluated to arbitrary levels of precision. Quantities that are evaluated are rounded to seven decimal digits and are adequate for computations in single precision (32 bits). The control points x_i are antisymmetric with respect to $(x = 0)$ and include $(x = 0)$ when n is even; the multipliers α_i are symmetric. These properties may be expressed by

$$x_{n-i} = -x_i; \qquad \alpha_{n-i} = \alpha_i; \qquad i = 0, 1, \ldots, \text{ Integer result of } (n/2)$$

To use Gaussian quadrature for an integral with respect to ξ between arbitrary limits a and b, we use the linear transformation

[6.23a] $$x = (2\xi - a - b)/(b - a)$$

to obtain

$$\int_a^b \phi(\xi)d\xi = 0.5 \int_{-1}^1 f(x)dx;$$

[6.23b] $$f(x) = (b - a)\phi(\{[b - a]x + a + b\}/2);$$

The application of Gaussian quadrature is illustrated in the following example.

EXAMPLE 6.3

The standardized normal distribution function that is used in statistics and given in Eq. [4.17] may be expressed as

$$\Phi(z) = 0.5 + I/\sqrt{2\pi}; \qquad I = \int_0^z e^{-u^2/2} du$$

Use a five-point Gaussian quadrature to evaluate I when z is equal to 1.5.

From Eq. [6.32b]:

$$I = 0.5 \int_{-1}^1 f(x)dx; \qquad f(x) = 1.5e^{-(0.75x + 0.75)^2/2}$$

Thus

$$I \simeq 0.5 \sum_{i=0}^4 \alpha_i f(x_i)$$

where α_i and x_i are obtained from Eq. [6.22e]. The result (rounded to 10 significant figures) is

$$I \simeq 1.085\ 853\ 299$$

For comparison, we give the "true" value of I as obtained from Eq. [4.19] and other estimates of I with five-point quadratures. All values are rounded to 10 significant figures.

"True" Value:	$I = 1.085\ 853\ 318$
Trapezoidal Rule:	$I \simeq 1.080\ 136\ 620$
Simpson's Rule:	$I \simeq 1.085\ 890\ 701$
Romberg:	$I \simeq 1.085\ 883\ 233$

We see, in this example, that the estimate from Gaussian quadrature is better than the estimates from any of the compound quadratures for the same number of control points.

6.2.5 Improper Integrals

The term *improper integral* is generally used to describe an integral that has one of the following properties:

- The integral has an upper limit of ∞, a lower limit of $(-\infty)$, or both.
- The integrand tends to a finite limiting value at a finite limit of the integral, but cannot be evaluated exactly at that limit. For example, the integrand function $[x/\tan(x)]$ tends to 1 at $(x = 0)$.
- There is an integrable singularity at a known location on the interval defined by the limits; that is, the integrand becomes infinite, but the integral is finite.

An integral whose value is either infinite or undefined is *not* classified as an improper integral; such an integral simply cannot be evaluated.

We begin our discussion of dealing with improper integrals by considering infinite limits. Let us rewrite the distribution function $\Phi(z)$ in Example 6.2 as

[6.24]
$$\Phi(z) = 1 - I/\sqrt{2\pi};\quad I = \int_{z}^{\infty} e^{-x^2/2}\,dx$$

The integral I in Eq. [6.24] has an integrand that tends to zero as the infinite upper limit is approached. The simplest way to deal with the infinite range is to replace the upper limit by a sufficiently large value, say L, and perform the integration from z to L. The difficulty with this approach is that we really do not know how large L should be; underestimation of L will cause an inaccurate result, and overestimation will cause us to expend needless effort.

Another approach is to split the integral into a set of integrals with smaller ranges of integration. The value of z in Example 6.3 was 1.5; we might therefore express I in Eq. [6.24] as

$$[6.25] \qquad I = \int_{1.5}^{4} e^{-x^2/2} dx + \int_{4}^{8} e^{-x^2/2} dx + \int_{8}^{16} e^{-x^2/2} dx + \cdots$$

The range of integration in the first integral is a reasonably small estimate over which we expect most of the contribution to I. Thereafter, the range is increased because the integrand is expected to be small in magnitude. The strategy for evaluating I is to add the integrals on the right-hand side of Eq. [6.25] until one of them is found to be sufficiently small. Following this strategy, we find that the first integral on the right-hand side of Eq. [6.25] has a value of 0.167381432 (rounded to nine decimal places), the second has a value of 0.000079388, and the third is insignificant. The value of I is then estimated by summing the first two integrals to obtain 0.167460820, which matches the true value to all nine decimal places.

Yet another approach for dealing with infinite limits is to transform the variable of integration so that the limits become finite. A typical transformation is to set x equal to $(1/t^m)$, where m is a positive integer. If we use this approach with m equal to 1 for the integral in Eq. [6.24], we obtain

$$[6.26] \qquad I = \int_{0}^{1/z} t^{-2} e^{-0.5/t^2} dt$$

We have achieved the objective of finite limits; however, the integral in this particular case now acquires a different property of improper integrals. The form in Eq. [6.26] has an integrand with a finite limiting value (of 0) at the lower limit where t is 0. Closed formulas such as the trapezoidal or Simpson's rule require the evaluation of the integrand at the endpoints of the range of integration. For well-behaved functions, it may be enough to replace the value of the integrand function by its limiting value at the appropriate endpoint. Another option is to avoid the endpoint completely by using one of the open Newton–Cotes formulas or by employing Gaussian quadrature.

The example given in Eq. [6.26] does not yield easily to either of the options that we have just described. The difficulty has nothing to do with the integration process; rather, it lies in the evaluation of the integrand *near* the limit ($t = 0$). Suppose that we are using 64-bit precision with underflow at about 10^{-308} to solve the problem. The exponential term in the integrand reaches the underflow value when $(-0.5/t^2)$ is approximately equal to $\ln(10^{-308})$, and the corresponding t value is about 0.027. For z equal to 1.5 in Eq. [6.26], the upper limit of integration is $(2/3)$; thus any closed quadrature requiring more than 25 equal intervals will require evaluation of the integrand for t values less than 0.027. Similarly, any Gaussian quadrature that has a control point less than about (-0.92) will require an evaluation at a t value less than 0.027.

The difficulty posed by the integrand in Eq. [6.26] is easily removed if we recognize that the difference of the integrand from zero is inconsequential when the exponential term is near the underflow limit. We may then replace the integral in Eq. [6.26] by

$$I = \int_0^{1/z} f(t)\, dt; \quad f(t) = \begin{cases} t^{-2}e^{-0.5/t^2}; & t \geq 0.03 \\ 0; & t < 0.03 \end{cases}$$

This last form is a proper integral, and it may be evaluated satisfactorily by any of the standard procedures.

The last type of improper integral to be considered is the one that contains an integrable singularity. Let us first look at the case where the singularity is at a finite limit of the integral. This case is exemplified by the expression for the period T of a simple pendulum of length ℓ under a gravitational acceleration g. The pendulum's angle θ with respect to the vertical is in the range $(-\theta_{max} \leq \theta \leq \theta_{max})$, where θ_{max} is assumed to be a positive angle no greater than $(\pi/2)$. For these conditions, T is given by

[6.27] $$T = I\sqrt{8\ell/g}; \quad I = \int_0^{\theta_{max}} \left[1/\sqrt{\cos\theta - \cos\theta_{max}} \right] d\theta$$

The integrand in Eq. [6.27] is clearly singular when θ is equal to θ_{max}. One way to remove the singularity is to find a suitable change in the variable of integration. For example, the change

$$\sin\beta = [\sin(\theta/2)]/[\sin(\theta_{max}/2)]$$

allows us to express I in the alternate form

$$I = \sqrt{2} \int_0^{\pi/2} \left[1/\sqrt{1 - \sin^2(\theta_{max}/2)\sin^2(\beta)} \right] d\beta$$

The new integral on the right hand side is known as an ***elliptic integral of the first kind*** and contains no singularities. It may therefore be evaluated by standard means.

The change of variable that we used for Eq. [6.27] is not obvious. Even with simpler integrands, finding an appropriate change of variable requires some experience. Let us concentrate, therefore, on numerical approaches to dealing with singularities by returning to the integral I as expressed in Eq. [6.27].

An approach to the integration is to rewrite the integral as

[6.28] $$I = I_1 + I_2$$

$$= \int_0^{\theta_{max}-\delta} \left[1/\sqrt{\cos\theta - \cos\theta_{max}} \right] d\theta + \int_{\theta_{max}-\delta}^{\theta_{max}} \left[1/\sqrt{\cos\theta - \cos\theta_{max}} \right] d\theta$$

where δ is a small, positive quantity. The first integral I_1 can be evaluated, but we are still left with the problem of what to do with the second integral I_2. We could either ignore it and run the risk of obtaining an inaccurate result, or find another way to evaluate it. The second option may be implemented for Eq. [6.28] by writing

$$\theta = \theta_{max} - \psi$$

This change of variable allows us to write I_2 as

$$I_2 = \int_0^\delta \left[1 / \sqrt{(\cos \theta_{max})(\cos \psi - 1) + (\sin \theta_{max})(\sin \psi)} \right] d\psi$$

Then by using the small-angle approximations for $(\cos \psi)$ and $(\sin \psi)$, we obtain the approximation

[6.29] $$I_2 \simeq \int_0^\delta \left[1 / \sqrt{\psi \sin \theta_{max}} \right] d\psi = 2 \sqrt{\delta / (\sin \theta_{max})}$$

The objection to evaluating I_2 in this fashion is the same as for our earlier approach; namely, it requires some experience to find a suitable change of variable. Nevertheless, our example is instructive because it allows us to see how the choice of δ affects the result of the integration.

The numerically obtained value of I_1 in Eq. [6.28] and the approximate value of I_2 from Eq. [6.29] are given to six decimal places in Table 6–2 for three values of δ. The true value of $(I_1 + I_2)$ is 2.384011 (to six decimal places). Even for relatively small values of δ, ignoring the contribution of I_2 will create an error in the second or third significant digit. The results for I_2 are themselves inexact because of the small-angle approximations used in deriving Eq. [6.29]. The errors in I_2 are approximately (-0.000104), (-0.000037), and (-0.000003) for δ values of 0.010, 0.005, and 0.001, respectively. These errors limit the accuracy of the overall integral even when the part I_1 is computed very accurately.

The next logical approach is to use Gauss–Legendre quadrature to avoid the limits of the integral. The errors that result from this approach may be large because they depend on the derivatives of the singular integrand. Another Gaussian quadrature known as **Gauss–Chebyshev quadrature** (Ref. 11) is usually effective for

Table 6–2 Variation of I_1 and I_2 from Eq. [6.28] and [6.29] with δ

δ	I_1	I_2
0.010	2.168993	0.214914
0.005	2.232007	0.151967
0.001	2.316046	0.067962

integrands that are singular at one of the limits of integration. This type of quadrature is applied according to

[6.30a] $$I = \int_{-1}^{1} \left[f(x)/\sqrt{1 - x^2} \right] dx \simeq [\pi/(n + 1)] \sum_{i=0}^{n} f(x_i)$$

where the control points x_0 through x_n are given by

[6.30b] $$x_i = \cos[(2i + 1)\pi/(2n + 2)]; \qquad i = 0, 1, \ldots, n$$

To apply the Gauss–Chebyshev quadrature of Eq. [6.30a, b] to the integral I in Eq. [6.27], we first use Eq. [6.23a, b] to rewrite it in the form

[6.31a] $$I = 0.5 \int_{-1}^{1} \left[\theta_{max}/\sqrt{\cos(\theta_{max}[1 + x]/2) - \cos \theta_{max}} \right] dx$$

For the integral on the right-hand side of Eq. [6.31a], the function corresponding to $f(x)$ in Eq. [6.30a] is

[6.31b] $$f(x) = \theta_{max} \sqrt{\{1 - x^2\}/\{\cos(\theta_{max}[1 + x]/2) - \cos \theta_{max}\}}$$

We may now proceed with the Gauss–Chebyshev quadrature, but we must exercise some care in evaluating the function $f(x)$ when x is close to 1. The numerator and denominator of the term in the square root each approaches zero as x approaches 1; furthermore, their approach to zero results from subtraction. The computation of $f(x)$ therefore requires a division with two operands that may be individually affected by subtractive cancellation.

When the integrand is singular at a location that is not a limit, the integral is split into two integrals with the location of the singularity as the common limit. Each of the two new integrals now has a singular integrand at one limit and may be treated in the same manner as we discussed earlier.

6.2.6   Multidimensional Integrals

An integral with more than one variable of integration is called a *multiple integral* or a *multidimensional integral*. In principle, multidimensional integrals may be thought of as nested one-dimensional integrals; in practice, they may be formidable, especially when the domain of integration is complicated.

Consider the (x, y) domain shown in Fig. 6–5. The *second moments of the area* of the domain (or its *area moments of inertia*) about the x and y axes are denoted by I_x and I_y, respectively. These moments are expressed by

$$I_x = \int_A y^2 dA; \qquad I_y = \int_A x^2 dA$$

NUMERICAL INTEGRATION **221**

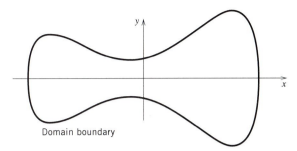

Figure 6–5 A nonconvex domain of integration in two dimensions.

where A is the area. If we replace dA by $(dx\ dy)$, the moments of inertia are represented by multidimensional integrals. The simple natures of the integrands make it possible to reduce the integrals to one dimension; however, the nonconvex shape of the domain makes it important to consider which variable is to be used for the first integration.

If we integrate with respect to x first, we may have to do it in two pieces because a line of constant y may intersect the boundary of the domain at four locations. This difficulty does not arise if we integrate with respect to y first; therefore, we may represent the moments of inertia by the integrals

$$I_x = \int_{x_a}^{x_b} \left[\int_{y_a(x)}^{y_b(x)} y^2 dy \right] dx; \qquad I_y = \int_{x_a}^{x_b} \left[\int_{y_a(x)}^{y_b(x)} x^2 dy \right] dx$$

The y limits of integration are functions of x, which, for the symmetric domain in Fig. 6–5, may be described by

$$-y_a(x) = y_b(x) = f(x); \qquad x_a \le x \le x_b$$

The limits x_a and x_b are the left and right extremities of the domain. When the inner integrals are performed, x is held constant; thus the one-dimensional forms of the moments of inertia are

$$I_x = \int_{x_a}^{x_b} 2[f(x)]^3/3\ dx; \qquad I_y = \int_{x_a}^{x_b} 2x^2 f(x)\ dx$$

It may not be possible to reduce the dimensions in a more general case. For example, suppose that the domain in Fig. 6–5 is the base of a three-dimensional solid of height $z(x, y)$. The volume V of the solid is then obtained from

$$V = \int_{x_a}^{x_b} \left[\int_{y_a(x)}^{y_b(x)} z(x, y)\ dy \right] dx$$

The concept of integrating with respect to y first is still in effect, but it may not be possible or easy to integrate $z(x, y)$ analytically with respect to y. We must then perform a full numerical integration in nested form. To do so, we will generally require two integration modules. The result of the y integration module is the value of the integrand function for the x integration module.

EXAMPLE   6.4

Establish the approximation for computing

$$I = \int_{y_1}^{y_n} \int_{x_1}^{x_m} z(x, y)\,dx\,dy$$

on the rectangular domain ($x_1 \leq x \leq x_m, y_1 \leq y \leq y_n$) by the trapezoidal rule. The control points are given by

$$x_i = x_1 + (i - 1)h; \quad y_j = y_1 + (j - 1)k$$

in which h and k are constant spacings.

Let $z_{i,j}$ denote $z(x_i, y_j)$. Then, denoting the integral of z with respect to x at a constant y value equal to y_i by u_j, we have from Eq. [6.12]

$$u_j = (h/2)z_{1,j} + h[z_{2,j} + z_{3,j} + \cdots + z_{m-1,j}] + (h/2)z_{m,j}$$

We then use Eq. [6.12] again to obtain the integral of u with respect to y as

$$I = (k/2)u_1 + k[u_2 + u_3 + \cdots + u_{n-1}] + (h/2)u_n$$

The entire integration may be expressed in matrix form by

$$I = (hk/4)\mathbf{r} \cdot \mathbf{Z} \cdot \mathbf{c}$$

where $\mathbf{r}$ is an m-component row vector and $\mathbf{c}$ is an n-component column vector given by

$$\mathbf{r} = [1 \quad 2 \quad \ldots \quad 2 \quad 1]; \quad \mathbf{c}^T = [1 \quad 2 \quad \ldots \quad 2 \quad 1]$$

and $\mathbf{Z}$ is the ($m \times n$) integrand matrix with elements $z_{i,j}$.

6.3   CLOSURE

The fundamentals of numerical calculus — differentiation and integration — are treated in the preceding sections. The approach that we have used for differentiation is based on Taylor series expansions about a base point. The order of accuracy for a given derivative depends on the number of expansions that are used and term cancellations that are possible when the expansion points are symmetric with respect to the base point. In general, a higher derivative will have a lower order of accuracy than a lower derivative for the same expansion points. Smaller interval spacings will give more accurate approximations, unless the spacings become so small that errors due to finite precision begin to dominate.

The workhorse methods of numerical integration are the trapezoidal and Simpson's rules, which are discussed in §6.2.1 and §6.2.2. These may be applied as summations for a prescribed set of control points, or they may be applied only to as many points as needed. Simpson's rule and Romberg integration in general are based on the trapezoidal rule pseudocode and require minimal extra effort for significant gains in accuracy. The Gauss–Legendre quadrature of §6.2.4 is a high-order method that is no more difficult to use than the trapezoidal rule. Another form of Gaussian quadrature, Gauss–Chebyshev quadrature, is described in our treatment of improper integrals. Good manipulative skills are helpful when dealing with improper or multidimensional integrals, as they allow us to recast these integrals in more desirable forms.

6.4 EXERCISES

1. The time t [s] for a car to accelerate from 40 mph to a speed v [mph] is given below for five values of v.

index, i:	1	2	3	4	5
v_i [mph]:	40	45	50	55	60
t_i [s]:	0.00	0.69	1.40	2.15	3.00

The car's acceleration a [ft/s^2] at a speed v [mph] may be expressed by

$$a = (22/15)(dv/dt) = (22/15)/(dt/dv)$$

Estimate a when v is 50 mph by the following techniques.

(a) Use a central-difference formula for $(dt/dv)_3$ with data at i equal to 2 and 4.
(b) Develop and use a central-difference formula for $(dt/dv)_3$ with data at i equal to 1, 2, 4, and 5.
(c) Develop and use a formula for $(dv/dt)_3$ with data at i equal to 2, 3, and 4.
(d) Develop and use a formula for $(dv/dt)_3$ with data at all values of i.

2. Use the data in Problem 1 to estimate a when v is 45 mph by the following techniques.

(a) Use a central-difference formula for $(dt/dv)_2$ with data at i equal to 1 and 3.
(b) Develop and use a third-order difference formula for $(dt/dv)_2$ with data at i equal to 1, 2, 3, and 4.
(c) Develop and use a fourth-order difference formula for $(dt/dv)_2$ with data at all values of i.

3. The altitude h [ft] attained by an aircraft climbing from sea level at its maximum rate of climb [ft/min] is shown at four times t [min].

index, i:	1	2	3	4
t_i [min]:	10	20	30	40
h_i [ft]:	5400	9000	11600	12800

Estimate the aircraft's rate of climb $(dh/dt)_3$ when it is at an altitude of 11600 ft by the following techniques.

(a) Use a backward-difference formula with data at i equal to 2 and 3.
(b) Use a forward-difference formula with data at i equal to 3 and 4.
(c) Use a central-difference formula with data at i equal to 2 and 4.
(d) Develop and use a third-order difference formula with data at all values of i.

4. Estimate the rate of climb dh/dt for the aircraft in Problem 3 at times t equal to 24 min and 37 min. To do so, develop the highest-order difference formula by expanding h at the three times closest to t about the time t.

5. How many function values $f(x_i)$, $f(x_{i+1})$, etc. are required to develop a third-order, forward-difference formula for d^3f/dx^3 at x_i? Develop such a formula, assuming that the x values are equally spaced.

6. How many function values $f(x_i)$, $f(x_{i-1})$, etc. are required to develop a first-order, backward-difference formula for d^4f/dx^4 at x_i? Develop such a formula, assuming that the x values are equally spaced.

7. Use the data in Problem 1 to estimate the distance x [ft] that the car travels in accelerating from 40 mph to 60 mph by applying the trapezoidal rule to

$$x = (22/15) \int_0^3 v \, dt$$

8. Use the idea that an integral is the "area under the curve" and a sketch of v versus t to show that the distance x in Problem 7 may be expressed by

$$x = (22/15) \left[t_5 v_5 - \int_{40}^{60} t \, dv \right]$$

9. Estimate x from the expression in Problem 8 and the data in Problem 1 by using

(a) The trapezoidal rule and Simpson's rule with five data points.
(b) The best Romberg integration that is available with the given data.

10. A circular shaft has a diameter d [m] that varies with axial position x [m] according to

$$d = 0.02(1 + x^2)/e^x; \qquad 0 \le x \le 3 \text{ m}$$

An axial load P of 30000 N is applied to one end of the shaft, whose modulus of elasticity E is (2×10^{11}) N/m². The axial elongation of the shaft is Δx [m] and is given by

$$\Delta x = (P/E) \int_0^3 (1/A) \, dx; \qquad A = \text{cross-sectional area} = \pi d^2/4$$

Estimate Δx by the trapezoidal rule and by Simpson's rule with 10 equal intervals.

11. Apply the pseudocode given for the trapezoidal rule in §6.2.1 to Problem 10. Use a convergence indicator ϵ of 0.1 for the integral of $(1/A)$. Report the results for Δx and for the number of refinements r.

12. Apply the pseudocode given for Romberg integration in §6.2.3 to Problem 10. Use a convergence indicator ϵ of 0.1 for the integral of $(1/A)$. Report the results for Δx and for the number of refinements r.

13. Solve Problem 10 by splitting the integral into two integrals with equal ranges of integration and by using a five-point Gaussian (Gauss–Legendre) quadrature for each of the two resulting integrals.

14. The error function erf(x) appears in many solutions of classical physical problems (for example, statistics, temperature distribution in a bar, and viscous flow velocity distribution above an accelerated surface). The error function is given by

$$\text{erf}(x) = (2/\sqrt{\pi}) \int_0^x \exp(-t^2) dt$$

Use three-point quadratures to estimate erf(0.5) by the trapezoidal rule, Simpson's rule, and Gaussian (Gauss–Legendre) quadrature.

15. Use five-point quadratures to estimate erf(0.8), defined in Problem 14, by the trapezoidal rule, Simpson's rule, and Gaussian (Gauss–Legendre) quadrature. Also use a Gaussian (Gauss–Legendre) four-point quadrature to estimate erf(0.8).

16. The width w (in the y direction) of the domain shown in Fig. 6–5 is given by

$$w/2 = (2x^2 + 0.2x + 0.08) \sqrt{1 - 4x^2}; \qquad -0.5 \le x \le 0.5$$

The area A of the domain and the x coordinate x_c of its centroid are given by

$$A = \int_{-0.5}^{0.5} w \, dx; \qquad x_c A = \int_{-0.5}^{0.5} xw \, dx$$

Use an appropriate numerical integration scheme to obtain accurate estimates of A and x_c.

17. Suppose that $(w/2)$ in Problem 16 is now the radius r of a solid body of revolution. The volume V and the x coordinate x_c of the body's centroid are given by

$$V = \pi \int_{-0.5}^{0.5} r^2 \, dx; \qquad x_c V = \pi \int_{-0.5}^{0.5} xr^2 \, dx$$

Use an appropriate numerical integration scheme to obtain accurate estimates of V and x_c.

18. The radius r of an ellipsoid varies with the axial position x according to

$$r^2 = 0.04(1 - 4x^2); \qquad -0.5 \le x \le 0.5$$

The surface area S of the ellipsoid is given by

$$S = 2\pi \int_{-0.5}^{0.5} r\,\sqrt{1 + (dr/dx)^2}\;dx = \pi \int_{-0.5}^{0.5} \sqrt{4r^2 + (d[r^2]/dx)^2}\;dx$$

The integrand in either integral is singular at the limits; however, the singularity may be removed by the transformation

$$x = -0.5\cos(\xi)$$

Rewrite S as an integral with respect to ξ and perform the integration by the trapezoidal rule or by Romberg integration.

19. Evaluate S in Problem 18 from one of the original integrals by Gauss–Chebyshev quadrature.

20. Repeat Problem 18 for the body of revolution described in Problem 17.

21. Estimate the surface area S of the body of revolution described in Problem 17 by the method described in Problem 19.

22. The boundary of the domain shown in Fig. 6–5 is given by

$$y^2 = (1 - 4x^2)(2x^2 + 0.2x + 0.08)^2; \qquad -0.5 \le x \le 0.5$$

The domain's area moments of inertia are given by

$$I_x = \int_A y^2\,dA; \quad I_y = \int_A x^2\,dA$$

where A is the area of the domain and dA is $(dx\ dy)$. Evaluate these moments of inertia.

23. The domain in Fig. 6–5 is described in Problem 22. It is the base of a solid whose height is

$$z(x,y) = (x + 1)e^{xy}$$

Evaluate the volume V of the solid from

$$V = \int_A z(x,y)\,dA$$

24. The ellipsoid given in Problem 18 may be described in a Cartesian (x, y, z) coordinate system by

$$y^2 + z^2 = 0.04(1 - 4x^2); \quad -0.5 \le x \le 0.5$$

The ellipsoid is cut by the plane

$$y = x - 0.2$$

Develop the integral for the volume V_r of the part of the ellipsoid to the right of the plane (that is, the part whose end is at x equal to 0.5). Estimate V_r.

7

Ordinary Differential Equations

An ***ordinary differential equation*** is one in which an ordinary derivative of a dependent variable y with respect to an independent variable x is related in a prescribed manner to x, y, and lower derivatives. That is, we have a relation of the form

[7.1] $$d^n y / dx^n = f(x, y, dy/dx, d^2 y/dx^2, \ldots, d^{n-1} y/dx^{n-1})$$

To solve an equation of this type, we also require a set of conditions. When all of the conditions are given at one value of x and the solution proceeds from that value of x, we have an ***initial-value*** problem. When the conditions are given at different values of x, we have a ***boundary-value*** problem.

Any ordinary differential equation (o.d.e.) can be replaced by a system of first-order differential equations (which involve only first derivatives). Our treatment of o.d.e.s will therefore begin with a single first-order equation.

The fundamental methods for initial-value problems are treated in Section 7.1. Special cases of systems of equations and stiff equations are discussed in Section 7.2 and Section 7.3, respectively. Methods for solving boundary-value problems are described in Section 7.4.

An o.d.e. whose solution exhibits ***exponential growth*** is used as a preliminary case study. Other discussion problems deal with the ***linear-lag*** equation, motion of a ***rocket,*** the ***temperature distribution*** in the wall of a cylindrical tube, and the ***deflection of a cantilevered beam.***

7.1   SINGLE FIRST-ORDER EQUATIONS WITH INITIAL VALUES

The single first-order o.d.e. with an initial value is a special case of Eq. [7.1]. It is described by

[7.2] $$dy/dx = f(x, y); \qquad y = y_0 \text{ at } x = x_0$$

The description in Eq. [7.2] consists of the differential equation itself and a given solution y_0 at an initial location x_0. We then seek the solution y as x ranges from its initial value to some other value. In cases where the right-hand side is a function of x only, the solution is obtained by analytic integration or by numerical integration as we have discussed in Chapter 6.

Basic concepts for an o.d.e. are presented along with the Euler method in §7.1.1. Other simple methods are discussed in §7.1.2, and general classes of methods are presented in §7.1.3 through §7.1.6. Adaptive techniques are discussed in §7.1.7.

7.1.1   Basic Concepts and the Euler Method

Let us consider, for convenience in discussion, a specific o.d.e. of the type given in Eq. [7.2]. This o.d.e. is

[7.3a] $$dy/dx = xy; \qquad y = 1 \text{ at } x = 0$$

whose analytic solution is

[7.3b] $$y = e^{x^2/2}$$

The general form of the solution in Eq. [7.3b] depends on the initial value of y. An arbitrary choice of the initial value, say y_0, would give a *family of solutions*

[7.4] $$y = y_0 \, e^{x^2/2}$$

During computations, numbers are generally not exactly represented. Errors due to inexact representation are *round-off errors*. They may occur even in specifying the initial value of y, and they cause us to move from one member of the family of solutions to another as the solution progresses. We introduce the quantity η to denote the computed value of the solution and to emphasize that it is different from the true solution y.

The algorithm to be developed for solving Eq. [7.3a] is one that uses computed solutions up to η_i at x_i to determine the solution η_{i+1} at a new location

[7.5] $$x_{i+1} = x_i + h$$

The interval from x_i to x_{i+1} is the integration step, and h is known as the *step size*. The simplest algorithm is the *Euler method*, which may be obtained by expressing the solution y_{i+1} at x_{i+1} in a Taylor series expansion as we did in Chapter 6. We write

$$y_{i+1} = y_i + h(dy/dx)_i + O(h^2)$$

and truncate the expansion immediately after the first-derivative term. The error resulting from the truncation is the *local truncation error* and is of the order of h^2. The Euler method is obtained by using the computed values η in place of y and $f(x_i, \eta_i)$ in place of the derivative; we therefore obtain

[7.6] $$\eta_{i+1} = \eta_i + hf(x_i, \eta_i)$$

Pseudocode 7–1 shows the Euler method for solving Eq. [7.3a].

The numerical solution of Eq. [7.3a] by the Euler method with n equal to 4 (so that the step size is 0.25) is shown by the straight-line segments in Fig. 7–1. The starting point of each segment is on a member of the family of solutions and is tangent to it. The true solution is shown by the solid curve in Fig. 7–1, and the other curves are the members of the family that dictate the solution for each step after the first. The effects of round-off errors cannot be seen because they are much smaller than the truncation errors.

The rapid growth of error that we see in Fig. 7–1 is the reason that the Euler method is rarely used. Nevertheless, its simplicity makes it a useful introductory method and simplifies discussion of certain concepts. Two of these concepts are consistency and convergence. A method is *consistent* with the differential equation if the approximation used by the method is equivalent to the equation when the step size h tends to zero. A method is *convergent* if the approximate solution (obtained with infinite precision so that there are no round-off errors) tends to the true solution as h tends to zero.

Pseudocode 7–1 The Euler method for Eq. [7.3a] from x = 0 to x = 1

Initialize: n	\ number of steps
$h \leftarrow 1/n$	\ step size
$\eta \leftarrow 1; \quad x \leftarrow 0$	\ initial conditions
For $[i = 1, 2, \ldots, n]$	\ solution loop
$\quad \eta \leftarrow \eta + hx\eta$	\ apply Eq. [7.6]
$\quad x \leftarrow x + h$	\ update the x value
Write: x, η	\ tabulate results
End For	

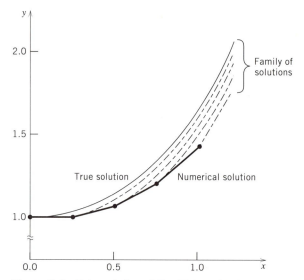

Figure 7–1   Euler solution of Eq. [7.3a] with four steps of size 0.25.

In the case of the Euler method, the derivative dy/dx is approximated by

$$(dy/dx)_i \simeq (y_{i+1} - y_i)/h$$

The method is clearly consistent since the limit of the right-hand side as h tends to zero is the definition of the derivative. The method is also convergent since the local truncation error approaches zero with h. It is not hard to imagine (without rigorous proof) that a method must be consistent if it is to be convergent. We can argue that a method that gives the true solution to an o.d.e. as h tends to zero can only do so if the method is equivalent to the o.d.e. under the same conditions.

A convergent method does not guarantee an accurate solution because convergence is based on a step size that approaches zero, and a practical solution is obtained with finite step sizes. In the latter case, both round-off and local truncation errors contribute to the global error at each step. In solving Eq. [7.3], we can reduce the local truncation error by reducing h; however, the number of steps is increased, and round-off errors become more significant. In practice, therefore, we must also be concerned about the stability of the method.

A loose definition of a *stable* method is one in which the global error does not grow in an unbounded fashion. A more formal definition involves an analysis that is usually applied to the o.d.e.

[7.7a] $dy/dx = \lambda y; \quad y = y_0 \text{ at } x = 0; \quad \lambda = \text{constant}$

The solution of Eq. [7.7a] is of course

[7.7b] $$y = y_0 \, e^{\lambda x}$$

With infinite precision, η_0 is equal to y_0, and application of the Euler method in Eq. [7.6] gives the approximate solution η_{i+1} at x_{i+1} equal to $(i + 1)h$ as

[7.8] $$\eta_{i+1} = \eta_i(1 + h\lambda) = \eta_0(1 + h\lambda)^{i+1} = y_0(1 + h\lambda)^{i+1}$$

If we assume that the solution is for positive x (and thus h) values, and we compare η_{i+1} in Eq. [7.8] as i tends to ∞ with the true solution in Eq. [7.7b], we see that they both become infinite when λ is positive and that they both remain at the initial value y_0 for the trivial case in which λ is zero.

When λ is positive and the effects of round-off errors are included in the analysis, it can be shown that the global error at any step is magnified by $(1 + h\lambda)$ in the next step of the solution. Thus, the Euler method is unstable for positive λ values, because the global error grows with each step of the solution.

The true solution in Eq. [7.7b] becomes zero at an infinite value of x when λ is negative; however, η_{i+1} can become zero with a negative value of λ only when $|1 + h\lambda|$ is less than 1. The Euler method is therefore said to be conditionally stable for negative λ. The interval $h\lambda$ that satisfies the condition for stability is the *interval of absolute stability* and is given by

[7.9] $$-2 < h\lambda < 0$$

As we can see from our preceding discussion, stability is important when we seek a solution over a large range of x values. Let us look more closely at what stability means in the case of Eq. [7.7a] when λ is positive. The exact solution of Eq. [7.7b] and the error in the numerical solution are both unbounded. The unbounded nature of the error is all that we mean when we say that the Euler method is unstable. A more meaningful property in the case of unbounded solutions is *relative stability*, which means that the growth rate of the error is less than the growth rate of the solution. A method that is neither stable nor relatively stable may still produce acceptable solutions over a sufficiently small range of x values.

The solution in Eq. [7.7b] is bounded for negative λ values; we therefore look for stability in the method to guarantee that the errors will also be bounded. For many o.d.e.s, the step size that is required to obtain an accurate solution is much smaller than the step size required for stability; stability is therefore not a major issue in such cases.

The concepts of consistency, convergence, and stability tend to be of background interest for most cases. All of the established methods are consistent and convergent; however, we would insist on establishing these properties if we were devising a new method. The lack of stability serves as a warning of potential trouble for problems with unbounded solutions. Stability is not a major concern when established methods are applied to o.d.e.s with bounded solutions; it does, however, play a major role in cases such as stiff o.d.e.s and partial differential equations.

A final concept is the **order** of a method. We note that the order of the Euler method is different from the order of the local truncation error. Recall from our derivation of Eq. [7.6] that the local truncation error is $O(h^2)$. The method itself is called a first-order method because the global error is $O(h)$. To see why this is so, recognize that we are performing n steps with error $O(h^2)$ for each step, so that the overall error is thus $O(nh^2)$. Because n is proportional to $1/h$, we have a method that is $O(h)$.

7.1.2   Some Simple Second-Order Methods

Four simple methods are now presented to illustrate some of the ideas that are used in the development of methods for o.d.e.s. These methods are all second-order (with third-order local truncation errors), and they are all consistent and convergent.

A natural extension of the Euler method for solving Eq. [7.2] is to include more terms of the Taylor series expansion. One of several ways to do this is through the **Taylor series method**, which uses the Taylor series directly. In this method, we expand y_{i+1} at x_{i+1} about x_i to obtain

[7.10] $$y_{i+1} = y_i + h(dy/dx)_i + (h^2/2!)(d^2y/dx^2)_i + \cdots$$

The first derivative on the right-hand side of Eq. [7.10] is, of course, given in Eq. [7.2] by $f(x, y)$. Higher derivatives are obtained by using the chain rule to differentiate $f(x, y)$ with respect to x. To express these higher derivatives, we shall use x and y subscripts to denote partial derivatives; for example,

$$f_x = \partial f/\partial x; \quad f_y = \partial f/\partial y; \quad f_{xx} = \partial^2 f/\partial x^2$$

Expressions for the second and third derivatives are given as the following examples.

[7.11a] $$d^2y/dx^2 = df/dx = f_x + f_y(dy/dx) = f_x + f_y f$$

[7.11b] $$d^3y/dx^3 = d^2f/dx^2$$
$$= f_{xx} + f_{xy}f + f_y(df/dx) + (f_{xy} + f_{yy}f)f$$
$$= f_y(d^2y/dx^2) + f_{xx} + 2f_{xy}f + f_{yy}f^2$$

The second-order Taylor series method is obtained by truncating the series after the term containing the second derivative and substituting the expressions for (dy/dx) and (d^2y/dx^2). The result in terms of the computed solution η is

[7.12] $\eta_{i+1} = \eta_i + hf(x_i, \eta_i) + [h^2/2][f_x(x_i, \eta_i) + f_y(x_i, \eta_i)f(x_i, \eta_i)]$

The Taylor series method is rarely used, even though it is conceptually simple to build a high-order method. Its disadvantage is the large number of functions that must be evaluated to represent the terms. For example, we must evaluate f, f_x, and f_y to apply Eq. [7.12]. To add the next term in the series, we must also evaluate the additional terms (f_{xx}, f_{xy}, and f_{yy}) that appear on the right-hand side of Eq. [7.11b].

A different Taylor series expansion about the point x_m equal to $(x_i + h/2)$ gives us

$$y_{i+1} = y_m + (h/2)(dy/dx)_m + (h^2/8)(d^2y/dx^2)_m + \cdots$$
$$y_i = y_m - (h/2)(dy/dx)_m + (h^2/8)(d^2y/dx^2)_m - \cdots$$

Subtraction yields

[7.13a] $y_{i+1} - y_i = h(dy/dx)_m + O(h^3)$

We now use similar expansions for the derivative (dy/dx) to obtain

$$(dy/dx)_{i+1} + (dy/dx)_i = 2(dy/dx)_m + O(h^2)$$

and rewrite Eq. [7.13a] as

[7.13b] $y_{i+1} - y_i = [h/2][(dy/dx)_i + (dy/dx)_{i+1}] + O(h^3)$

The second-order method that is derived from Eq. [7.13b] is given by

[7.14] $\eta_{i+1} = \eta_i + [h/2][f(x_i, \eta_i) + f(x_{i+1}, \eta_{i+1})]$

The method given by Eq. [7.14] is known as an ***implicit method*** because the required solution η_{i+1} is implicitly contained on the right-hand side. Implicit methods generally have better stability characteristics than explicit methods; however, they may be difficult to implement when the derivative $f(x, y)$ is complicated.

To avoid possible difficulties in solving the equation for the implicit method, we can look for an explicit way of expressing the term $f(x_{i+1}, \eta_{i+1})$ in Eq. [7.14]. For example, we can use the Euler method of §7.1.1 to predict η_{i+1} as

[7.15a] $(\eta_{i+1})_p = \eta_i + hf(x_i, \eta_i)$

and replace Eq. [7.14] by

[7.15b] $\eta_{i+1} = \eta_i + [h/2][f(x_i, \eta_i) + f(x_{i+1}, (\eta_{i+1})_p)]$

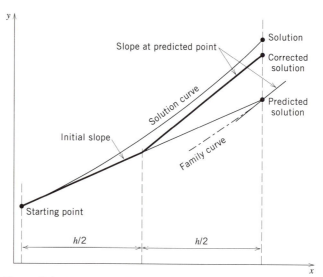

Figure 7–2 Illustration of the predictor–corrector method of Eq. [7.15a, b].

The combination of Eq. [7.15a, b] is known as a ***predictor–corrector method*** because a predicted value $(\eta_{i+1})_p$ from Eq. [7.15a] is used to replace (η_{i+1}) on the right-hand side of Eq. [7.14]. The result is Eq. [7.15b], which is the corrector for the method. A step of the predictor–corrector method is shown in Fig. 7–2.

The illustration in Fig. 7–2 shows how the initial slope is extended for a full step h to obtain the predicted solution according to Eq. [7.15a]. The predicted solution is on another member of the family of solutions, and it is the slope of this family member at the new location that is used in the corrector. Obtaining the corrected solution from Eq. [7.15b] is equivalent to advancing half a step with the initial slope and the other half with the new slope. The difference between the predictor–corrector method and the implicit method is in how the new slope is computed.

The equations for the predictor–corrector method can be made part of a class of equations whose forms are

[7.16a] $$k_1 = f(x_i, \eta_i); \quad k_2 = f(x_i + \alpha h, \eta_i + \beta h k_1)$$

[7.16b] $$\eta_{i+1} = \eta_i + h(\gamma_1 k_1 + \gamma_2 k_2)$$

Let us expand k_2 about (x_i, η_i) by a Taylor series for two variables (see Appendix C) as

$$k_2 = f + \alpha h f_x + \beta h k_1 f_y + (\alpha^2 h^2/2)f_{xx} + \alpha\beta h^2 k_1 f_{xy} + (\beta^2 h^2[k_1]^2/2)f_{yy} + \cdots$$

We use this expansion and Eq. [7.16a] to rewrite Eq. [7.16b] in terms of values at point i as

[7.17a] $\quad \eta_{i+1} = \eta_i + (\gamma_1 + \gamma_2)hf + \gamma_2 h^2(\alpha f_x + \beta f f_y)$
$$+ \gamma_2 h^3[(\alpha^2/2)f_{xx} + \alpha\beta f f_{xy} + (\beta^2 f^2/2)f_{yy}] + \cdots$$

We may also use a direct Taylor series expansion with the expressions in Eq. [7.11a, b] as we did earlier to obtain Eq. [7.12]. The direct expansion is

[7.17b] $\quad \eta_{i+1} = \eta_i + hf + (h^2/2)(f_x + f f_y)$
$$+ (h^3/6)(f_y[f_x + f f_y] + f_{xx} + 2f f_{xy} + f^2 f_{yy}) + \cdots$$

Finally, we match the coefficients of f and its derivatives in Eq. [7.17a, b] through the terms of order h^2 to obtain the relations

[7.18] $\qquad\qquad \gamma_1 + \gamma_2 = 1; \qquad \gamma_2\alpha = \gamma_2\beta = 1/2$

Values of α, β, γ_1, and γ_2 that satisfy Eq. [7.18] are used in Eq. [7.16a, b] to form the class of methods known as ***two-stage Runge–Kutta methods***. One of the four values (usually γ_2) must be chosen before we can apply Eq. [7.18] to determine the others. The specified value is usually chosen to meet some assumed condition for the leading error term, which has order h^3. Some typical values and their associated methods are

Modified Euler (second order):

[7.19a] $\qquad\qquad \gamma_2 = 1/2; \quad \alpha = \beta = 1; \quad \gamma_1 = 1/2$

Midpoint (second order):

[7.19b] $\qquad\qquad \gamma_2 = 1; \quad \alpha = \beta = 1/2; \quad \gamma_1 = 0$

The Runge–Kutta methods of this section are called two-stage because they use two evaluations, k_1 and k_2, of the derivative $f(x, y)$. The names given to various two-stage methods are not unique; indeed, a particular method may have more than one name, and the same name may be applied to different methods. The modified Euler method is the same as the predictor–corrector method we discussed earlier. If we combine Eq. [7.19b] and [7.16a, b], we obtain

$$k_1 = f(x_i, \eta_i)$$
$$\eta_{i+1} = \eta_i + hf(x_i + h/2, \eta_i + hk_1/2)$$

The method is called the midpoint method because the derivative used to advance the solution is an approximation to the derivative at the midpoint of the step.

Table 7–1 Errors in the Solution of Eq. [7.3a] by Various Methods

	Error				
x	Euler	Taylor	Implicit	Mod. Euler	Midpoint
0.1	−0.00501	−0.00001	0.00001	−0.00001	−0.00001
0.2	−0.01020	−0.00008	0.00005	−0.00003	−0.00005
0.3	−0.01583	−0.00020	0.00012	−0.00004	−0.00012
0.4	−0.02218	−0.00038	0.00022	−0.00006	−0.00022
0.5	−0.02960	−0.00064	0.00037	−0.00010	−0.00037
0.6	−0.03849	−0.00101	0.00057	−0.00015	−0.00058
0.7	−0.04937	−0.00151	0.00085	−0.00023	−0.00087
0.8	−0.06290	−0.00218	0.00122	−0.00035	−0.00127
0.9	−0.07994	−0.00308	0.00173	−0.00055	−0.00182
1.0	−0.10161	−0.00430	0.00241	−0.00084	−0.00257

It is interesting to note that when dy/dx is described in terms of x only (so that dy/dx is equal to $f(x)$), the modified Euler method is the same as the trapezoidal rule of Chapter 6. The midpoint rule corresponds to the lowest-order, open Newton–Cotes integration formula of the same name.

The results of using the first-order Euler method and the second-order Taylor series, implicit, modified Euler, and midpoint methods to solve Eq. [7.3a] with ten steps of size 0.1 are summarized in Table 7–1 by the errors $(\eta - y)$ at each step. Although the errors are shown only to five decimal places, the calculations that produced them were in 64-bit precision.

We see in Table 7–1 that the errors in the first-order Euler method are significantly worse than those for the four second-order methods, as expected. The computed solutions for the second-order methods all fall short of the true solution, except for the implicit method. The modified Euler gives the best results for the particular problem we solved; this does not mean that the modified Euler method is generally the best of the four.

In developing the second-order methods, we have introduced different ways of using the Taylor series, the concept of implicit and predictor–corrector methods, and a generalized set of explicit, second-order methods. The generalized methods belong to the class of Runge–Kutta methods, which we shall now consider in more detail.

7.1.3 Runge–Kutta Methods

A **_Runge–Kutta method_** (Ref. 12) refers to **_one_** of a class of methods rather than to a specific method. Two of these methods, the modified Euler and the midpoint methods, were introduced in §7.1.2. Runge–Kutta methods are **_one-step_** methods; that is, we use information about the solution at one location x_i to advance the solution to the next location x_{i+1}. The advancement is through a weighted average of r derivatives k_1 through k_r, and is performed according to

[7.20a]
$$\eta_{i+1} = \eta_i + h \sum_{j=1}^{r} \gamma_j k_j$$

in which the weights γ_j satisfy

[7.20b]
$$\sum_{j=1}^{r} \gamma_j = 1$$

A Runge–Kutta method that uses r derivatives is known as an **r-stage** method.

The derivatives k_j that are used in the r-stage method for solving Eq. [7.2] are compactly expressed by

[7.20c] $\quad k_1 = f(x_i, \eta_i);$

$$k_j = f\left\{ [x_j + h\alpha_{j-1}], \left[\eta_i + h \sum_{m=1}^{j-1} \beta_{j-1,m} k_m \right] \right\}; \quad j = 2, 3 \ldots, r;$$

The two-stage methods described earlier in Eq. [7.16a, b] would now be written as

$$k_1 = f(x_i, \eta_i); \quad k_2 = f([x_i + \alpha_1 h], [\eta_i + h\beta_{1,1} k_1]);$$
$$\eta_{i+1} = \eta_i + h(\gamma_1 k_1 + \gamma_2 k_2)$$

The values of α, β, and γ that are used to specify a method are chosen to match a Taylor series expansion of the kind given in Eq. [7.17b]. The order of the method depends on the choices that are made. The maximum order that can be attained for r-stage Runge–Kutta methods is h^r for r up to 4. Thereafter, matters become more complicated; for example, a fifth-order method requires six stages. The derivation of Runge–Kutta methods with very high orders is really a moot issue because such methods are very inefficient.

In addition to the modified Euler and midpoint methods, another popular two-stage, second-order method is the **Heun method** whose coefficients are given by

Heun (second order):

[7.21] $\qquad \gamma_2 = 1 - \gamma_1 = 3/4; \qquad \alpha_1 = \beta_{1,1} = 2/3$

We again point out that names used for Runge–Kutta methods are not unique; for example, the modified Euler method is sometimes called the Heun method.

A third-order method ascribed to Kutta is the **Kutta method** with coefficients

Kutta (third order):

[7.22] $\qquad \gamma_3 = \gamma_1 = 1 - \gamma_2 = 1/6; \qquad \alpha_1 = \beta_{1,1} = 1/2;$
$\qquad \alpha_2 = -\beta_{2,1} = \beta_{2,2}/2 = 1$

The full form of the third-order Kutta method is

$$k_1 = f(x_i, \eta_i); \qquad k_2 = f(x_i + h/2, \eta_i + hk_1/2);$$
$$k_3 = f(x_i + h, \eta_i - hk_1 + 2hk_2);$$
$$\eta_{i+1} = \eta_i + (h/6)(k_1 + 4k_2 + k_3)$$

By far the most popular Runge–Kutta method is the *classical Runge–Kutta method* whose full form is described as follows.

Classical Runge–Kutta (fourth order):

[7.23] $k_1 = f(x_i, \eta_i); \quad k_2 = f(x_i + h/2, \eta_i + hk_1/2);$
$$k_3 = f(x_i + h/2, \eta_i + hk_2/2); \quad k_4 = f(x_i + h, \eta_i + hk_3)$$
$$\eta_{i+1} = \eta_i + (h/6)(k_1 + 2k_2 + 2k_3 + k_4)$$

The classical Runge–Kutta method is equivalent to Simpson's rule in Chapter 6 when the derivative dy/dx is described in terms of x only.

Stability analyses based on the solution of Eq. [7.7a] give the following conditions for the intervals of absolute stability for Runge–Kutta methods of order r up to 4.

[7.24a] $|1 + h\lambda| < 1;$ $r = 1$
[7.24b] $|1 + h\lambda + (h\lambda)^2/2| < 1;$ $r = 2$
[7.24c] $|1 + h\lambda + (h\lambda)^2/2 + (h\lambda)^3/6| < 1;$ $r = 3$
[7.24d] $|1 + h\lambda + (h\lambda)^2/2 + (h\lambda)^3/6 + (h\lambda)^4/24| < 1;$ $r = 4$

The intervals are $(-2,0)$ for r equal to 1 and 2 and are approximately $(-2.51,0)$ for r equal to 3 and $(-2.78,0)$ for r equal to 4.

Runge–Kutta methods are easy to use. Pseudocode 7–2 shows a general r-stage method; however, equivalent codes for specific methods are slightly easier to write and are slightly more efficient.

To illustrate the effect of the order of a method, we consider the o.d.e.

[7.25a] $dy/dx = f(x, y) = x - y; \qquad y = 0$ at $x = 0$

This is a simple form of the type of equations that are used to describe the behavior of a *control system device*. The variable y is the output of the device, and x represents time. The quantity x on the right-hand side is in reality a ramp function of time. The response of the device is given by

[7.25b] $y = x + e^{-x} - 1$

Pseudocode 7–2   A general Runge–Kutta module

\ Pseudocode for a general r-stage Runge–Kutta method. The step size h, the
\ number of stages, the current solution η and the corresponding location x, and
\ arrays for the coefficients of the method are supplied. The updated values of η
\ and x are returned.

Module RungeKutta($h, r, \eta, x, \alpha, \beta, \gamma$):

Declare: k \ r-element array for derivatives at each stage

$S \leftarrow 0$ \ initialize sum of $\gamma_j k_j$

For $[j = 1, 2, \cdots, r]$ \ stage loop

    If $[j = 1]$ **then**

        $k_1 \leftarrow f(x, \eta)$

    Else

        $T \leftarrow 0$ \ initialize βk sum

        For $[m = 1, 2, \cdots, j - 1]$ \ loop for βk sum

           $T \leftarrow T + \beta_{j-1,m} k_m$

        End For

        $k_j \leftarrow f(x + \alpha_{j-1}h, \eta + hT)$

    End If

    $S \leftarrow S + \gamma_j k_j$

End For

$\eta \leftarrow \eta + hS$ \ updated solution

$x \leftarrow x + h$ \ updated x value

End Module RungeKutta

At large x (or time) values, the exponential term approaches zero and the output y lags behind the input by an almost constant value of 1.

The step sizes for stability are obtained by setting λ in Eq. [7.24a] through [7.24d] equal to f_y; thus, the appropriate λ value is (-1) for Eq. [7.25a]. A step size less than 2 will satisfy the stability requirements for all of the methods. If we use such a large step size to solve Eq. [7.25a] for x values up to 4, we will obtain very inaccurate results. We need a much smaller step size than is allowed by stability requirements to achieve good accuracy.

The solutions of Eq. [7.25a] obtained by the Euler, modified Euler, and classical methods with ten steps of size 0.4 are compared with the analytical solution of Eq. [7.25b] in Table 7–2. It is not surprising that accuracy increases with the order of the method. If we examine the computed solutions more closely, we see that the errors grow during the first few steps, but then decay as we continue the solution for larger x values. This decay in the error is the signature of a stable method.

As we noted earlier, Runge–Kutta methods are easy to use. In addition, because they are one-step methods, changing the step size to advance from one x location

Table 7–2 Solutions of Eq. [7.25a] for x Values from 0 to 4

| x | Solution | | | |
	Analytical	Euler	Mod. Euler	Classical
0.4	0.070320	0.000000	0.080000	0.070400
0.8	0.249329	0.160000	0.262400	0.249436
1.2	0.501194	0.416000	0.514432	0.501302
1.6	0.801897	0.729600	0.813814	0.801993
2.0	1.135335	1.077760	1.145393	1.135416
2.4	1.490718	1.446656	1.498867	1.490783
2.8	1.860810	1.827994	1.867230	1.860861
3.2	2.240762	2.216796	2.245716	2.240801
3.6	2.627324	2.610078	2.631087	2.627353
4.0	3.018316	3.006047	3.021139	3.018337

to the next is a trivial matter. On the other hand, high-order Runge–Kutta methods require several evaluations of the derivative $f(x, y)$ for each step of the solution, and it is generally a difficult task to estimate the error at each step. Alternatives to the Runge–Kutta methods are multistep methods; these include the Adams methods, which we shall now consider.

7.1.4   The Adams Methods

The **Adams methods** (Ref. 12) use information at multiple steps of the solution to obtain the solution at the next x location. For example, a three-step method uses information at x_i, x_{i-1}, and x_{i-2} to generate a solution at the new point x_{i+1}. The Adams methods form two main classes — the **explicit Adams–Bashforth methods** and the **implicit Adams–Moulton methods**. These may be combined to form **Adams–Bashforth–Moulton predictor–corrector methods**.

Let us rewrite the o.d.e. of Eq. [7.2] in the form

[7.26] $$dy/dx = f(x, y) = \psi(x)$$

The idea behind the n-step Adams–Bashforth method is to use an interpolating polynomial based on information at the current location x_i and at previous locations $(x_{i-1}, x_{i-2}, \ldots x_{i-n+1})$ to express $\psi(x)$. The solution at x_{i+1} is then found from

[7.27] $$y_{i+1} - y_i = \int_{x_i}^{x_{i+1}} \psi(x)dx$$

We illustrate the derivation of the Adams–Bashforth methods by considering the two-step case. We assume that we have a constant step size h from point $(i - 1)$ to point i and from point i to point $(i + 1)$, and that a solution exists up to point i.

We may use a Lagrange interpolating polynomial from Chapter 5 with information at the two steps x_i and x_{i-1} to write

$$\psi(x) \simeq [(x_i - x)\psi(x_{i-1}) + (x - x_{i-1})\psi(x_i)]/h$$

Integration of $\psi(x)$ as required by Eq. [7.27] yields

[7.28] $$y_{i+1} = y_i + (h/2)[3\psi(x_i) - \psi(x_{i-1})]$$

We then use the relation between $\psi(x)$ and $f(x, y)$ from Eq. [7.26] to write the two-step Adams–Bashforth method as

[7.29] $$\eta_{i+1} = \eta_i + (h/2)[3f(x_i, \eta_i) - f(x_{i-1}, \eta_{i-1})]$$

The local truncation error is found by comparing the Taylor series expansion of Eq. [7.28] about x_i with the expansion for y_{i+1}. The expansion of Eq. [7.28] is

$$y_{i+1} = y_i + (3h/2)\psi - (h/2)[\psi - h(d\psi/dx) + (h^2/2)(d^2\psi/dx^2) - \cdots]$$

and the direct expansion of y_{i+1} is

$$y_{i+1} = y_i + h(dy/dx) + (h^2/2)(d^2y/dx^2) + (h^3/6)(d^3y/dx^3) + \cdots$$

Recognizing that ψ is (dy/dx), we see that the terms in the two expansions match up to those containing h^2. The leading error term is therefore $O(h^3)$ and is

$$[-h^3/4 - h^3/6](d^3y/dx^3) = (-5h^3/12)(d^3y/dx^3)$$

The n-step Adams–Bashforth methods have the general form

[7.30] $$\eta_{i+1} = \eta_i + \beta h \sum_{j=1}^{n} \alpha_j f(x_{i-j+1}, \eta_{i-j+1})$$

The coefficients in Eq. [7.30] are summarized in Table 7–3 for n values up to four. All of these methods are explicit and convergent and are of order n. Note that the method for n equal to 1 is the same as the Euler method.

The development of the Adams–Moulton methods is similar to that of the preceding Adams-Bashforth methods. The difference is that the interpolating polynomial now includes the data point $\psi(x_{i+1})$. As a result, the Adams–Moulton methods are implicit, and their general form is

[7.31] $$\eta_{i+1} = \eta_i + \beta h \sum_{j=0}^{n} \alpha_j f(x_{i-j+1}, \eta_{i-j+1})$$

Table 7–3   Coefficients for the Adams–Bashforth Methods

n	β	α_1	α_2	α_3	α_4
1	1	1			
2	1/2	3	-1		
3	1/12	23	-16	5	
4	1/24	55	-59	37	-9

Note that the summation now starts from 0 instead of 1 and therefore contains the term $f(x_{i+1}, \eta_{i+1})$ on the right-hand side. The Adams–Moulton methods are of order $(n + 1)$. Their coefficients are summarized in Table 7–4 for n values up to three. The case for n equal to 0 is known as the ***implicit Euler method***, and the case for n equal to 1 is the method given by Eq. [7.14] in §7.1.2.

Implementation of the Adams–Moulton methods requires us to be able to extract the solution η_{i+1}, which appears on both sides of Eq. [7.31]. Root-solving may be required to do this in some cases. An alternative is to form a predictor–corrector method by combining the Adams–Bashforth and Adams–Moulton methods. The former is used to obtain a predicted value $(\eta_{i+1})_p$, which is then used to replace η_{i+1} in the right-hand side of the Adams–Moulton equation.

In general, the value of n for the Adams–Bashforth predictor is usually one more than the value of n for the Adams–Moulton corrector. For example, the scheme with n equal to 2 for the predictor and n equal to 1 for the corrector is

[7.32]
$$(\eta_{i+1})_p = \eta_i + (h/2)[3f(x_i, \eta_i) - f(x_{i-1}, \eta_{i-1})];$$
$$\eta_{i+1} = \eta_i + (h/2)[f(x_{i+1}, (\eta_{i+1})_p) + f(x_i, \eta_i)]$$

The predictor–corrector method has the same order as its explicit and implicit parent methods. The advantages of the predictor–corrector formulation are that it simplifies the Adams–Moulton part of the method and is expected to have smaller errors than the Adams–Bashforth part. The expectation of smaller errors is due to the fact that the magnitude of the local truncation error is smaller for an Adams–Moulton method than it is for the corresponding Adams–Bashforth method of the same order.

The efficiency of the Adams methods relative to that of the Runge–Kutta methods can be discerned by comparing the four-step Adams-Bashforth method with a four-stage Runge–Kutta method. The Runge–Kutta method always requires four new

Table 7–4   Coefficients for the Adams–Moulton Methods

n	β	α_0	α_1	α_2	α_3
0	1	1			
1	1/2	1	1		
2	1/12	5	8	-1	
3	1/24	9	19	-5	1

Pseudocode 7–3   The Adams predictor–corrector step of Eq. [7.32]

Module AdamsPC2(h, x, η, f_{old}) :

$f_{cur} \leftarrow f(x, \eta)$ \qquad \ derivative at current step
$x \leftarrow x + h$ \qquad \ update x
$\eta_p \leftarrow \eta + (h/2)(3f_{cur} - f_{old})$ \qquad \ predictor
$\eta \leftarrow \eta + (h/2)(f(x, \eta_p) + f_{cur})$ \qquad \ corrector; update η
$f_{old} \leftarrow f_{cur}$ \qquad \ update f_{old} to prepare for next step

End Module AdamsPC2

---

evaluations of $f(x, y)$ to advance the solution. The Adams-Bashforth method also requires four values of $f(x, y)$, but only one is new since three of them (prior to the current stage of the solution) already exist.

Coding of an Adams method is relatively straightforward. The important points are that derivatives at previous steps of the solution must be saved to obtain efficiency, and that a different method must be used to start the solution process. For example, the method given by Eq. [7.32] to advance the solution from x_i to x_{i+1} requires one saved derivative $f(x_{i-1}, \eta_{i-1})$, which we shall denote by f_{old}. One step of the second-order Adams predictor–corrector method is then given by Pseudocode 7–3.

The module cannot be invoked until we have found a solution at the first step after the initial condition. A compatible starter method is the modified Euler method of Eq. [7.16a, b] and Eq. [7.19a]. The starter method is used only once to obtain the first new solution. The value f_{old} at this stage of the solution process is the derivative at the initial values of x and y.

The pseudocode for Eq. [7.32] is in a pure predictor–corrector form. It can be modified to implement the implicit Adams–Moulton method from which the corrector is derived. The modification is used when the solution of the implicit equation is difficult to obtain analytically. It uses the predictor as the starting point for an iterative solution scheme that is akin to the fixed-point iteration method of §3.1.2. Because the predictor is usually a good approximation to the required solution, only a very few iterations are required for convergence. The modification consists of replacing the corrector in the preceding pseudocode by the segment in Pseudocode 7–4.

Pseudocode 7–4   Segment to modify AdamsPC2 to obtain an implicit solution

---

Repeat

$\quad \eta_{old} \leftarrow \eta_p$
$\quad \eta_p \leftarrow \eta + (h/2)(f(x, \eta_{old}) + f_{cur})$
Until $[|\eta_p - \eta_{old}| \leq$ small positive value]

$\eta \leftarrow \eta_p$

Table 7–5   Second-Order Adams and Runge–Kutta Solutions of Eq. [7.25a]

	Solution			
x	Analytical	Mod. Euler	Pred.–Corr.	Implicit
0.4	0.070320	0.080000	0.080000	0.080000
0.8	0.249329	0.262400	0.249600	0.253333
1.2	0.501194	0.514432	0.496512	0.502222
1.6	0.801897	0.813814	0.795505	0.801481
2.0	1.135335	1.145393	1.128903	1.134321
2.4	1.490718	1.498867	1.484990	1.489547
2.8	1.860810	1.867230	1.856037	1.859698
3.2	2.240762	2.245716	2.236947	2.239799
3.6	2.627324	2.631087	2.624360	2.626533
4.0	3.018316	3.021139	3.016061	3.017688

The problem described by Eq. [7.25a] is solved by the Adams predictor corrector method of Eq. [7.32] and by the second-order Adams–Moulton method with ten equal steps of size equal to 0.4. The modified Euler method is used as a starter in both cases. Results are given in Table 7–5 along with previous solutions from Table 7–2. The modified Euler and predictor–corrector methods require evaluations of two new derivatives per step. The implicit method requires only one if the resulting equation can be solved easily as in this case. We see, therefore, the potential of the Adams methods to produce solutions that are comparable to Runge–Kutta solutions at the same or less computational cost. We caution, however, that this potential is not always realized.

As we stated earlier, the local truncation error for an Adams–Moulton method is expected to be smaller than that for the Adams–Bashforth method of the same order. The local truncation error (l.t.e.) may be estimated from

$$[7.33] \qquad \text{l.t.e.} = \eta_{i+1} - \eta_i - \beta h \sum_{j=m}^{n} \alpha_j f(x_{i-j+1}, \eta_{i-j+1})$$

in which m is 1 for Adams–Bashforth methods and m is 0 for Adams–Moulton methods.

The stability range for the Adams methods is greatest for the Adams–Moulton methods and least for the Adams–Bashforth methods. Stability is determined from the o.d.e. $(dy/dx = \lambda x)$ that we introduced earlier in Eq. [7.7a]. Occurrences of $f(x_{i-j+1}, \eta_{i-j+1})$ in the Adams methods are replaced by $(\lambda \eta_{i-j+1})$, and we seek intervals $h\lambda$ that do not amplify errors as the step number i tends to ∞. Table 7–6 shows the intervals of absolute stability for the Adams methods.

7.1.5   The Milne–Simpson and Hamming Methods

The *Milne method* (Ref. 13) for solving an o.d.e. of the type given in Eq. [7.2] belongs to a class of methods that is derived from direct integration of (dy/dx) over a

Table 7–6   Intervals of Absolute Stability $(-s, 0)$ for the Adams Methods

Order	s		
	Bashforth	Moulton	Pred.–Corr.
1	–	∞	2.0
2	1.00	∞	2.0
3	0.55	6	1.8
4	0.30	3	1.3

range of equally spaced control points. This approach is different from the integration of an interpolating polynomial over one step as used to derive the Adams methods. The Milne method is derived from the five-point, open Newton–Cotes formula known as *Milne's rule* and given in Appendix C. It is based on the relation

$$y_{i+1} - y_{i-3} = \int_{y_{i-3}}^{y_{i+1}} dy = \int_{x_{i-3}}^{x_{i+1}} (dy/dx)dx$$

We replace y by the computed value η, replace (dy/dx) by $f(x, y)$, and use Milne's rule for the last integral to obtain the explicit, fourth-order Milne method

[7.34] $\eta_{i+1} = \eta_{i-3} + (4h/3)[2f(x_i, \eta_i) - f(x_{i-1}, \eta_{i-1}) + 2f(x_{i-3}, \eta_{i-3})]$

Because Eq. [7.34] is an extrapolation formula, which tends to produce inaccurate results, the Milne method is rarely used by itself; it is instead used as the predictor in a predictor–corrector method. A fourth-order method based on Simpson's rule (see Chapter 6) with control points x_{i-1}, x_i, and x_{i+1} is given by

[7.35] $\eta_{i+1} = \eta_{i-1} + (h/3)[f(x_{i-1}, \eta_{i-1}) + 4f(x_i, \eta_i) + f(x_{i+1}, \eta_{i+1})]$

This method is an implicit method when used alone; if, however, the Milne method is used to predict the value of η_{i+1} to be used on the right-hand side of Eq. [7.35], the resulting method is the *Milne–Simpson predictor–corrector method*.

The Milne–Simpson method is unstable; it is therefore unpopular, although it can produce accurate solutions over short enough ranges of x and although there are techniques for overcoming the instability. A popular alternative to the Milne–Simpson method is the *Hamming method* in which the Simpson corrector is replaced. Hamming's corrector belongs to a general class that includes the Adams–Moulton methods and is chosen as a compromise between stability and accuracy. The corrector for Hamming's method is

[7.36] $\eta_{i+1} = (9\eta_i - \eta_{i-2})/8$
   $+ (3h/8)[f(x_{i+1}, \eta_{i+1}) + 2f(x_i, \eta_i) - f(x_{i-1}, \eta_{i-1})]$

The Hamming corrector may be used directly as an implicit method, or we may use the Milne predictor to start an iterative process for the solution of Eq. [7.36]. The iteration process was described in our discussion of the Adams methods in §7.1.4; other discussions pertaining to the multistep nature of the Adams methods are also applicable here. The interval of absolute stability for $(h\lambda)$ is $(-0.75,0)$ for Eq. [7.36] when λ in Eq. [7.7a] is negative. The method is also relatively stable for positive λ when $(h\lambda)$ is less than 0.4.

7.1.6   The Gragg Method

The *Gragg method* (Ref. 8, 11) is based on a second-order multistep method that belongs to the same class as the Milne method. The basis method is the *multistep midpoint method* whose properties allow us to use the same concepts as for Romberg integration in Chapter 6.

The multistep midpoint method is obtained from the open Newton–Cotes formula known as the midpoint rule. It is expressed by

[7.37]
$$\eta_{i+1} = \eta_{i-1} + 2hf(x_i, \eta_i)$$

As is usual with multistep methods, another method must be used to start the solution. Although we can use any starter method, the desired behavior of the error for the Gragg method requires the Euler method as a starter. The potentially large error due to the Euler method and the extrapolatory nature of the multistep midpoint method can produce severe errors. For example, the solution of Eq. [7.25a] with a step size of 0.4 for x values up to 4 produces the behavior shown in Fig. 7–3.

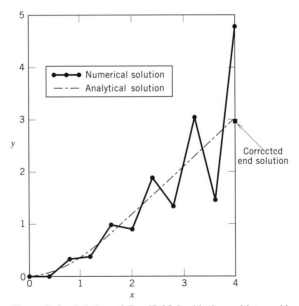

Figure 7–3   Solution of Eq. [7.25a] with the multistep midpoint method.

Despite the oscillations, the solution at the final x value can be corrected by smoothing. We denote the final x value by x_f, the last two uncorrected solutions by η_{f-1} and η_f, and the correction to η_f by η_c. The corrected value is obtained from

[7.38]
$$\eta_c = [\eta_f + \eta_{f-1} + hf(x_f, \eta_f)]/2$$

When the correction is used, the method is known as the ***modified midpoint method***. The correction for the solution of Eq. [7.25a] is also shown in Fig. 7–3.

A special result about the modified midpoint method is due to Gragg and states that the error $(\eta_f - y_f)$ is a series in even powers of the step size h. This is exactly the same type of result used in Chapter 6 to obtain Romberg integration from the trapezoidal rule. The Gragg method is to combine corrected solutions obtained with $(m, 2m, 4m, \ldots)$ steps as in Romberg integration to build higher-order methods. In particular, if we combine the solutions $(\eta_c)_m$ from m steps and $(\eta_c)_{2m}$ from $2m$ steps according to

[7.39]
$$\eta_g = [4(\eta_c)_{2m} - (\eta_c)_m]/3$$

the result η_g is the solution from a fourth-order Gragg method.

The fourth-order solution of Eq. [7.39] requires $(3m + 3)$ evaluations of the derivative $f(x, y)$; a four-stage, fourth-order Runge–Kutta method with $2m$ steps would require $8m$ evaluations. Although the Gragg method is more efficient for the same number of steps, it may not produce the same accuracy. For example, the solution of Eq. [7.25a] by the Gragg method is considerably worse than the fourth-order Runge–Kutta solution. A major cause of inaccuracy is the use of the Euler method as a starter, as we can see in Fig. 7–3. For cases in which the initial Euler solution is accurate, the Gragg method will match the Runge–Kutta method for accuracy.

Our treatment of the Gragg method is intended to introduce an approach to obtaining high-order solutions of o.d.e.s that is somewhat different from those used in Runge–Kutta and predictor-corrector methods. A variation on the approach due to Bulirsch and Stoer produces a powerful method for solving o.d.e.s. We shall not discuss this method, because it is beyond our scope; however, we note that it makes use of rational function extrapolation to obtain accurate solutions with very large step sizes.

7.1.7 Adaptive Methods

Our discussions of o.d.e. solvers have assumed that step sizes are constant throughout the solution. We now turn our attention to varying the step size so that we can attain better efficiency. If we know the error in a step of a method and how it depends on the step size h, we can set h to the largest value that will keep the error below some prescribed tolerance. A method that allows us to achieve this goal is called an ***adaptive method*** because the step size is adapted to the behavior of the error.

One very popular adaptive method is from the class of ***Runge–Kutta–Fehlberg methods*** (Ref. 14). We shall focus our attention on a six-stage member of this class. The same six stages (that is, the same set of derivatives k_1 through k_6 defined in Eq. [7.20c]) are used to produce fourth-order ***and*** fifth-order Runge–Kutta methods.

One six-stage Runge–Kutta–Fehlberg (RKF6) method is as follows.

[7.40]

$$k_1 = f(x_i, \eta_i); \qquad k_2 = f(x_i + h/4, \eta_i + hk_1/4);$$

$$k_3 = f(x_i + 3h/8, \eta_i + h[3k_1/32 + 9k_2/32]);$$

$$k_4 = f(x_i + 12h/13, \eta_i + h[1932k_1/2197 - 7200k_2/2197 + 7296k_3/2197]);$$

$$k_5 = f(x_i + h, \eta_i + h[439k_1/216 - 8k_2 + 3680k_3/513 - 845k_4/4104]);$$

$$k_6 = f(x_i + h/2, \eta_i + h[-8k_1/27 + 2k_2$$
$$- 3544k_3/2565 + 1859k_4/4104 - 11k_5/40]);$$

$$\eta_{i+1} = \eta_i + h(25k_1/216 + 1408k_3/2565 + 2197k_4/4104 - k_5/5);$$

$$\zeta_{i+1} = \eta_i + h(16k_1/135 + 6656k_3/12825 + 28561k_4/56430 - 9k_5/50 + 2k_6/55)$$

The set of expressions except the last one for ζ_{i+1} represents a standard fourth-order, six-stage method. The expression for ζ_{i+1} is a fifth-order solution based on the current ***fourth-order*** solution η_i at x_i.

The fourth-order solution is the one that is reported; the fifth-order value is used solely to estimate the error in the fourth-order value. The magnitude E of the error in η_{i+1} is estimated from

$$[7.41] \quad E \simeq |\zeta_{i+1} - \eta_{i+1}|$$
$$= h |k_1/360 - 128k_3/4275 - 2197k_4/75240 + k_5/50 + 2k_6/55|$$

We can, of course, evaluate E from Eq. [7.41] without actually computing ζ_{i+1}.

A simple strategy for changing the step size consists of halving or doubling the step size depending on the value of (E/h). To implement the strategy, we first set a tolerance ϵ for the error per unit step size. If E/h for a particular step exceeds ϵ, we repeat the step with a new h equal to half the previous value. If E/h does not exceed ϵ, we accept the solution for η_{i+1}.

If E/h does not exceed ϵ, we must decide whether we should retain the same step size or double it for the next step of the solution. In doubling the step size, the error in the new fourth-order solution will be $O(16h^4)$. Step doubling is therefore used only if E/h is smaller than $\epsilon/16$.

The results of applying the six-stage RKF6 method to Eq. [7.25a] with an initial step size of 0.4 and a tolerance ϵ equal to 10^{-5} are shown in Table 7–7. The value of

Table 7–7   Runge–Kutta–Fehlberg Solutions of Eq. [7.25a]

h	x	y	E/h
0.4	0.0	0.000000	$3.77(10^{-5})$
0.2	0.2	0.018730	$2.21(10^{-6})$
0.2	0.4	0.070319	$1.81(10^{-6})$
0.2	0.6	0.148811	$1.48(10^{-6})$
0.2	0.8	0.249328	$1.21(10^{-6})$
0.2	1.0	0.367878	$9.91(10^{-7})$
0.2	1.2	0.501193	$8.11(10^{-7})$
0.2	1.4	0.646596	$6.64(10^{-7})$
0.2	1.6	0.801896	$5.44(10^{-7})$
0.4	2.0	1.135331	$7.62(10^{-6})$
0.4	2.4	1.490713	$5.11(10^{-6})$
0.4	2.8	1.860805	$3.42(10^{-6})$
0.4	3.2	2.240758	$2.29(10^{-6})$
0.4	3.6	2.627320	$1.54(10^{-6})$
0.4	4.0	3.018313	$1.03(10^{-6})$

E/h exceeded ϵ initially; thus the first step is repeated with a step size of 0.2. This step size is maintained until x reaches 1.6. The value of E/h is found to be less than $\epsilon/16$ at that point, and the step size is doubled to 0.4. Fifteen steps, including the repeated step, are required to obtain solutions for x values up to 4. The results are much more accurate than those given for the classical Runge–Kutta method (with ten steps) in Table 7–2, and we are assured of our level of accuracy.

The RKF6 method suffers from the same inefficiency as the ordinary Runge–Kutta methods; however, it can be implemented very easily because it is a one-step method, which does not rely on a history of previous solutions. Multistep methods such as the Adams methods of §7.1.4 can also be used in an adaptive way. The error that occurs in a step is computed easily from Eq. [7.33]. The difficulty with these methods is that step halving or step doubling requires a history that may not exist. One approach to easing this difficulty is simply to use the starter method to build an appropriate history every time the step size is changed.

7.2   SYSTEMS OF FIRST-ORDER EQUATIONS

High-order o.d.e.s (that is, those with derivatives higher than the first) can be reduced to a system of first-order equations. As an example, we consider the case of the *motion of a rocket* under a constant thrust.

The rocket has an initial mass of 300 kg, which *includes* 180 kg of propellant. The propellant is consumed at a rate of 10 kg/s and is expelled at a certain speed so that the thrust is maintained at a constant value of 5000 N (as long as the propellant lasts). The rocket is initially at rest at an altitude h equal to 0. It is then launched vertically upwards. In its upward flight, it is subjected to a drag force of $(0.1v^2)$

N when v is measured in m/s. The gravitational acceleration g is assumed to be constant at 9.81 m/s^2.

The variables for the rocket problem are its mass, its altitude h [meters], its upward speed v [meter/second], and time t [seconds]. The mass is obtained by subtracting the mass of the propellant that is consumed from the initial mass. All other quantities are related through Newton's Second Law of Motion by

[7.42] $dh/dt = v; \qquad dv/dt = (5000 - 0.1v^2)/(300 - 10t) - g$

The system of two first-order o.d.e.s given in Eq. [7.42] is a reduced version of the single, second-order equation

$$d^2h/dt^2 = (5000 - 0.1[dh/dt]^2)/(300 - 10t) - g$$

The general approach for reducing a high-order equation of the form

$$d^ny/dx^n = f(x, y, dy/dx, d^2y/dx^2, \ldots, d^{n-1}y/dx^{n-1})$$

is to write y and each of its derivatives except the highest in terms of the next higher derivative. That is, we represent the derivatives by

$$dy^i/dx^i = u_i$$

and rewrite the high-order o.d.e. as the system of n equations

$$dy/dx = u_1$$
$$d(u_i)/dx = u_{i+1}; \quad i = 1, 2, \ldots, n - 2$$
$$d(u_{n-1})/dx = f(x, y, u_1, u_2, \ldots, u_{n-1})$$

Let us now return to the rocket problem that was posed in Eq. [7.42] and write the initial conditions

[7.43] $h = 0$ and $v = 0$ at $t = 0$

We wish to determine the altitude and velocity during the 18 s that thrust is available from the rocket's engine. We can make the system of equations look like the o.d.e. in Eq. [7.2] by using vector representations (see Chapter 2). Each vector will have two components since there are two dependent variables. For example, we may write (with vectors in boldface type)

[7.44a] $\mathbf{y} = \begin{bmatrix} y_1 \\ y_2 \end{bmatrix} = \begin{bmatrix} h \\ v \end{bmatrix};$

[7.44b] $dy/dt = \mathbf{f}(t, \mathbf{y}) = \begin{bmatrix} f_1 \\ f_2 \end{bmatrix} = \begin{bmatrix} v \\ (5000 - 0.1v^2)/(300 - 10t) - g \end{bmatrix}$

The methods applied to the scalar o.d.e. in Eq. [7.2] can now be applied to the vector form in Eq. [7.44b].

To illustrate the applications of our earlier methods, we shall use the Taylor series method of Eq. [7.12] and the modified Euler method of Eq. [7.16a, b] and Eq. [7.19a]. In both cases we shall use 36 steps of size Δt equal to 0.5 s.

For the second-order Taylor series method of Eq. [7.12], we need the partial derivatives $\mathbf{f}_t$ and $\mathbf{f}_y$. The derivative $\mathbf{f}_t$ is simply the vector

[7.45a] $\mathbf{f}_t = \begin{bmatrix} (f_1)_t \\ (f_2)_t \end{bmatrix} = \begin{bmatrix} 0 \\ 10(5000 - 0.1v^2)/(300 - 10t)^2 \end{bmatrix}$

The derivative $\mathbf{f}_y$ is the Jacobian (see Appendix C) and is given by

[7.45b] $\mathbf{f}_y = \begin{bmatrix} \partial f_1/\partial y_1 & \partial f_1/\partial y_2 \\ \partial f_2/\partial y_1 & \partial f_2/\partial y_2 \end{bmatrix} = \begin{bmatrix} (f_1)_h & (f_1)_v \\ (f_2)_h & (f_2)_v \end{bmatrix}$

$= \begin{bmatrix} 0 & 1 \\ 0 & -0.2v/(300 - 10t) \end{bmatrix}$

The vector equivalent of Eq. [7.12] is then

[7.46] $\boldsymbol{\eta}_{i+1} = \boldsymbol{\eta}_i + \Delta t \mathbf{f}(t_i, \boldsymbol{\eta}_i) + [\Delta t^2/2][\mathbf{f}_t(t_i, \boldsymbol{\eta}_i) + \mathbf{f}_y(t_i, \boldsymbol{\eta}_i) \cdot \mathbf{f}(t_i, \boldsymbol{\eta}_i)]$

The vector $\boldsymbol{\eta}$ is our notation for the computed solution of $\mathbf{y}$, and the other vectors in Eq. [7.46] are given by Eq. [7.44b] and Eq. [7.45a, b].

The formulation of the vector problem for the more common o.d.e. methods such as the modified Euler method is easier to develop than for the Taylor series methods because there are no partial derivatives to consider. We can write the vector forms for the modified Euler method by substituting vectors for the appropriate scalars in the scalar method. The forms for the modified Euler method are

[7.47] $\mathbf{k}_1 = \mathbf{f}(t_i, \boldsymbol{\eta}_i); \quad \mathbf{k}_2 = \mathbf{f}(t_i + \Delta t, \boldsymbol{\eta}_i + \Delta t \mathbf{k}_1);$
$\boldsymbol{\eta}_{i+1} = \boldsymbol{\eta}_i + (\Delta t/2)(\mathbf{k}_1 + \mathbf{k}_2)$

Implementation of Eq. [7.46] or [7.47] on a computer will generally require us to use the components of the vectors rather than the direct vectors. We can not compute the components simultaneously if we work with components on a sequential machine. We must therefore be very careful with our "bookkeeping" so that we do not inadvertently update a quantity before we are completely finished with it.

Table 7–8 Solutions for the Rocket Problem with Δt Equal to 0.5 s

	Taylor series		Modified Euler	
t [s]	h [m]	y [m/s]	h [m]	y [m/s]
2.0	14.409	14.8270	14.410	14.8272
4.0	60.680	31.8439	60.680	31.8425
6.0	143.037	50.8583	143.032	50.8529
8.0	265.108	71.4609	265.088	71.4487
10.0	429.468	93.0034	429.417	92.9816
12.0	637.169	114.6287	637.068	114.5959
14.0	887.421	135.3709	887.249	135.3277
16.0	1177.518	154.3173	1177.257	154.2667
18.0	1503.130	170.7901	1502.772	170.7371

The solutions of the rocket problem by the methods given in Eq. [7.46] and [7.47] are shown in Table 7–8. The results for the Taylor series method are slightly larger, in general, than those for the modified Euler method.

We note that there are alternative approaches to solving the rocket problem. For example, we can solve the scalar, first-order o.d.e. in (dv/dt) for v as a tabulated function of time. The results for v can then be integrated numerically to obtain values of h. The numerical integration would, of course, be constrained by the times that are used to obtain the solution for v.

7.3 STIFF EQUATIONS

Certain differential equations display a quality known as ***stiffness***. Stiff equations are those whose solutions contain significantly different scales for the independent variable. When the larger scale is the one of interest, but the smaller scale dictates the step size of a method on the basis of stability, the equation is stiff.

We have used Eq. [7.25a] to illustrate several of the methods that we have discussed for x values from 0 to 4. Suppose now that we are interested in the solution at an x value as large as 25. We see from the analytic solution in Eq. [7.25b] that the exponential term will decay relatively quickly, and that the solution tends to the linear function $(x - 1)$ for large values of x.

If we are interested in the solution at large x values, we do not wish to expend a great deal of effort at the smaller x values. However, if we use too large a step size in an attempt to get to our solution quickly, we may encounter instability problems. To illustrate this concept, Eq. [7.25a] is solved by the Euler method, the modified Euler method, and the second-order Adams–Moulton method which is also known as the ***implicit trapezoidal method***. Twelve steps of size $(25/12)$ were used in all three cases. This step size is slightly larger than the stability limit of 2 for the Euler and modified Euler methods.

The solutions for the three methods are shown in Fig. 7–4. As expected, the Euler and modified Euler methods exhibit unstable behavior. The implicit method, which is stable for all step sizes, shows noticeable inaccuracy in the first step of

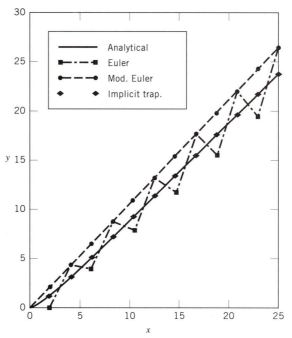

Figure 7–4 Solutions of Eq. [7.25a] with a Step Size of (25/12).

the solution but gives accurate results at larger values of x. The early error in the implicit method is caused by the large step size while the exponential term still has a strong influence. After the decay of that term, the stability of the implicit method allows those initial errors to die out.

The test equation in this case is mildly stiff. If the right-hand side of Eq. [7.25a] is changed to $(x - cy)$ with a large value for c, the stability limits for the Euler and modified Euler methods would impose much smaller step sizes and make the solution process for large x values considerably more costly.

The problem of stiffness is usually associated with higher-order o.d.e.s. Let us examine the linear, second-order o.d.e.

[7.48] $$d^2y/dx^2 + b(dy/dx) + cy = g(x)$$

The form of Eq. [7.48] with positive constants b and c is used to model classic problems such as spring–mass–damper mechanical circuits and resistor–inductor–capacitor electrical circuits.

The solution of Eq. [7.48] consists of the homogeneous solution (for zero right-hand side) and a particular solution. The homogeneous solution is expressed in the form

$$y_h = C e^{\lambda x}$$

When we substitute the solution into Eq. [7.48] with a zero right-hand side, we obtain the characteristic equation

[7.49] $$\lambda^2 + b\lambda + c = 0$$

The roots λ_1 and λ_2 of Eq. [7.49] are the characteristic values or eigenvalues. They are used to represent the homogeneous solution y_h as

$$y_h = C_1 e^{\lambda_1 x} + C_2 e^{\lambda_2 x}$$

in which C_1 and C_2 depend on the initial conditions for the problem.

Suppose now that b is equal to 10 and c is equal to 9. The roots of Eq. [7.49] are (-9) and (-1), so that the homogeneous solution is

$$y_h = C_1 e^{-9x} + C_2 e^{-x}$$

If the initial conditions are such that C_1 is zero, we have a solution that decays as e^{-x}. However, the **other** exponential term dictates the step size for stability. Since the eigenvalue for that term is 9 times larger in magnitude than the one for the term of interest, we would be forced to use a smaller than desirable step size with any conditionally stable method.

The way to deal with stiff equations is to find a stable method. The implicit trapezoidal method is unconditionally stable for decaying solutions; however, its accuracy is limited by its order. The higher-order Adams–Moulton methods are likely to be more accurate; but, as seen from Table 7–6, their intervals of stability become smaller as the order is increased.

Other implicit methods with better stability characteristics have been developed by Gear. These **Gear methods** (Ref. 12) have the general implicit form

[7.50] $$\eta_{i+1} = \gamma \left[\beta h f(x + h, \eta_{i+1}) + \sum_{j=1}^{n} (\alpha_j \eta_{i+j-1}) \right]$$

Coefficients for the Gear methods are given in Table 7–9 for order k up to four. The first-order method is equivalent to the first-order Adams–Moulton or implicit Euler method. The higher-order Gear methods are somewhat difficult to implement because of their multistep nature. Added complications arise when we are solving a system of nonlinear o.d.e.s; then the implicit nature of the method requires us to solve a nonlinear system of equations for the new solution vector. Two possible approaches for nonlinear problems are the use of either iterative refinement or the Newton–Raphson method. We discussed iterative refinement in our treatment of the Adams predictor–corrector methods, and we discussed the Newton–Raphson method for a system of nonlinear equations in Chapter 3.

Table 7–9 Coefficients for the Gear Methods of Order k

k	γ	β	α_1	α_2	α_3	α_4
1	1	1	1			
2	1/3	2	4	-1		
3	1/11	6	18	-9	2	
4	1/25	12	48	-36	16	-3

7.4 BOUNDARY-VALUE PROBLEMS

Ordinary differential equations with known conditions at more than one value of the independent variable are known as ***boundary-value problems***. We shall consider two methods for dealing with such problems. The shooting method is described in §7.4.1, and the finite-difference method is described in §7.4.2.

7.4.1 The Shooting Method

The idea of the ***shooting method*** is analogous to shooting at a target with a rifle. One set of conditions consists of the location of the rifle and the direction in which it is aimed. Another condition is the location of the target. We know the locations of the rifle and the target, but we do not know with certainty the direction in which we should aim. We therefore estimate the direction and then adjust it as needed when we see the results of taking the shot.

An example to illustrate the shooting method deals with the ***temperature distribution*** in a tube. The inner and outer walls of a cylindrical tube are concentric. The inner wall has a radius of 1 in. and is maintained at 80°C; the outer wall has a radius of 3 in. and is maintained at 40°C. Let us denote the temperature by T and the radial position by r. The model for the temperature distribution between the inner and outer walls is

[7.51] $d^2T/dr^2 + (dT/dr)/r = 0;$ $T = 80$ at $r = 1;$ $T = 40$ at $r = 3$

The temperatures at the inner and outer walls correspond to the respective locations of the rifle and the target in our analogy. The missing value of (dT/dr) at the inner wall corresponds to the aim of the rifle. We are seeking the correct value of (dT/dr).

The shooting method is, in reality, a root-solving problem. Let q be any choice we make for the initial value of (dT/dr), and let $T_{out}(q)$ be the temperature at the outer wall for that choice. Then the value of (dT/dr) that satisfies the boundary conditions in Eq. [7.51] is the root of the equation

$$g(q) = T_{out}(q) - 40 = 0$$

Any appropriate root-solving method from Chapter 3 may be used to solve for q. Since we are not dealing with an analytic function, the best choice is the secant method of §3.1.6. To start the secant method, we must obtain values of $g(q)$ for two q values q_a and q_b. A crude estimate of (dT/dr) is obtained by assuming that the drop of 40°C from the inner wall to the outer wall is linear in r; thus, we guess that (dT/dr) is $(-20$ °C/in.). Our starting values of q are shifted from this value by a small amount; that is, we choose

$$q_a = -19 \text{ °C/in.;} \qquad q_b = -21 \text{ °C/in.}$$

The modified Euler method with a step size of 0.05 in. is used to solve the o.d.e. in Eq. [7.51]. The results for $g(q)$ are

$$g(q_a) = 19.133595; \qquad g(q_b) = 16.937131$$

The improved estimate of q from the secant method is

$$\text{improved } q = q_b - [q_b - q_a]g(q_b)/[g(q_b) - g(q_a)] = -36.422182 \text{ °C/in.}$$

It happens that the function $g(q)$ is linear in q for our example; therefore, the improved value of q is the one that satisfies the boundary conditions of the problem. A new solution of the o.d.e. with this value of q gives the required temperature distribution. The solution is shown in Fig. 7–5.

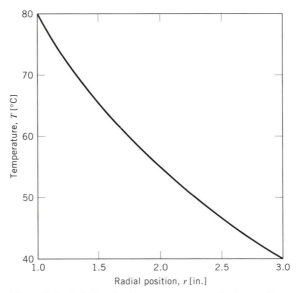

Figure 7–5 Solution for the temperature distribution problem.

7.4.2   The Finite-Difference Method

For o.d.e.s in which more than one boundary condition is to be satisfied at a location other than an initial one, the shooting method of §7.4.1 requires us to solve a system of equations. An alternative to the shooting method in such cases is to model derivatives in the o.d.e. by difference formulas and to solve the resulting system of linear algebraic equations. This approach is categorized as the *finite-difference method*.

A suitable problem for the finite-difference method is one of determining the *deflection of a cantilevered* beam under a distributed load. A beam of length 12 ft and constant cross section has one end fixed and the other free. The modulus of elasticity E and the cross-sectional area moment of inertia I are such that (EI) is equal to 2000 kip·ft². The beam is under a distributed load $q(x)$ expressed in kip/ft by

$$q(x) = (x/12)^2$$

where $(x = 0)$ is the fixed end of the beam and $(x = 12)$ is the free end. A schematic diagram for the problem is given in Fig. 7–6. The model for the downward deflection y of the beam is

[7.52] $d^4y/dx^4 = q(x)/[EI];$

$y = dy/dx = 0$ at $x = 0$; $d^2y/dx^2 = d^3y/dx^3 = 0$ at $x = 12$

The methods of §6.1.1 may be used to derive a second-order difference formula for the fourth derivative. Let us denote control points with equal spacing h of 2 ft by

[7.53] $x_i = (2i)$ ft; $i = 1, 2, \ldots, 6$

along the beam, and let us denote the deflection at x_i by y_i. The difference formula is

[7.54] $(d^4y/dx^4)_i \simeq (y_{i-2} - 4y_{i-1} + 6y_i - 4y_{i+1} + y_{i+2})/h^4$

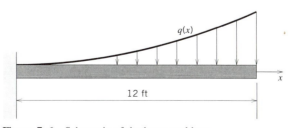

Figure 7–6   Schematic of the beam problem.

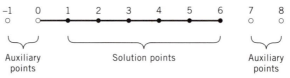

Figure 7–7   Finite-difference grid for the beam problem.

If we apply Eq. [7.54] to our problem, it is clear that control points other than the six that we have defined will be needed to fully express (d^4y/dx^4) for i values of 1, 2, 5, and 6. We have labelled these extra control points as auxiliary points on the grid diagram of Fig. 7–7.

The approximation in Eq. [7.54] is directly applicable at points 3 and 4 on the grid. The linear equations that approximate the beam model at these points are

[7.55a] $y_{i-2} - 4y_{i-1} + 6y_i - 4y_{i+1} + y_{i+2} = h^4(x_i/12)^2/2000; \quad i = 3, 4$

It is also applicable when i is equal to 2, because the deflection y_0 at the auxiliary point 0 is zero as specified by the boundary condition in Eq. [7.52]. The equation when i is equal to 2 is therefore

[7.55b] $-4y_1 + 6y_2 - 4y_3 + y_4 = h^4(x_2/12)^2/2000$

The other condition $(dy/dx = 0)$ at the fixed end is written in difference form to the second order by using the central-difference formula of Eq. [6.6]. The difference form is

$$0 = (dy/dx)_0 \simeq (y_1 - y_{-1})/(2h)$$

We therefore replace y_{-1} by y_1 and y_0 by 0 when we write the equation at point 1. The result is

[7.55c] $7y_1 - 4y_2 + y_3 = h^4(x_1/12)^2/2000$

A similar technique is used at the free end of the beam to express y values at the auxiliary points 7 and 8 in terms of those at solution points. Second-order difference formulas for (d^2y/dx^2) and (d^3y/dx^3) at point 6 in terms of y_4 through y_8 are set to zero to give us

$$0 = (d^2y/dx^2)_6 \simeq (y_5 - 2y_6 + y_7)/h^2$$
$$0 = (d^3y/dx^3)_6 \simeq (-y_4 + 2y_5 - 2y_7 + y_8)/h^3$$

These equations are solved to obtain y_7 and y_8 as

$$y_7 = 2y_6 - y_5; \qquad y_8 = 4y_6 - 4y_5 + y_4$$

Finally, the solutions are substituted into the difference equations for points 5 and 6 to produce

[7.55d]
$$y_3 - 4y_4 + 5y_5 - 2y_6 = h^4(x_5/12)^2/2000$$

[7.55e]
$$2y_4 - 4y_5 + 2y_6 = h^4(x_6/12)^2/2000$$

The original problem is now in the form of a system of linear equations given by Eq. [7.55a] through [7.55e]. The matrix form of the system is

[7.56]
$$
\begin{bmatrix}
7 & -4 & 1 & 0 & 0 & 0 \\
-4 & 6 & -4 & 1 & 0 & 0 \\
1 & -4 & 6 & -4 & 1 & 0 \\
0 & 1 & -4 & 6 & -4 & 1 \\
0 & 0 & 1 & -4 & 5 & -2 \\
0 & 0 & 0 & 2 & -4 & 2
\end{bmatrix}
\cdot
\begin{bmatrix}
y_1 \\ y_2 \\ y_3 \\ y_4 \\ y_5 \\ y_6
\end{bmatrix}
=
\begin{bmatrix}
0.032/144 \\ 0.128/144 \\ 0.288/144 \\ 0.512/144 \\ 0.800/144 \\ 1.152/144
\end{bmatrix}
$$

The matrix in Eq. [7.56] is pentadiagonal (with at most two nonzero elements on either side of the diagonal element). For larger matrices that would be obtained with smaller step sizes, it may be worthwhile to use the concepts for developing the Thomas algorithm to develop a pentadiagonal matrix solver. The size of the current matrix is small enough so that any direct solver may be used without significant loss of efficiency. The solution of Eq. [7.56] is given in ft to five decimal places by

$$\mathbf{y} = [0.03700 \; 0.13178 \; 0.26833 \; 0.43156 \; 0.60833 \; 0.78911]^T$$

7.5   CLOSURE

Methods for solving ordinary differential equations are designed to proceed from given initial conditions. Major methods for initial-value problems are the Runge–Kutta methods and the multistep Adams–Bashforth and Adams–Moulton methods.

The Runge–Kutta methods that we have presented are explicit and easy to use. They can be used effectively when efficiency and stability are not major concerns. The multistep Adams–Bashforth methods are also explicit and provide efficient alternatives to the Runge–Kutta methods at the expense of more complicated coding. The implicit, multistep Adams–Moulton methods have smaller truncation errors and better stability characteristics than the Adams–Bashforth methods. Because their implicit nature makes them more difficult to apply directly to nonlinear o.d.e.s than their explicit counterparts, we may form predictor–corrector methods by combining the Bashforth form as a predictor and the Moulton form as a corrector.

Predictor–corrector formulations may be used in a pure form to preserve efficiency. By this, we mean that the corrector equation is applied only once. Refinement by iterating on the corrector is in effect a method of solving the implicit equation at a cost of one derivative evaluation per iteration. If the predictor provides a good approximation to the implicit solution, only a very few iterations are needed to achieve reasonable accuracy.

Another multistep formulation is the Milne–Simpson method. Although this is an unstable method, the concepts used in deriving it are also applicable to the modified midpoint and Gragg methods. The Gragg method in turn leads to other powerful methods that are beyond the scope of our discussion. The Milne predictor is also used in the Hamming method, which was devised as a compromise between accuracy and stability.

Adaptive methods are used to vary the step size so that it is as large as possible without generating errors beyond specified tolerances. The Runge–Kutta–Fehlberg method is one of the most popular adaptive methods because it is so easy to use. The local error is estimated by using the same stages to produce results of different order. Adaptive versions of multistep methods may also be implemented. The local error is easier to compute than for Runge–Kutta methods; however, the method must be restarted every time there is a change in step size.

Any of the methods may be applied to systems of o.d.e.s. Such systems may be formulated directly, or they may be obtained by reducing higher-order o.d.e.s. These systems look the same as their scalar counterparts when they are expressed in vector notation. Good bookkeeping is essential because we are generally forced to work with vector components rather than directly with the vectors.

Stability is not a major concern for many of the o.d.e.s that we are likely to encounter, because the step size for accuracy is often much smaller than is required for stability. Stiff equations with wide variation in the scale of the independent variable do occur, however. Such equations require methods that have good stability characteristics.

Boundary-value problems may be solved by the methods for initial-value problems by assuming values for the required initial conditions and iterating until the boundary conditions are satisfied. This approach is the shooting method. With finite-difference methods, we recast the problem as a system of linear algebraic equations. The latter approach is recommended when more than one initial condition is missing.

7.6 EXERCISES

1. The equation for the upward velocity v of a rocket from Eq. [7.42] is

$$dv/dt = (5000 - 0.1v^2)/(300 - 10t) - g; \quad v = 0 \text{ at } t = 0$$

where v is in meters/second, t is the time in seconds, and g is 9.81 m/s^2.

Use time steps of 0.2 s to solve for v by the Euler method, and produce a table of (t, v) values in time increments of 2 s up to 18 s.

2. A circular arc of radius r equal to 8 ft is in the vertical plane. The position of an object on the arc is given by the angle θ between the vertical and the radial line through the

object. The object is given an initial speed of 20 ft/s at the bottom of the arc where θ is equal to 0. It then slides up the arc under the influence of gravity and friction. The gravitational acceleration g is 32.2 ft/s^2, and the dynamic friction coefficient μ is equal to 0.2. The equation of motion relating the speed v of the object to the angular position θ is

$$d(v^2)/d\theta = -2[gr(\sin\theta + \mu\ \cos\theta) + \mu v^2)$$

Note that the dependent variable is the square of the speed.

Solve the equation to produce a table of (θ, v) values in θ increments of 0.1 radian until v becomes smaller than 5 ft/s. Use the Euler method and the modified Euler method with step sizes of 0.1 radian.

3. Repeat Problem 2 with the second-order Taylor series method of Eq. [7.12].

4. Repeat Problem 2 with the implicit method of Eq. [7.14].

5. The problem posed in Eq [7.25a] is

$$dy/dx = x - y; \qquad y = 0 \text{ at } x = 0$$

The problem is to be solved with ten equal step sizes for x values up to 4. Solve the problem by the second-order Taylor series method of Eq. [7.12].

6. Repeat Problem 5 with the implicit method of Eq. [7.14].

7. A rectangular open channel has a mild slope S_0 equal to 0.0008 and a width w equal to 15 m. The depth y of the flow in the channel under certain conditions is given by

$$dy/dx = (S_0 - S)/(1 - F^2)$$

where x and y are measured in meters, x is in the direction of the channel, S is the energy slope, and F is the Froude number. S and F are in turn given by

$$S^3 = (nV)^6/R^4; \qquad F^2 = V^2/(gy)$$

in which n is the friction coefficient equal to 0.015, g is the gravitational acceleration equal to 9.81 m/s^2, V is the flow velocity, and R is the hydraulic radius. These last two quantities are expressed in terms of a volume flow rate Q of 20 m^3/s by

$$V = Q/(wy); \qquad R = wy/(w + 2y)$$

The initial condition for the flow is

$$y = 0.8 \text{ m at } x = 0$$

Solve the differential equation from $(x = 0)$ to $(x = 150$ m$)$ to produce a table of (x, y) values in 5 m increments of x. Use the modified Euler method with a step size h of 1 m.

8. Repeat Problem 2 with the classical Runge–Kutta method.

9. Repeat Problem 7 with the classical Runge–Kutta method.

10. Solve Problem 1 by the second-order Adams–Bashforth method. Use the modified Euler method as a starter.

11. Solve Problem 1 by the third-order Adams–Bashforth method. Use the Kutta method of Eq. [7.22] as a starter.

12. Solve Problem 2 by the second-order Adams–Moulton method. Use the implicit method of Eq. [7.14] as a starter.

13. Solve Problem 7 by the Adams predictor–corrector method of Eq. [7.32] with iterative correction. Use the modified Euler method as a starter.

14. Solve Problem 1 by using a fourth-order Adams–Bashforth predictor and a fourth-order Adams–Moulton corrector, using the classical Runge–Kutta method as a starter. Do not use iterative correction.

15. Solve Problem 2 by Hamming's method from Eq. [7.36], using the classical Runge–Kutta method as a starter. Do not use iterative correction.

16. Solve Problem 7 with Milne's method from Eq. [7.34] as a predictor and Hamming's method from Eq. [7.36] as a corrector. Use the classical Runge–Kutta method as a starter. Do not use iterative correction.

17. Solve Problem 7 by the Gragg method to find y at x equal to 100 m. Use step sizes of 1 m, 5 m, and 10 m in three separate solutions.

18. Solve Problem 7 for x up to 100 m by the six-stage Runge–Kutta–Fehlberg method. Start with a step size of 1 m and set a tolerance of 10^{-5} for the fourth-order local truncation error. Reduce the size of the last step if necessary so that the solution ends at x equal to 100 m. Your results should indicate where step doubling or step halving takes place.

19. A three-stage Runge–Kutta–Fehlberg method is given by

$$k_1 = f(x_i, \eta_i); \qquad k_2 = f(x_i + h, \eta_i + hk_1);$$
$$k_3 = f(x_i + h/2, h[k_1 + k_2]/4);$$
$$\eta_{i+1} = \eta_i + h(k_1 + k_2)/2; \qquad \zeta_{i+1} = \eta_i + h(k_1 + k_2 + 4k_3)/6$$

The solutions η_{i+1} and ζ_{i+1} have second-order and third-order local truncation errors, respectively. The criterion for step doubling must be modified from the one used with the six-stage method to be consistent with the order of the local truncation error for the three-stage method.

Repeat Problem 18 with this method, but now use 10^{-5} as the tolerance for the second-order local truncation error.

20. A rocket is designed to be launched vertically upwards from a pole of Mars. The atmospheric drag and the effects of the planet's rotation are neglected; however, the gravitational acceleration g in km/s^2 varies with altitude h in km according to

$$g = 0.0037\{3394/(h + 3394)\}^2$$

The equation of motion for the rocket is

$$d^2h/dt^2 = qu/(m_0 - qt) - g$$

where q is the propellant consumption rate of 600 kg/s, t is time in seconds, u is the axial speed of 2 km/s at which propellant is ejected relative to the rocket, and m_0 is the rocket's initial total mass of 198000 kg. The initial mass of the propellant is 96000 kg and is included in m_0. We wish to obtain a history of the rocket's altitude h (in km) and speed v (in km/s) as a function of time t, for as long as the propellant lasts. The initial conditions for the problem are

$$h = 0 \text{ and } v = dh/dt = 0 \text{ at } t = 0$$

Express the o.d.e. as a first-order system and use any method to obtain an accurate solution. First obtain a solution with g held constant at 0.0037 km/s^2 and verify the accuracy of your method by comparing the results to the analytic solution. Then apply the method to the case in which g varies with h. Consider what the most accurate way of computing g is.

21. The mass m supported by one wheel of a car rides on a suspension consisting of a spring and a damper (or shock absorber). When the wheel is displaced from equilibrium by the road, the motion of the supported mass depends on the spring stiffness k and damping coefficient b. A schematic of a suspension system and a sudden road elevation are shown in Fig. 7–P21.

   The two differential equations and initial conditions for this problem are

$$dy/dt = v; \quad y = 0 \text{ when } t = 0$$
$$dv/dt = [k(y_w - y) + b(v_w - v)]/m; \quad v = 0 \text{ when } t = 0$$

in which t is time in seconds, y and v are, respectively, the displacement and velocity of the mass, y_w and v_w are, respectively, the displacement and velocity of the wheel, k is equal to 2146 lb/ft, m is equal to 30.9 slugs, and b is given by ($b^2 = km$).

   Time t is zero when the wheel hits the "bump." The values of y_w and v_w are then constants equal to 0.25 ft and 0 ft/s, respectively.

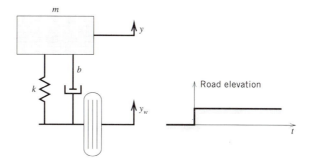

Figure 7–P21

Solve the problem accurately by any method for t values up to 2 s, and compare the numerical solution for y with the analytical solution

$$y = y_w - y_w e^{-\beta t}[\cos(\omega t) + (\beta/\omega)\sin(\omega t)];$$

$$\beta = b/(2m); \qquad \omega^2 = k/m - \beta^2$$

Provide a table of $(t, y, y_{analytic}, v)$ values.

22. Solve the two differential equations in Problem 21 accurately for y and v for t values up to 4 s when the wheel displacement is given by

$$y_w = 0.1 \, \sin(7.5t)[1 - \cos(\pi t/2)]$$

The arguments of the sin and cos functions are in radians, and v_w is (dy_w/dt). If possible, show y and y_w versus t on the same plot.

23. Two first-order reactions (such as those that occur in radioactive decay) may be modeled as

$$x_1 \xrightarrow{\kappa_1} x_2 \xrightarrow{\kappa_2} x_3$$

in which the x_i values represent concentrations of three species, and the κ_j values are the reaction rates. A prototypical system of equations for such reactions is

$$d(x_1)/dt = -\kappa_1 x_1; \qquad d(x_2)/dt = \kappa_1 x_1 - \kappa_2 x_2; \qquad d(x_3)/dt = \kappa_2 x_2$$

Solve the system accurately by any method for

$$x_1 = 1, x_2 = x_3 = 0 \text{ at } t = 0; \qquad \kappa_1 = 3, \kappa_2 = 1$$

End the solution when x_1 reaches a value of 0.2, and show the three x_i values versus t on the same plot, if possible.

24. Repeat Problem 23 with κ_1 equal to 10 and κ_2 equal to 1.

25. One end of a rod that extends from $(x = 0)$ to $(x = 1)$ is held at a temperature T_0 of 100 degrees. The other of the rod is at $(x = 1)$ and is insulated, so that dT/dx is zero. The temperature distribution in the rod is obtained from

$$d^2T/dx^2 = 2.3(T - T_a); \qquad T = T_0 \text{ at } x = 0, \, dT/dx = 0 \text{ at } x = 1$$

in which the ambient temperature T_a is 20 degrees.

Use the shooting method to obtain the temperature distribution in the rod at x increments of 0.2. Use the modified Euler method with a step size of 0.2 for this problem.

26. Repeat Problem 25 with the finite-difference method. Use an x spacing of 0.2.

27. The Blasius solution for the boundary layer on a flat plate is formulated in terms of a function $f(\eta)$ as

$$f(d^2f/d\eta^2) + 2(d^3f/d\eta^3) = 0;$$

$$f = 0, df/d\eta = 0 \text{ at } \eta = 0; \quad df/d\eta = 1 \text{ at } \eta = \infty$$

The infinite value of η must be replaced by a large value η_{max}. This value is assumed to be the point at which $(df/d\eta)$ becomes essentially constant. Use the shooting method to solve this problem. Assume that a suitable value for η_{max} will be approximately equal to 10.

28. The model for small deflections y of a simply supported beam under a uniform load is given by

$$EI(d^2y/dx^2) = qx(x - L)/2; \quad y = 0 \text{ at } x = 0 \text{ and at } x = L$$

The beam extends from $(x = 0)$ to $(x = L)$, where L is 12 ft, EI is 2000 kip·ft^2, and q is 0.7 kip/ft. Solve the problem for deflections at every 2 ft by the shooting method. Use a second-order o.d.e. solver with a step size of 2 ft.

29. Repeat Problem 28 with the finite-difference method. Use a step size of 2 ft.

PART IV

ADVANCED TOPICS

8

Matrix Eigenproblems

The system of linear algebraic equations ($\mathbf{A} \cdot \mathbf{x} = \mathbf{f}$), which we considered in Chapter 2, was to be solved for the vector $\mathbf{x}$ when the vector $\mathbf{f}$ and the ***nonsingular*** matrix $\mathbf{A}$ were specified. We now consider the system of equations

[8.1]
$$\mathbf{A} \cdot \mathbf{x} = \lambda \mathbf{x}$$

in which λ is a scalar quantity. We may use the identity matrix $\mathbf{I}$ to rewrite Eq. [8.1] as

[8.2]
$$(\mathbf{A} - \lambda\mathbf{I}) \cdot \mathbf{x} = \mathbf{0}$$

and look for values of λ such that Eq. [8.2] is satisfied when $\mathbf{x}$ is a nonzero vector. Such a problem is known as a ***matrix eigenproblem*** in which λ is an ***eigenvalue*** and $\mathbf{x}$ is an ***eigenvector***. An example of a matrix eigenproblem was given in Section 3.3.

The name eigenproblem comes from the meaning "own" of the German word *eigen*. The eigenvalues are a problem's ***own*** or ***characteristic*** values. Characteristic values are sought for problems such as those involving ordinary differential equations and buckling of columns. Other problems involving systems of equations yield the classic matrix form that is given in Eq. [8.1].

To obtain a nontrivial solution $\mathbf{x}$ for Eq. [8.1] or [8.2], the matrix $(\mathbf{A} - \lambda\mathbf{I})$ must be singular; that is, its determinant must be zero. The expansion of the determinant as a polynomial in λ gives the ***characteristic polynomial***, whose roots are the eigenvalues. Each eigenvalue is associated with an infinite number of vectors $\mathbf{x}$, which are multiples of each other. Any of these vectors is an eigenvector.

There are several methods for computing eigenvalues and eigenvectors for a system. The choice of method depends on what we are seeking; for example, it depends on if we are looking for the full set of eigenvalues and eigenvectors, or for eigenvalues only, or for the dominant eigenvalue and its associated eigenvector. These methods range from the relatively straightforward to the complicated, and they all require at least a working knowledge of matrix algebra. The basic concepts of Chapter 2 are therefore prerequisites for this chapter.

We shall consider three basic approaches to eigenproblems. The first is based on the Faddeev–Leverrier method and allows computation of all eigenvalues and eigenvectors. This approach is discussed in Section 8.1. The power method for computing the dominant eigenvalue and its associated eigenvector is discussed in Section 8.2. Finally, the Jacobi method for symmetric matrices is considered in Section 8.3.

Three applied problems are used as case studies. A **spring–mass system** and an analogous **capacitor–inductor filter** are used to illustrate the Faddeev–Leverrier method as well as variants of the power method. The problem of finding the **maximum principal stress** from a given stress tensor is used for the power method and for the Jacobi method.

8.1 THE FADDEEV–LEVERRIER METHOD

The **Faddeev–Leverrier method** (Ref. 5, 15) is a procedure for determining the coefficients of the characteristic polynomial. It also contains mechanisms for computing the eigenvectors after the eigenvalues have been found, and for computing the inverse of a matrix.

If $\mathbf{A}$ is an $(n \times n)$ matrix, the determinant $|\lambda \mathbf{I} - \mathbf{A}|$ is an n-th degree polynomial in λ and may be expressed as

[8.3] $$|\lambda \mathbf{I} - \mathbf{A}| = \lambda^n + \alpha_{n-1} \lambda^{n-1} + \cdots + \alpha_1 \lambda + \alpha_0$$

To use the Faddeev–Leverrier method, we also need the **trace** (or spur) of a matrix. The trace of an $(n \times n)$ matrix $\mathbf{A}$ is defined by

[8.4] $$\text{tr}\,(\mathbf{A}) = \sum_{i=1}^{n} a_{ii} = \text{sum of the diagonal elements}$$

The Faddeev–Leverrier algorithm begins with a matrix $\mathbf{D}_1$ set equal to the identity matrix $\mathbf{I}$. The coefficients of the characteristic polynomial are then found from the following sequence of computations.

[8.5a] $$\mathbf{D}_1 = \mathbf{I}; \quad \alpha_{n-1} = -\text{tr}\,(\mathbf{A} \cdot \mathbf{D}_1)$$

[8.5b] $$\mathbf{D}_i = \mathbf{A} \cdot \mathbf{D}_{i-1} + (\alpha_{n-i+1})\mathbf{I};$$
$$\alpha_{n-i} = (-1/i)\,\text{tr}\,(\mathbf{A} \cdot \mathbf{D}_i); \quad i = 2, 3, \ldots, n$$

Once we have used Eq. [8.5a, b] to determine the coefficients of the characteristic polynomial in Eq. [8.3], we may use a root-solving method such as Bairstow's method in Chapter 3 to determine the eigenvalues λ.

Bairstow's method is capable of finding both real and complex eigenvalues. Other methods such as the Newton-Raphson or bisection method may also be used for real eigenvalues if we have sufficient information to establish good initial estimates. We shall consider how we might obtain such information later in this section; at the moment, however, we shall illustrate how eigenproblems arise by formulating two problems. We shall then demonstrate the application of Eq. [8.5a, b].

A spring–mass system is shown in Fig. 8–1. The three masses are $2m$ and $2m$ and m, and the spring stiffnesses are k and k and $2k$ and $2k$. The displacements of the masses from their equilibrium positions are denoted by x_1, x_2, and x_3. These cause the springs to expand or to compress. The resulting spring force is the stiffness multiplied by the expansion or compression in a direction opposite to the expansion or compression. The free-body diagrams for the three masses are also given in Fig. 8–1 and show the spring forces.

We now apply Newton's Second Law of Motion to each of the three masses. The results, with t denoting time, are

$$2m(d^2x_1/dt^2) = -2kx_1 + kx_2; \quad 2m(d^2x_2/dt^2) = kx_1 - 3kx_2 + 2kx_3;$$
$$m(d^2x_3/dt^2) = 2kx_2 - 4kx_3$$

A natural frequency ω is one for which each displacement may be written as

$$x_i = B_iC \, \cos(\omega t + \phi)$$

in which B_iC is an amplitude and ϕ is a phase angle. Therefore, we see that $[d^2(x_i)/dt^2]$ is equal to $(-\omega^2 x_i)$. We introduce λ as the quantity $(2m\omega^2/k)$, and

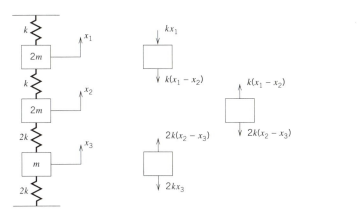

Figure 8–1 A spring-mass system and its free-body diagrams.

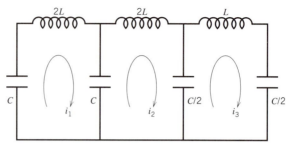

Figure 8–2   An electrical analog of the system in Fig. 8–1.

we use **x** to denote the displacement vector with components x_1, x_2, and x_3. These relations allow us to rewrite the equations of motion as

[8.6]
$$(\mathbf{A} - \lambda\mathbf{I}) \cdot \mathbf{x} - 0; \qquad \mathbf{A} = \begin{bmatrix} 2 & -1 & 0 \\ -1 & 3 & -2 \\ 0 & -4 & 8 \end{bmatrix}$$

The same situation is obtained with the capacitor–inductor filter of Fig. 8–2. This circuit is the electrical analog of our mechanical circuit. Equations for the loop currents i_1, i_2, and i_3, are obtained by setting the voltage drop in each loop to zero and differentiating with respect to time.

The voltage drops across an inductor L' and across a capacitor C' are, respectively,

$$\text{Drop across } L' = L(di/dt); \quad \text{Drop across } C' = (1/C)\int (i \; dt)$$

in which i is the current and t denotes time. The equations for the currents in Fig. 8–2 are therefore

$$i_1/C + 2L(d^2i_1/dt^2) + (i_1 - i_2)/C = 0$$

$$(i_2 - i_1)/C + 2L(d^2i_2/dt^2) + 2(i_2 - i_3)/C = 0$$

$$2(i_3 - i_2)/C + L(d^2i_3/dt^2) + 2i_3/C = 0$$

The assumed solution for i_j is $[B_j Q \; \cos(\omega t + \phi)]$, and **x** is the current vector with components i_1, i_2, and i_3. When we introduce λ as $(2LC\omega^2)$, we obtain exactly the same eigenproblem given by Eq. [8.6].

The application of Eq. [8.5a, b] to the eigenproblem in Eq. [8.6] is now detailed. We have the following sequence of computations beginning with $\mathbf{D}_1$ equal to **I**.

[8.7a]
$$(\mathbf{A} \cdot \mathbf{D}_1) = \begin{bmatrix} 2 & -1 & 0 \\ -1 & 3 & -2 \\ 0 & -4 & 8 \end{bmatrix}; \quad \alpha_2 = -\text{tr}(\mathbf{A} \cdot \mathbf{D}_1) = -13$$

[8.7b] $\mathbf{D}_2 = (\mathbf{A} \cdot \mathbf{D}_1) + \alpha_2 \mathbf{I} = \begin{bmatrix} -11 & -1 & 0 \\ -1 & -10 & -2 \\ 0 & -4 & -5 \end{bmatrix}$;

$(\mathbf{A} \cdot \mathbf{D}_2) = \begin{bmatrix} -21 & 8 & 2 \\ 8 & -21 & 4 \\ 4 & 8 & -32 \end{bmatrix}$; $\alpha_1 = (-1/2)\,\text{tr}(\mathbf{A} \cdot \mathbf{D}_2) = 37$

[8.7c] $\mathbf{D}_3 = (\mathbf{A} \cdot \mathbf{D}_2) + \alpha_1 \mathbf{I} = \begin{bmatrix} 16 & 8 & 2 \\ 8 & 16 & 4 \\ 4 & 8 & 5 \end{bmatrix}$;

$(\mathbf{A} \cdot \mathbf{D}_3) = \begin{bmatrix} 24 & 0 & 0 \\ 0 & 24 & 0 \\ 0 & 0 & 24 \end{bmatrix}$; $\alpha_0 = (-1/3)\,\text{tr}(\mathbf{A} \cdot \mathbf{D}_3) = -24$

The characteristic equation for Eq. [8.6] is thus obtained from Eq. [8.7a, b, c] to be

[8.8] $|\lambda \mathbf{I} - \mathbf{A}| = \lambda^3 - 13\lambda^2 + 37\lambda - 24 = 0$

This result may be verified by an actual expansion of the determinant.

For the general $(n \times n)$ matrix $\mathbf{A}$, the accuracy of the computations may be discerned by looking at the matrix $(\mathbf{A} \cdot \mathbf{D}_n)$. The theoretical result for this matrix is

$$\mathbf{A} \cdot \mathbf{D}_n = -\alpha_0 \mathbf{I}$$

The accuracy of the computation is therefore indicated by how much the off-diagonal elements of the final matrix differ from zero, and by how much the diagonal elements differ from each other.

We also note that the inverse of the matrix $\mathbf{A}$, if it is nonsingular, is given by

[8.9] $$\mathbf{A}^{-1} = (-1/\alpha_0)\mathbf{D}_n$$

A singular matrix is indicated by a zero value of α_0. The inverse of the matrix $\mathbf{A}$ in our case study is obtained from the results of Eq. [8.7c].

A pseudocode for the Faddeev–Leverrier method is built on three other modules, which are described as follows.

- **Module Multiply**$(n, \mathbf{A}, \mathbf{B}, \mathbf{C})$: Forms the product $\mathbf{C} = \mathbf{A} \cdot \mathbf{B}$ of two $(n \times n)$ matrices.
- **Module Trace**$(n, \mathbf{A}, \text{tr})$: Computes the trace tr of an $(n \times n)$ matrix $\mathbf{A}$.
- **Module Copy**$(n, \mathbf{A}, \mathbf{B})$: Copies an $(n \times n)$ matrix $\mathbf{A}$ to $\mathbf{B}$.

These auxiliary modules are simple to code and are used in the Faddeev–Leverrier module of Pseudocode 8–1.

We now look at the determination of the eigenvalues once the characteristic equation has been found. A root-solving method such as Bairstow's method in Section

---

Pseudocode 8–1   Faddeev–Leverrier module for coefficients α of the characteristic equation

---

Module FadLev(n, A, α)   \ A is an ($n \times n$) matrix;
                                         \ α is an array with subscripts from 0 to ($n - 1$)

Declare: D, Tmp   \ ($n \times n$) arrays representing matrices

Initialize: D   \ set D to the identity matrix I

For [$i = 1, 2, \ldots, n$]

    Multiply(n, A, D, Tmp)   \ form **Tmp** $=$ **A** $\cdot$ **D** at step i

    Trace(n, A, α_{n-i})

    $\alpha_{n-i} \leftarrow -\alpha_{n-i}/i$

    Copy(n, Tmp, D)

    If [$i < n$] then

        For [$j = 1, 2, \ldots, n$]

            $D_{jj} \leftarrow D_{jj} + \alpha_{n-i}$ \ **D** matrix for next step

        End For

    End If

End For

End Module FadLev

3.3 may be used to find all of the eigenvalues, including complex conjugate pairs. We may also use other root-solving methods to find the eigenvalues one at a time.

We may be able to obtain reasonable estimates of the eigenvalues from properties of polynomials and from Gerschgorin's theorem. Some properties of polynomials are as follows.

- An n-th degree polynomial $p(\lambda)$ has n roots (some of which may be repeated). At least one root is real if n is odd, and complex roots always occur in complex conjugate pairs.

- *Descartes' rule of signs* states that the number of positive, real roots is either equal to the number of sign changes in the coefficients or is fewer by an even integer. The number of negative, real roots is estimated in the same manner by transforming the polynomial $p(\lambda) = 0$ to $p(-\lambda) = 0$.

The coefficients of the polynomial of Eq. [8.8] undergo three sign changes; there is at least one positive root, and possibly three. The coefficients of the transformed polynomial $p(-\lambda)$ do not change sign; therefore, no negative root exists.

Gerschgorin's theorem is used to locate the eigenvalues of an ($n \times n$) matrix **A** in the complex z plane. Let r_i and c_j be defined by

$$r_i = \left(\sum_{j=1}^{n} |a_{ij}| \right) - |a_{ii}|; \quad c_j = \left(\sum_{i=1}^{n} |a_{ij}| \right) - |a_{jj}|$$

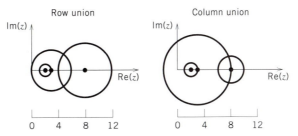

Figure 8–3   Gerschgorin unions for the matrix **A** of Eq. [8.6].

Gerschgorin's theorem states that the eigenvalues of **A** lie in the union of circles with centers a_{ii} on the real axis and radii r_i. Alternately, the eigenvalues lie in the union of circles with centers a_{jj} on the real axis and radii c_j. A related theorem states that a union of m circles that forms a connected set and is isolated from any other set contains exactly m eigenvalues, including repeated eigenvalues.

The row and column unions for the matrix **A** in Eq. [8.6] are shown in Fig. 8–3. In either case, the only connected set is formed by the union of three circles; thus, there is no way to isolate a particular eigenvalue. By looking at the intersection of the row and column unions, however, we can see that any real eigenvalue must be on the interval from 0 to 10. Indeed, the eigenvalues for our problem are given in descending order by

[8.10] $\lambda_1 = 9.298408; \quad \lambda_2 = 2.769686; \quad \lambda_3 = 0.931906$

The eigenvectors corresponding to a given eigenvalue may be obtained from the matrices $\mathbf{D}_i$ of Eq. [8.5a, b] of the Faddeev–Leverrier method. If eigenvectors are required, the basic pseudocode for the method should be modified to store a column of each $\mathbf{D}_i$ (the same column in each case) before they are overwritten by a later matrix. An eigenvector is any column of the matrix **C** given by

[8.11] $$\mathbf{C} = \sum_{i=1}^{n}(\lambda^{n-i})\mathbf{D}_i$$

We may choose any column because the columns are multiples of each other, and an eigenvector is any one of an infinite number of vectors that are multiples of each other. The results for the third column of the **C** matrices for each of the eigenvalues in Eq. [8.10] are as follows.

[8.12a] $\lambda = \lambda_1: \quad \mathbf{x} = \begin{bmatrix} 2.000000 \\ -14.596816 \\ 44.968351 \end{bmatrix}$

[8.12b] $\qquad \lambda = \lambda_2: \quad \mathbf{x} = \begin{bmatrix} 2.000000 \\ -1.539372 \\ -1.177269 \end{bmatrix}$

[8.12c] $\qquad \lambda = \lambda_3: \quad \mathbf{x} = \begin{bmatrix} 2.000000 \\ 2.136188 \\ 1.208919 \end{bmatrix}$

An alternative method of computing an eigenvalue $\mathbf{x}$ is to set any one of its components, say the last one, to 1. This value is used to reduce the $(n \times n)$ system $(\mathbf{A} \cdot \mathbf{x} = \lambda \mathbf{x})$ to a system of $(n - 1)$ equations for the other components of $\mathbf{x}$. The result that we would obtain by this procedure is a vector that is a multiple of one of the vectors in Eq. [8.12a, b, c].

A specific eigenvector for a problem depends on the conditions of the problem. For example, Eq. [8.10] and Eq. [8.12a, b, c] give a complete set of solutions for the problem posed in Eq. [8.6]. Each solution gives the conditions under which all of the mass displacements in our mechanical circuit or all of the currents in our electrical circuit oscillate in phase at the same frequency. The frequency is related to λ as we stated in the formulations of these problems, and the amplitudes of the oscillations are proportional to the eigenvectors. If one of the amplitudes is given, we may then obtain a specific eigenvector that represents all of the amplitudes.

The Faddeev–Leverrier method is aimed at computing the coefficients of the characteristic equation for an eigenproblem. The eigenvalues are the roots of that equation and must be found by a separate mechanism. Once they have been found, the auxiliary matrices $\mathbf{D}_i$ of the Faddeev–Leverrier method may be used to compute the eigenvectors.

The operational count for the Faddeev–Leverrier method is on the order of n^4 for an $(n \times n)$ matrix. These are costly computations if we are interested only in a limited solution of the eigenproblem. The power method in Section 8.2 provides an alternate approach when only the dominant eigenvalue and its associated eigenvector are required.

8.2 THE POWER METHOD

The *power method* (Ref. 4) is used to find the distinct eigenvalue with the largest magnitude. It is an iterative method that strongly resembles the fixed-point iteration method of Chapter 3 for solving nonlinear equations.

The application of the method to Eq. [8.1] begins with an initial estimate of the eigenvector $\mathbf{x}$. This estimate is usually chosen so that the infinity norm of the vector is 1. This choice is made easily by setting the last component to 1 and all others to zero. The basic iteration for the power method is used to improve $\mathbf{x}$ according to

[8.13] $\qquad\qquad\qquad\qquad \mathbf{x}_{\text{imp}} = \mathbf{A} \cdot \mathbf{x}$

The improved vector $\mathbf{x}_{\text{imp}}$ is then scaled by the component λ with the largest magnitude according to

[8.14]
$$(\mathbf{x}_{\text{imp}})_{\text{scaled}} = (\mathbf{x}_{\text{imp}})/\lambda$$

If the infinity norm of $[\mathbf{x} - (\mathbf{x}_{\text{imp}})_{\text{scaled}}]$ is greater than a specified tolerance and the iteration number has not reached a preset limit, $\mathbf{x}$ is reset to $(\mathbf{x}_{\text{imp}})_{\text{scaled}}$, and iterations are continued. The iteration limit is imposed to prevent infinite looping in cases where the method may fail to converge. If the method converges, the final value of λ is the required eigenvalue, and the final value of $(\mathbf{x}_{\text{imp}})_{\text{scaled}}$ is the corresponding eigenvector.

To illustrate the power method, we consider a three-dimensional coordinate system (y_1, y_2, y_3) and we let a matrix $\mathbf{A}$ denote a stress or inertia tensor associated with the three directions. For example, we consider the stress tensor

[8.15]
$$\mathbf{A} = \begin{bmatrix} 55 & 14 & 10 \\ 14 & 60 & 20 \\ 10 & 20 & 40 \end{bmatrix}$$

There is a principal set of directions for which the stress tensor has zero shear stresses; that is, only the diagonal elements are nonzero. The corresponding eigenproblem is

$$\mathbf{A} \cdot \mathbf{x} = \lambda \mathbf{x}$$

in which the eigenvalues λ are the principal normal stresses, and the eigenvector $\mathbf{x}$ is proportional to the direction cosines of the corresponding principal axis. In such problems, we are often interested only in the maximum stress; that is, the maximum eigenvalue.

To apply the power method to the problem associated with Eq. [8.15], we begin with an initial vector

$$\mathbf{x} = \begin{bmatrix} 0 \\ 0 \\ 1 \end{bmatrix}$$

which is premultiplied by $\mathbf{A}$ to give

$$\mathbf{x}_{\text{imp}} = \begin{bmatrix} 10 \\ 20 \\ 40 \end{bmatrix}$$

The scale factor λ is seen to be 40, and the scaled vector is $(\mathbf{x}_{\text{imp}})/40$.

A second iteration begins by resetting $\mathbf{x}$ to $(\mathbf{x}_{imp})/40$ and premultiplying by $\mathbf{A}$ to produce

$$\mathbf{x}_{imp} = \begin{bmatrix} 30.75 \\ 53.50 \\ 52.50 \end{bmatrix}$$

with a scale factor λ equal to 53.5. The process is continued for a total of 20 iterations until a tolerance of 10^{-6} is reached. The eigenvalue and eigenvector at this stage are

$$\lambda = 82.9231; \quad \mathbf{x} = \begin{bmatrix} 0.729074 \\ 1.000000 \\ 0.635805 \end{bmatrix}$$

The direction cosines of the principal axis are proportional to the eigenvector, and the sum of their squares is equal to 1. The direction cosines are therefore 0.524011, 0.718738, and 0.456977. The maximum stress is the value found for λ.

The major disadvantage of the power method is that its convergence rate is slow when the eigenvalues with the two largest magnitudes are almost equal to each other. Let these eigenvalues be λ_1 and λ_2 with $|\lambda_1|$ greater than $|\lambda_2|$. The convergence of the power method is proportional to $|\lambda_2/\lambda_1|$. A matrix without a distinct dominant eigenvalue (for example, one with a dominant complex conjugate pair of eigenvalues) would fail to converge.

A similar method for finding the eigenvalue with the smallest magnitude is the *inverse power method*. The eigenvalues of $\mathbf{A}^{-1}$ are the reciprocals of the eigenvalues of $\mathbf{A}$. If we apply the power method to $\mathbf{A}^{-1}$ to find its distinct eigenvalue μ with the largest magnitude, the corresponding eigenvalue of $\mathbf{A}$ is $(1/\mu)$ and is the one with the smallest magnitude.

The inverse power method is essentially the same as the power method; however, Eq. [8.13] is replaced by

[8.16] $$\mathbf{x}_{imp} = \mathbf{A}^{-1} \cdot \mathbf{x}$$

We, of course, do not invert the matrix $\mathbf{A}$ to apply Eq. [8.16]; rather, we use a method such as the LU decomposition method of Chapter 2 to solve for $\mathbf{x}_{imp}$. The procedure for using Eq. [8.16] is illustrated in Example 8.1.

EXAMPLE 8.1
The matrix $\mathbf{A}$ of Eq. [8.6] is factored as follows:

$$\mathbf{A}^* = \mathbf{L} \cdot \mathbf{U}; \quad \mathbf{A}^* = \begin{bmatrix} 2 & -1 & 0 \\ 0 & -4 & 8 \\ -1 & 3 & -2 \end{bmatrix}$$

$$\mathbf{L} = \begin{bmatrix} 1 & 0 & 0 \\ 0 & 1 & 0 \\ -1/2 & -5/8 & 1 \end{bmatrix}; \quad \mathbf{U} = \begin{bmatrix} 2 & -1 & 0 \\ 0 & -4 & 8 \\ 0 & 0 & 3 \end{bmatrix}$$

Note that $\mathbf{A}^*$ is obtained by interchanging rows 2 and 3 of $\mathbf{A}$.

Show how $\mathbf{x}_{\text{imp}}$ is determined from Eq. [8.16] when the inverse power method is used to compute the smallest eigenvalue of $\mathbf{A}$.

We write:

$$\mathbf{L} \cdot (\mathbf{U} \cdot \mathbf{x}_{\text{imp}}) = \mathbf{L} \cdot \mathbf{g} = \begin{bmatrix} x_1 \\ x_3 \\ x_2 \end{bmatrix}$$

We use $\mathbf{L}$ and forward substitution to determine the intermediate vector $\mathbf{g}$ as

$$g_1 = x_1; \quad g_2 = x_3; \quad g_3 = x_2 + x_1/2 + 5x_2/8$$

Then we use $\mathbf{U}$ and back substitution to obtain $\mathbf{x}_{\text{imp}}$ as

$$(x_{\text{imp}})_3 = g_3/3; \quad (x_{\text{imp}})_2 = -[g_2 - 8(x_{\text{imp}})_3]/4; \quad (x_{\text{imp}})_1 = [g_1 + (x_{\text{imp}})_2]/2$$

The basic power method cannot be used to find complex eigenvalues. Variations of the method do allow this possibility. The basis for the variation is that the eigenvalues of the matrix $(\mathbf{A} - b\mathbf{I})$ are equal to $(\lambda - b)$, where λ represents the eigenvalues of $\mathbf{A}$. Complex choices of b allow the computation of complex eigenvalues. Other choices allow the computation of eigenvalues of $\mathbf{A}$ that are not dominant. The technique of using the power method with a matrix $\mathbf{A}$ that is modified by an amount $b\mathbf{I}$ is known as the ***power method with shift***.

EXAMPLE 8.2

The dominant eigenvalue λ_{max} of the matrix $\mathbf{A}$ in Eq. [8.6] was found to be 9.2984. What shift b must be used so that the power method can be applied to determine the smallest eigenvalue λ_{min} of $\mathbf{A}$?

From the Gerschgorin unions in Fig. 8–3, we know that $\mathbf{A}$ has no negative eigenvalues. If we choose $b \geq \lambda_{\text{max}}$, the dominant eigenvalue μ_{max} of $(\mathbf{A} - b\mathbf{I})$ will be its most negative eigenvalue and will be equal to $(\lambda_{\text{min}} - b)$.

To maximize the convergence rate, the ratio of μ_{max} to the intermediate eigenvalue should be as large as possible. Thus, we do not want a shift b that is any larger than necessary. The choice of shift is therefore

$$b = \lambda_{\text{max}} = 9.2984$$

8.3   THE JACOBI METHOD FOR SYMMETRIC MATRICES

The *Jacobi method* (Ref. 4, 16) for obtaining eigenvalues and eigenvectors consists of a sequence of transformations that reduce a *symmetric* matrix $\mathbf{A}$ to a diagonal matrix with the same eigenvalues. The eigenvalues of the diagonal matrix are equal to the diagonal elements themselves.

Each transformation is designed to drive off-diagonal elements a_{pq} and a_{qp} to zero. The transformation has the form

[8.17]
$$\mathbf{A}_{k+1} = \mathbf{R} \cdot \mathbf{A}_k \cdot \mathbf{R}^{\mathbf{T}}$$

where $\mathbf{A}_k$ and $\mathbf{A}_{k+1}$ are two successive matrices in the sequence of transformations, and $\mathbf{R}$ is a matrix with zero-valued elements *except* for

[8.18] $r_{pp} = r_{qq} = c = \cos\theta; \quad r_{pq} = -r_{qp} = s = \sin\theta;$
           $r_{ii} = 1; \; i \neq p, i \neq q$

For example, to drive elements a_{13} and a_{31} of a (4×4) matrix to zero, we use

$$\mathbf{R} = \begin{bmatrix} c & 0 & s & 0 \\ 0 & 1 & 0 & 0 \\ -s & 0 & c & 0 \\ 0 & 0 & 0 & 1 \end{bmatrix}$$

The matrix $\mathbf{R}$ and its transpose are *orthogonal* because they are inverses of each other. The transformation of Eq. [8.17] is called a *plane rotation* because $\mathbf{R}$ is of the form used to rotate two-dimensional coordinates, and $\mathbf{R}$ itself is called a *rotation matrix*. The value of the angle θ that causes the elements a_{pq} and a_{qp} to become zero is given by

[8.19]
$$\cot 2\theta = (a_{pp} - a_{qq})/(2a_{pq})$$

The transformation of Eq. [8.17] is a *similarity transformation*; that is, the eigenvalues are invariant under it. In applying the transformation, we have two matrix multiplications. If a matrix $\mathbf{U}$ is premultiplied by $\mathbf{R}$ to produce $\mathbf{V}$, then $\mathbf{U}$ and $\mathbf{V}$ are the same except for the elements in rows p and q. Elements in these rows are given by

[8.20a] $v_{pk} = cu_{pk} + su_{qk}; \quad v_{qk} = -su_{pk} + cu_{qk}; \quad k = \text{column}$

When $\mathbf{V}$ is postmultiplied by $\mathbf{R}^{\mathbf{T}}$ to produce $\mathbf{W}$, then $\mathbf{V}$ and $\mathbf{W}$ are the same except for the elements in columns p and q. Elements in these columns are given by

[8.20b] $w_{kp} = cv_{kp} + sv_{kq}; \quad w_{kq} = -sv_{kp} + cv_{kq}; \quad k = \text{row}$

Of course, w_{pq} and w_{qp} are zero valued as intended after both multiplications.

In the classical Jacobi method, the elements a_{pq} and a_{qp} that are chosen to become zero are those with the largest magnitude. In the more commonly used *cyclic* Jacobi method, they are chosen in a systematic fashion from among the off-diagonal elements. The systematic approach eliminates the need to search for the element with the largest magnitude. It proceeds from row 1 to row $(n - 1)$ of an $(n \times n)$ matrix, and from column $(i + 1)$ to column n for each of the rows. Only elements above the diagonal are chosen, and this is all that is necessary because the matrix is symmetric. If a particular element is already very close to zero, no transformation is performed.

The Jacobi method is an iterative procedure. As each rotation is done, elements that were previously driven to zero may become nonzero again. Thus, the iteration is carried out until *all* of the off-diagonal elements become very small in magnitude.

The eigenvalues λ_j are the elements a_{jj} of the final matrix. The eigenvector $\mathbf{x}_j$ is the j-th column of the matrix $\mathbf{X}$ formed by the product of the transposes of $\mathbf{R}$; that is,

[8.21] $$\mathbf{X} = (\mathbf{R}_1)^{\mathbf{T}} \cdot (\mathbf{R}_2)^{\mathbf{T}} \cdots$$

Codes for the Jacobi method are given in Pseudocode 8–2. The procedure shown in Module Angle is an efficient and accurate way of computing θ from Eq. [8.19].

The first rotation for the symmetric matrix $\mathbf{A}$ in Eq. [8.15] is aimed at reducing the elements ($a_{12} = a_{21} = 14$) to zero. The cosine and sine of the rotation angle θ are

$$c = 0.641954; \quad s = 0.766743$$

It should be noted that there are many angles differing by integer multiples of $(\pi/2)$ that satisfy Eq. [8.19]. The angle corresponding to the results given for c and s is $(\pi/2)$ more than the angle that would be obtained from direct inversion of Eq. [8.19]. When c and s are used to perform the first rotation of $\mathbf{A}$, the matrix is overwritten with the new elements. The result (affecting only rows 1 and 2 and columns 1 and 2) is

$$\mathbf{A} = \begin{bmatrix} 71.721463 & 0.000000 & 21.754404 \\ 0.000000 & 43.278537 & 5.171643 \\ 21.754404 & 5.171643 & 40.000000 \end{bmatrix}$$

A cycle or sweep consists of transforming $\mathbf{A}$ as necessary for each of the elements above the diagonal. Convergence to a diagonal form is obtained in the third sweep

Pseudocode 8–2   The cyclic Jacobi method

\ Main module for the Jacobi method. **A** and **X** are $(n \times n)$ arrays. **X** is set to
\ the identity matrix **I** for input to the module, and ϵ is a small, positive value.

Module Jacobi$(n, \mathbf{A}, \mathbf{X}, \epsilon)$:
Repeat
    Done $\leftarrow$ 1   \ assume convergence has occurred
    For $[p = 1, 2, \ldots, (n - 1)]$
        For $[q = p + 1, p + 2, \ldots, n]$ \ loops for choosing a_{pq}
            If $[|a_{pq}| > \epsilon]$ **then**
                Done $\leftarrow$ 0   \ no convergence yet
                Angle$(a_{pp}, a_{qq}, a_{pq}, c, s)$ \ to find c and s
                Rotate$(n, p, q, c, s, \mathbf{A}, \mathbf{X})$ \ modify **A** and **X**
            End If
        End For   \ with counter q
    End For   \ with counter p
Until [Done $= 1$]
Write: Diagonal elements of **A**, columns of **X**
End Module Jacobi

\ Module to compute c and s for the rotation matrix

Module Angle$(a_{pp}, a_{qq}, a_{pq}, c, s)$:
$\alpha \leftarrow [(a_{pp} - a_{qq})^2 + 4(a_{pq})^2]^{0.5}$
$c \leftarrow [\{\alpha + (a_{pp} - a_{qq})\}/\{2\alpha\}]^{0.5}$; $s \leftarrow a_{pq}/(c\alpha)$
End Module Angle

\ Module to find new elements of **A** and **X** after a rotation.

Module Rotate$(n, p, q, c, s, \mathbf{A}, \mathbf{X})$:
For $[k = 1, 2, \ldots, n]$ \premultiplication loop
    $t1 \leftarrow a_{pk}$ \temporary storage of old element value
    $a_{pk} \leftarrow c(t1) + s\,a_{qk}$
    $a_{qk} \leftarrow s(t1) + c\,a_{qk}$
End For

For $[k = 1, 2, \ldots, n]$ \ postmultiplication loop
    $t1 \leftarrow a_{kp}$; $t2 \leftarrow x_{kp}$ \ temporary storage of old element values
    $a_{kp} \leftarrow c(t1) + s\,a_{kq}$; $x_{kp} \leftarrow c(t2) + s\,x_{kq}$
    $a_{kq} \leftarrow -s(t1) + c\,a_{kq}$; $x_{kq} \leftarrow -s(t2) + c\,x_{kq}$
End For
End Module Rotate

or cycle with an ϵ value of 10^{-6}. The final matrices **A** and **X** are (to six decimal places)

$$\mathbf{A} = \begin{bmatrix} 82.923141 & 0.000000 & 0.000000 \\ 0.000000 & 44.500577 & 0.000000 \\ 0.000000 & 0.000000 & 27.576282 \end{bmatrix}$$

$$\mathbf{X} = \begin{bmatrix} 0.524012 & -0.849939 & -0.054899 \\ 0.718737 & 0.475866 & -0.506921 \\ 0.456977 & 0.226175 & 0.860242 \end{bmatrix}$$

The largest eigenvalue is 82.923141, and the corresponding eigenvalue is the first column of **X**. This vector is proportional to the one obtained earlier by the power method. The eigenvalues of **A** are the principal stresses for the problem formulated in Section 8.2. The eigenvalues are the direction cosines of the principal directions, and they should be mutually orthogonal. We can verify orthogonality easily by checking that the dot or inner product of any two eigenvectors is zero.

The Jacobi method is not as efficient as other methods such as the ***Givens or Householder methods***, which reduce the original matrix to tridiagonal form; however, it is very reliable, and it is relatively simple to apply. The Householder method can also be used to reduce nonsymmetric matrices to ***upper Hessenberg matrices***, which are upper triangular except for one subdiagonal. Algorithms such as the ***QR algorithm*** may then be used to deal with the tridiagonal or Hessenberg forms that are obtained. We shall not discuss these other methods since they are much more complicated than the Jacobi method.

8.4 CLOSURE

We have presented three types of methods for computing the eigenvalues and eigenvectors of a matrix, and we have given physical examples of how the standard matrix eigenproblem arises. The Faddeev–Leverrier method is the most general of the three. Its major objective is to produce the characteristic equation, which must then be solved by a separate method to obtain the eigenvalues.

Bairstow's method for polynomials can be used for solving the general problem, and will determine both real and complex eigenvalues. Other simpler methods may be used for small matrices, especially when the properties of polynomials and Gerschgorin's theorem can help us to locate the solutions within a relatively small region. We may then return to the Faddeev–Leverrier computations to obtain the eigenvectors, or we may solve a reduced system of equations separately.

Many eigenproblems do not require computation of all of the eigenvalues. The power method and the inverse power method give us mechanisms for computing the dominant and least dominant eigenvalues along with their associated eigenvectors. The convergence rate of the power method is poor when the two largest eigenvalues in magnitude are nearly equal. The technique of shifting the matrix by an amount

$(-b\mathbf{I})$ can help us to overcome this disadvantage, and it can also be used to find intermediate eigenvalues by the power method.

A technique for symmetric matrices, which occur frequently, is the Jacobi method. It belongs to a class of methods that are used to reduce the original matrix to a simpler form. The Jacobi method uses similarity transformations based on plane rotations to reduce the matrix to a diagonal form. The rotation matrices are used at the same time to form the matrix whose columns contain the eigenvectors for the problem. The disadvantage of the Jacobi method, other than the requirement for symmetry, is that it is iterative and may take many rotations to converge to a diagonal form. Nevertheless, it is reliable, is easy to use, and converges with uniform accuracy.

Because matrices are generally large structures, they require a great number of computations, which can cause the accumulation of round-off errors. We should therefore use as high a precision as possible, except for the very smallest of matrices. Even with high precision, we should seek ways to verify that our computations are accurate and that our convergence criteria, if required, are not so stringent as to be impossible to attain. Verification of accuracy for the Faddeev–Leverrier and Jacobi methods may be done easily by examining the final matrices.

8.5   EXERCISES

1.   The moments of inertia for a thin plate about the x and y axes of a Cartesian coordinate system are

$$I_{xx} = 0.20 \text{ kg} \cdot \text{m}^2; \quad I_{yy} = 0.12 \text{ kg} \cdot \text{m}^2$$

The products of inertia are

$$I_{xy} = I_{yx} = -0.14 \text{ kg} \cdot \text{m}^2$$

The inertia tensor is denoted by the matrix

$$\mathbf{A} = \begin{bmatrix} I_{xx} & -I_{xy} \\ -I_{yx} & I_{yy} \end{bmatrix}$$

The principal moments of inertia are the eigenvalues of $\mathbf{A}$, and the direction cosines of the corresponding principal axes are proportional to the eigenvectors. The two direction cosines c_1 and c_2 for an axis satisfy the relation

$$(c_1)^2 + (c_2)^2 = 1$$

Use the Faddeev–Leverrier method to find the characteristic equation for the eigenproblem. Then obtain the principal moments of inertia and the direction cosines of the corresponding principal axes.

2. Use the power method to determine the larger of the two principal moments of inertia in Problem 1.

3. Repeat Problem 1 with the Jacobi method.

4. The following matrix is associated with the spring–mass system of Fig. 3–7.

$$\mathbf{A} = \begin{bmatrix} 2 & 0 & -1 & 0 \\ 0 & 1 & 0 & 0 \\ -2 & 0 & 3 & 0 \\ 0 & 0 & 0 & 1 \end{bmatrix}$$

Find the eigenvalues of the matrix by the Faddeev–Leverrier method.

5. Use the power method to find the eigenvalue of **A** in Problem 4 with the largest magnitude.

6. A tridiagonal matrix has five diagonal elements, each of which is equal to 2. Each of the nonzero off-diagonal elements is equal to (-1). This matrix is related to the problem of transverse vibrations of five equal masses that are equally spaced on a taut string. Find the eigenvalues of the matrix by the Faddeev–Leverrier method.

7. Use the power method to find the largest eigenvalue in magnitude of the matrix in Problem 6.

8. Use the inverse power method to find the smallest eigenvalue in magnitude of the matrix in Problem 6.

9. Use the Jacobi method to find the eigenvalues of the matrix in Problem 6.

10. The system of first-order ordinary differential equations

$$d\mathbf{x}/dt = \mathbf{A} \cdot \mathbf{x}$$

has a general solution

$$\mathbf{x} = k_1 \mathbf{c_1} e^{\lambda_1 t} + k_2 \mathbf{c_2} e^{\lambda_2 t} + \cdots$$

in which λ_i are the eigenvalues of **A** and $\mathbf{c_i}$ are the eigenvectors.
     Consider the system with

$$\mathbf{A} = \begin{bmatrix} -5 & 1 & 1 & 0 \\ 1 & -8 & 2 & 0 \\ 1 & 2 & -6 & 1 \\ 0 & 0 & 1 & -3 \end{bmatrix}$$

Find the characteristic equation for **A** by the Faddeev–Leverrier method.

11. Use the Jacobi method to find the eigenvalues and eigenvectors of **A** in Problem 10.

12. The maximum eigenvalue of the matrix **A** in Eq. [8.15] is approximately equal to 82.923. What value of b must be used so that the maximum eigenvalue of the matrix $(\mathbf{A} + b\mathbf{I})$ is equal to 100? Verify your result by applying the power method and by applying the Jacobi method.

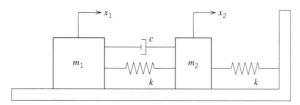

Figure 8–P13

13. The equations of motion for the spring-mass-damper system shown in Fig. 8–P13 are given by

$$m_1(d^2x_1/dt^2) + c[(dx_1/dt) - (dx_2/dt)] + k[x_1 - x_2] = 0$$
$$m_2(d^2x_2/dt^2) + c[(dx_2/dt) - (dx_1/dt)] + k[2x_2 - x_1] = 0$$

By writing

$$x_3 = dx_1/dt; \quad x_4 = dx_2/dt$$

the equations may be rewritten in the form

$$\mathbf{B} \cdot d\mathbf{x}/dt = \mathbf{C} \cdot \mathbf{x}$$

where

$$\mathbf{B} = \begin{bmatrix} 1 & 0 & 0 & 0 \\ 0 & 1 & 0 & 0 \\ 0 & 0 & m_1 & 0 \\ 0 & 0 & 0 & m_2 \end{bmatrix}; \quad \mathbf{C} = \begin{bmatrix} 0 & 0 & 1 & 0 \\ 0 & 0 & 0 & 1 \\ -k & k & -c & c \\ k & -2k & c & -c \end{bmatrix}; \quad \mathbf{x} = \begin{bmatrix} x_1 \\ x_2 \\ x_3 \\ x_4 \end{bmatrix}$$

By setting $d\mathbf{x}/dt$ equal to $\lambda\mathbf{x}$, we obtain the eigenvalue problem

$$\mathbf{A} \cdot \mathbf{x} = \lambda\mathbf{x}; \quad \mathbf{A} = \mathbf{B}^{-1} \cdot \mathbf{C}$$

Use the Faddeev–Leverrier method to obtain the characteristic equation for the matrix **A** when $m_1 = 4m_2 = 0.2$ slug, $k = 50$ lb/ft, and $c = 2$ lb · s/ft. Then compute the (possibly complex) eigenvalues λ of the matrix **A**.

9

Introduction to Partial Differential Equations

Phenomena that vary either with more than one space dimension or in both time and space are often modeled as *partial differential equations*, which contain partial derivatives of one or more dependent variables. The study of numerical methods for such equations is extensive; therefore, we can barely cover some of the simpler cases in the space of one chapter.

Our coverage is intended to provide an introduction to methods for partial differential equations (p.d.e.s) and will focus on the classical problems of parabolic, elliptic, and hyperbolic equations with relatively simple domains. The methods that we shall consider are *finite-difference* methods (Ref. 6, 17) in which derivatives of a function are replaced by difference equations. The finite-difference concept was introduced in Chapter 6 and was used for boundary-value problems in Chapter 7.

We shall present some preliminary concepts in Section 9.1, and follow with discussions of methods for the three types of p.d.e.s in Sections 9.2, 9.3, and 9.4. Section 9.5 deals with generalized, orthogonal coordinate systems.

9.1 PRELIMINARY CONCEPTS

The major classes of physical problems may be categorized as equilibrium, eigen-value, or propagation problems. In the *equilibrium* problem, the variable u to be found on some domain D is typically expressed in terms of an elliptic operator and has boundary conditions specified on the boundary of D. We can think of equilibrium problems as those in which u responds instantaneously to the imposed boundary conditions. Examples of equilibrium problems are those for steady fluid flows, steady-state temperature distributions, and equilibrium stresses.

Eigenvalue problems are those in which the governing equation has two parts that are expressed in terms of elliptic operators and related by an eigenvalue. The boundary conditions have a similar construction. Some examples are problems for structural stability and for natural frequencies of vibrations.

Propagation problems are generally called *initial-boundary-value* problems (or simply *initial-value* problems). The governing equation is typically a parabolic or hyperbolic p.d.e., and they contain initial and (sometimes) boundary conditions. We can think of a propagation problem as one in which the solution "marches" out from an initial state and is modified by boundary conditions at later times. Parabolic problems show damped responses to imposed boundary conditions; hyperbolic problems allow oscillations. Examples of propagation problems are those for unsteady heat conduction and viscous flows.

We have used the terms parabolic, elliptic, and hyperbolic without really defining them. One pragmatic means of definition is in terms of what the solutions do, as we have just described in the preceding paragraphs. For a more formal look at these terms, we consider a quasilinear p.d.e. in two dimensions p and q, and we use subscripts to denote partial differentiation. The equation is

[9.1] $$a u_{pp} + 2b u_{pq} + c u_{qq} = f$$

The quantities a, b, c, and f in Eq. [9.1] are all functions of u, p, q, u_p, and u_q. We wish to find conditions for which a knowledge of u, u_p, and u_q uniquely determines the second partial derivatives that satisfy the p.d.e. We write

$$d(u_p) = u_{pp}dp + u_{pq}dq; \qquad d(u_q) = u_{pq}dp + u_{qq}dq$$

so that we have

[9.2] $$\begin{bmatrix} a & 2b & c \\ dp & dq & 0 \\ 0 & dp & dq \end{bmatrix} \cdot \begin{bmatrix} u_{pp} \\ u_{pq} \\ u_{qq} \end{bmatrix} = \begin{bmatrix} f \\ d(u_p) \\ d(u_q) \end{bmatrix}$$

If the matrix in Eq. [9.2] is nonsingular, the second derivatives are unique; if the matrix is singular, however, its determinant is zero, and the characteristic equation is

$$a(dq)^2 - 2b\, dp\, dq + c(dp)^2 = 0$$

We may recast the characteristic equation as a quadratic in (dp/dq) by

[9.3] $$a(dp/dq)^2 - 2b\,(dp/dq) + c = 0$$

The roots of the characteristic equation are used to classify the p.d.e. of Eq. [9.1] according to the following relations for $(b^2 - ac)$.

[9.4a] **Hyperbolic:** $b^2 - ac > 0$; real and distinct roots

[9.4b] **Parabolic:** $b^2 - ac = 0$; real and identical roots

[9.4c] **Elliptic:** $b^2 - ac < 0$; complex conjugate roots

The values of (dp/dq) from the characteristic equation define paths along which the second derivatives cannot be uniquely determined. There are no such paths for the elliptic problem on a real domain; thus every point on the domain affects every other point. For parabolic and hyperbolic equations, the existence of real paths on the domain restricts the way in which information at one point may affect the solution at another point.

We will be dealing with equations that differ from the quasilinear form of Eq. [9.1]. As we have stated earlier, we can generally classify the equations on the basis of the behavior of the solution instead of relying on the characteristics.

In dealing with p.d.e.s with finite differences, we will make use of the Taylor series for several variables $(x_1, x_2, \ldots, x_n)$. Let f be a function of these variables; then the Taylor series expansion is given by

$$f(x_1 + \Delta x_1, x_2 + \Delta x_2, \ldots, x_n + \Delta x_n) =$$
[9.5a] $$[1 + \mathscr{D} + (\mathscr{D}^2/2!) + (\mathscr{D}^3/3!) + \cdots]f(x_1, x_2, \ldots, x_n)$$

The operator $\mathscr{D}$ is given by

[9.5b] $$\mathscr{D} = \sum_{i=1}^{n}[\Delta x_i(\partial/\partial x_i)]$$

For example, the operator $\mathscr{D}$ for n equal to 2 is

$$\mathscr{D} = \Delta x_1(\partial/\partial x_1) + \Delta x_2(\partial/\partial x_2)$$

so that

$$\mathscr{D}f = \Delta x_1(\partial f/\partial x_1) + \Delta x_2(\partial f/\partial x_2)$$

For terms such as $\mathscr{D}^2 f$, the operator is "squared" in algebraic fashion so that

$$\mathscr{D}^2 f = (\Delta x_1)^2 (\partial^2 f / \partial[x_1]^2) + 2(\Delta x_1)(\Delta x_2)(\partial^2 f / [\partial x_1 \partial x_2]) + (\Delta x_2)^2 (\partial^2 f / \partial[x_2]^2)$$

The Taylor series expansion may be used as it was in Chapter 6 to obtain finite-difference approximations for partial derivatives. Let us consider the common case in which a quantity u depends on two variables x and y. We look at equally spaced points given by

[9.6] $$x_i = i\Delta x; \qquad y_j = j\Delta y$$

We denote $u(x_i, y_j)$ by $u_{i,j}$ and write some of the more common central-difference approximations for derivatives of $u_{i,j}$ as follows.

[9.7] $$\begin{aligned}
u_x &= (u_{i+1,j} - u_{i-1,j})/(2\Delta x); \\
u_y &= (u_{i,j+1} - u_{i,j-1})/(2\Delta y); \\
u_{xx} &= (u_{i-1,j} - 2u_{i,j} + u_{i+1,j})/(\Delta x)^2; \\
u_{yy} &= (u_{i,j-1} - 2u_{i,j} + u_{i,j+1})/(\Delta y)^2 \\
u_{xy} &= (u_{i+1,j+1} + u_{i-1,j-1} - u_{i+1,j-1} - u_{i-1,j+1})/(4\Delta x \Delta y);
\end{aligned}$$

Forward-difference and backward-difference formulas (except for u_{xy}) are similar to those given in Chapter 6.

Two common operators are the **Laplacian** and the **biharmonic** operators, which are written, respectively, as ∇^2 and ∇^4. The symbol ∇ is the **nabla** or **del** operator. The quantities $\nabla^2 u$ and $\nabla^4 u$ are defined in (x, y) coordinates by

[9.8] $$\nabla^2 u = u_{xx} + u_{yy}; \qquad \nabla^4 u = u_{xxxx} + 2u_{xxyy} + u_{yyyy}$$

A finite-difference approximation for $\nabla^2 u$ is obtained easily from the forms in Eq.[9.7]. The approximation may also be shown pictorially as a **computational molecule**. Molecules for $\nabla^2 u$ and $\nabla^4 u$ with $(\Delta x = \Delta y = \Delta)$ are shown in Fig. 9–1. Both molecules represent second-order approximations. Other second-order approximations are possible, as are other approximations of different order.

9.2   METHODS FOR PARABOLIC EQUATIONS

We begin our discussion of methods for p.d.e.s with parabolic equations because they have much in common with the ordinary differential equations of Chapter 7. A general form for a parabolic p.d.e. is given in terms of time t and coordinates (x, y, z) by

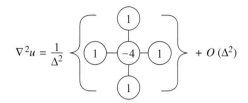

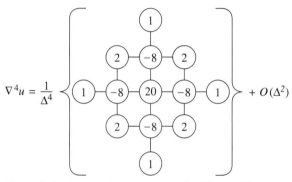

Figure 9–1 Computational molecules for $\nabla^2 u$ and $\nabla^4 u$.

[9.9]
$$u_t = (a_1 u_x)_x + (a_2 u_y)_y + (a_3 u_z)_z$$
$$+ b_1 u_x + b_2 u_y + b_3 u_z - cu;$$
$$a_1 > 0, a_2 > 0, a_3 > 0, c \geq 0$$

The quantities $a_1, a_2, a_3, b_1, b_2, b_3$, and c are functions of x, y, z, and t.

The prototype equation in one space dimension is a special case of Eq. [9.9] and is expressed by

[9.10a]   **P.D.E.:** $\qquad u_t = u_{xx}; \qquad 0 \leq x \leq 1, t \geq 0$

In addition to the p.d.e., we must also specify initial conditions (i.c.) and boundary conditions (b.c.). These are given in the general form

[9.10b]   **I.C.:** $\qquad u(x, 0) = f(x); \qquad 0 \leq x \leq 1$

[9.10c]   **B.C.:** $\qquad a u_x + p u = g(t); \qquad x = 0, t \geq 0$
$$b u_x + q u = h(t); \qquad x = 1, t \geq 0$$

If $(a = b = 0)$ in Eq. [9.10c], we have a ***Dirichlet or 1st problem***. If $(p = q = 0)$, we have a ***Neumann or 2nd problem***. Otherwise, we have a ***mixed or 3rd problem***. We shall consider the Dirichlet problem with p and q equal to 1 in our development of methods for parabolic p.d.e.s; that is, we shall use the boundary conditions

[9.10d]   **B.C.:** $u(0, t) = g(t);$ $u(1, t) = h(t);$ $t \geq 0.$

We shall begin our presentation of methods for parabolic equations with an analysis of a time-explicit method in §9.2.1. The implicit Crank–Nicolson method, the multi-level DuFort–Frankel method, and a predictor–corrector method are discussed in §9.2.2 through §9.2.4. Discussion of some special cases is presented in §9.2.5.

9.2.1   Analysis of an Explicit Method

An explicit method for the problem of Eq. [9.10a, b, d] uses a forward-difference approximation in time and a central-difference approximation in space. Let time t and position x be denoted by

[9.11] $t_n = n\Delta t, n \geq 0;$ $x_i = i\Delta x, \ 0 \leq i \leq m$

The approximations for u_t and u_{xx} are

[9.12a] $(u_t)_{i,n} \simeq (u_{i,n+1} - u_{i,n})/\Delta t$

[9.12b] $(u_{xx})_{i,n} \simeq (u_{i-1,n} - 2u_{i,n} + u_{i+1,n})/(\Delta x)^2$

The time approximation is first order and the space approximation is second order. When these approximations are substituted into the p.d.e. of Eq. [9.10a], we obtain the explicit scheme

[9.13] $u_{i,n+1} = r(u_{i-1,n}) + (1 - 2r)u_{i,n} + r(u_{i+1,n});$ $r = \Delta t/(\Delta x)^2$

The practice of changing notation as we did in Chapter 7 to distinguish between exact and computed solutions has been dropped; we now rely on context for this distinction.

The solution by the explicit scheme begins with setting n equal to 0 so that all u values at this level are given by the initial conditions. We then seek solutions for u at points 1 through $(m - 1)$ at the next time level; the solutions at points 0 and m are given by the boundary conditions. The problem and the explicit scheme are shown schematically in Fig. 9–2.

The roles of the boundary conditions and the initial condition are illustrated in the following Pseudocode 9–1 for the explicit method. A one-dimensional array (in x) is used to store the u values. Array elements are overwritten as u is updated. We must wait for one count of i before doing so; otherwise, we will destroy the value $u_{i-1,n}$ before we are finished with it.

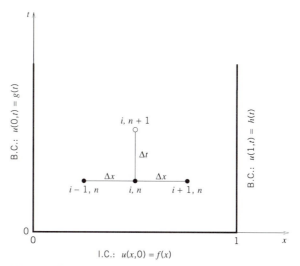

Figure 9–2 Schematic for the explicit method.

Pseudocode 9–1 Implementation of the explicit method

Declare: u \ with subscripts i from 0 to m
Enter: $\Delta t, K$ \ K = maximum value of time counter n
$\Delta x \leftarrow 1/m; \quad\quad r \leftarrow \Delta t / (\Delta x)^2$
For $[i = 0, 1, \ldots, m]$
  $u_i \leftarrow f(i\Delta x)$ \ initial conditions
End For

For $[np1 = 1, 2, \ldots, K]$ \ time loop for solution; $np1 = n + 1$
  unew $\leftarrow g(np1[\Delta t])$ \ left b.c. at new time level
  For $[i = 1, 2, \ldots, M - 1]$ \ M corresponds to the right boundary
   $t1 \leftarrow ru_{i-1}; \quad\quad t2 \leftarrow ru_{i+1}$
   $u_{i-1} \leftarrow$ unew   \ saved value for previous i
   unew $\leftarrow t1 + t2 + (1 - 2r)u_i$
  End For   \ with counter i
  $u_{m-1} \leftarrow$ unew   \ last value from i loop
  $u_m \leftarrow h(np1[\Delta t])$ \ right b.c. at new time level
  Write: u values
End For   \ with counter $np1$

The local truncation error for the explicit scheme is

$$\tau_e = (u_{i,n+1} - u_{i,n})/\Delta t - (u_{i-1,n} - 2u_{i,n} + u_{i+1,n})/(\Delta x)^2 - (u_t - u_{xx})$$

By expanding u values about (x_i, t_n) in a Taylor series, the leading terms of the local truncation error are contained in

[9.14] $$\tau_e = [\Delta t/2]u_{tt} - [(\Delta x)^2/12]u_{xxxx} + [(\Delta t)^2/6]u_{ttt} + \cdots$$

As Δt and Δx tend to zero, we see that the local truncation error also tends to zero. The explicit scheme is therefore **consistent** with the p.d.e.

Let us denote the exact solution of the p.d.e. by u_{pde}, the computed solution by $u_{computed}$, and the exact solution of the finite-difference equation (that would be obtained with infinite machine precision) by u_{fde}. The error in the computed solution is

$$\text{Error} = u_{computed} - u_{pde} = (u_{computed} - u_{fde}) + (u_{fde} - u_{pde})$$

The error term $(u_{computed} - u_{fde})$ is the one associated with the **stability** of a method. We denote this term by E and use Eq. [9.13] to write it at point $(i, n+1)$ as

[9.15] $$E_{i,n+1} = rE_{i-1,n} + (1 - 2r)E_{i,n} + rE_{i+1,n} + \text{higher-order terms}$$

To implement a **Fourier stability analysis** for a method, we represent the error E by a Fourier series. Note that by doing so for a general form, such as the one given by Eq. [9.15], we do not account for any effects due to the boundary conditions. We assume that the initial errors at $(t = 0)$ are given by the finite Fourier series

[9.16] $$E_{i,0} = \sum_{k=1}^{K} A_k \exp(j\lambda_k x_i)$$

in which j is the unit imaginary number, A_k values are generally complex, and λ_k are real wave numbers. The exponential of an imaginary number is given by Euler's formula

$$\exp(j\theta) = \cos\theta + j\sin\theta$$

When the problem is linear, we can apply our analysis to one Fourier component at a time. Let us assume that a time-varying component has the form $[\exp(\alpha t)\exp(j\lambda x)]$, which is reduced to the form in Eq. [9.16] when t is zero. To prevent growth of the error with time, we require

$$|\exp(\alpha[t + \Delta t])\exp(j\lambda x)|/|\exp(\alpha t)\exp(j\lambda x)| \le 1$$

To meet this condition, the time step Δt must satisfy

[9.17] $$|\exp(\alpha \Delta t)| \le 1$$

By substituting $[\exp(\alpha t)\exp(j\lambda x)]$ for E in Eq. [9.15], we obtain

$$\exp(\alpha[n + 1]\Delta t)\exp(j\lambda i\Delta x) =$$
$$\exp(\alpha n\Delta t)\{r\exp(j\lambda[i - 1]\Delta x) + (1 - 2r)\exp(j\lambda i\Delta x) + r\exp(j\lambda[i + 1]\Delta x)\}$$

Simplification yields

$$\exp(\alpha \Delta t) = (1 - 2r) + r[\exp(-j\lambda \Delta x) + \exp(j\lambda \Delta x)]$$
$$= (1 - 2r) + 2r\cos(\lambda \Delta x) = 1 - 4r\sin^2(\lambda \Delta x/2)$$

Therefore, the stability criterion of Eq. [9.17] is now recast as

$$-1 \le 1 - 4r\sin^2(\lambda \Delta x/2) \le 1$$

The inequality on the right is satisfied for nonnegative values of r, and the one on the left is satisfied when r is at most $(1/2)$. Thus, the stability condition for the explicit scheme is

[9.18] $$(0 \le) \, r = \Delta t/(\Delta x)^2 \le 1/2$$

An obvious modification for improving the accuracy of the explicit method is to replace the approximation for u_t with the central-difference formula

$$(u_t)_{i,n} \simeq (u_{i,n+1} - u_{i,n-1})/(2\Delta t)$$

The result is an explicit method that spans three time levels. Unfortunately, this new method is unconditionally unstable. The central-difference idea can, however, be applied in other ways. One of these leads to the Crank–Nicolson method, which is now discussed.

9.2.2 The Crank–Nicolson Method

The *Crank–Nicolson method* is derived from central-difference approximations for u_t and u_{xx} about the point $(x_i, t_n + \Delta t/2)$. The approximation for u_{xx} is obtained by averaging u values at the time levels t_n and t_{n+1}. The base approximations are

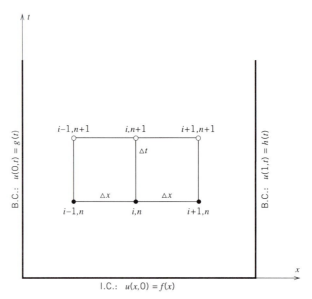

Figure 9–3 Schematic for the Crank–Nicolson method.

[9.19a] $$(u_t)_{i,p} \simeq (u_{i,n+1} - u_{i,n})/\Delta t; \qquad p = n + 1/2$$

[9.19b] $$(u_{xx})_{i,p} \simeq (u_{i-1,n+1} - 2u_{i,n+1} + u_{i+1,n+1}$$
$$+ u_{i-1,n} - 2u_{i,n} + u_{i+1,n})/[2(\Delta x)^2]; \qquad p = n + 1/2$$

The method is shown schematically in Fig. 9–3 for our prototype problem.

The local truncation error for the Crank–Nicolson (C–N) method is obtained by comparing the approximation to the exact derivatives at the point (i, n), **not** at the base point of the expansions. The local truncation error is

[9.20] $$\tau_{\mathrm{cn}} = (\Delta t/2)(u_{tt} - u_{xxt}) - [(\Delta x)^2/12]u_{xxxx} + \cdots$$

The method is also consistent and unconditionally stable for linear p.d.e.s.

To implement the C–N method, the formulas of Eq. [9.19a, b] are rearranged in the form

[9.21] $$(-r/2)u_{i-1,n+1} + (1 + r)u_{i,n+1} + (-r/2)u_{i+1,n+1}$$
$$= (r/2)u_{i-1,n} + (1 - r)u_{i,n} + (r/2)u_{i+1,n}; \qquad r - \Delta t/(\Delta x)^2$$

The matrix formulation corresponding to Eq. [9.21] for our prototype problem of Eq.[9.10a, b, d] is

[9.22a] $$\mathbf{A} \cdot \mathbf{u}_{n+1} = \mathbf{v}$$

where the solution vector $\mathbf{u}_{n+1}$ has components $u_{0,n+1}$ through $u_{m,n+1}$. The tridiagonal matrix $\mathbf{A}$ and the right-hand vector $\mathbf{v}$ are as follows. The u values in the initial vector $\mathbf{v}$ are obtained from the initial condition.

$$
[9.22b] \qquad \mathbf{A} = \begin{bmatrix}
1 & 0 & & & & \\
-r/2 & (1+r) & -r/2 & & & \\
& \cdot & \cdot & \cdot & & \\
& & \cdot & \cdot & \cdot & \\
& & & \cdot & \cdot & \cdot \\
& & & -r/2 & (1+r) & -r/2 \\
& & & & 0 & 1
\end{bmatrix}
$$

$$
[9.22c] \qquad \mathbf{v} = \begin{bmatrix}
g(t_{n+1}) \\
(1-r)u_{1,n} + (r/2)(u_{0,n} + u_{2,n}) \\
\vdots \\
(1-r)u_{m-1,n} + (r/2)(u_{m-2,n} + u_{m,n}) \\
h(t_{n+1})
\end{bmatrix}
$$

The solution of Eq. [9.22 a,b,c] may be obtained by using the Thomas algorithm of Chapter 2, *once only*, to obtain $\mathbf{L}$ and $\mathbf{U}$ factors, which are used at all time levels.

The C–N method has obvious advantages over the explicit method of Eq. [9.13]; it is unconditionally stable for linear p.d.e.s and it is higher order in time. It has another, more subtle, advantage in that the influence of the boundary conditions is felt at all spatial points at a given time level. To see that all points feel the effect of the boundary, we have only to look at the matrix formulation of the C–N scheme. If the boundary condition terms in the vector $\mathbf{v}$ are changed, the solution for time level $(n + 1)$ will also be changed.

For the explicit scheme, on the other hand, a solution at $(i, n + 1)$ depends only on earlier solutions at $(i - 1, n)$, (i, n), and $(i + 1, n)$. This situation is depicted in Fig. 9–4. The shaded area is called the ***domain of dependence***. The point at the top of the domain depends only on those other points in the domain. In other words, the influence of the boundary conditions experiences a time lag when we use the explicit scheme.

9.2.3   The DuFort–Frankel Method

Recall that we looked at an unstable explicit scheme at the end of §9.1.1. In that scheme, the derivative u_t was expressed by the central-difference approximation

$$
[9.23a] \qquad (u_t)_{i,n} \simeq (u_{i,n+1} - u_{i,n-1})/(2\Delta t)
$$

Let us rewrite our earlier approximation for u_{xx} by using an average for $u_{i,n}$ to get

$$
[9.23b] \qquad (u_{xx})_{i,n} \simeq (u_{i-1,n} - [u_{i,n-1} + u_{i,n+1}] + u_{i+1,n})/(\Delta x)^2
$$

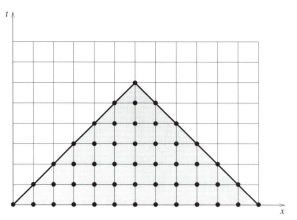

Figure 9–4   Domain of dependence for the explicit method of Eq. [9.13].

The method formed by combining Eq. [9.23a,b] is the ***DuFort–Frankel method***, which is explicit and three-level (in time). The method is unconditionally stable for linear p.d.e.s, and it has a local truncation error

[9.24] $$\tau_{\mathrm{df}} = (\Delta t/\Delta x)^2 u_{tt} + [(\Delta t)^2/6]u_{tt} - [(\Delta x)^2/12]u_{xxxx} + \cdots$$

It suffers from the same boundary-condition time lag as the explicit method, and it requires a special starting procedure because of its three-level nature.

    The DuFort–Frankel method is only conditionally consistent. For example, if Δt and Δx tend to zero in such a way that $(\Delta t/\Delta x)$ is equal to a constant value β, the DuFort–Frankel scheme is actually consistent with the hyperbolic p.d.e.

$$u_t = u_{xx} - \beta^2 u_{tt}$$

Thus, even though the method is unconditionally stable, we must still keep $(\Delta t/\Delta x)$ small to suppress its hyperbolic behavior.

9.2.4 Predictor–Corrector Methods

Predictor–corrector methods are useful when we extend our definition of parabolic equations to nonlinear forms. Consider, for example, the nonlinear prototype equation

[9.25] $$u_{xx} = f(x, t, u, u_x)u_t + g(x, t, u, u_x)$$

The form given in Eq. [9.25] includes the nonlinear diffusion equation and Burger's equation, which are given, respectively, by

$$[\kappa(u)u_x]_x = \alpha(u)u_t; \qquad u_{xx} = (u_t + uu_x)/\nu$$

Let us define the approximations to u_x and u_{xx} by the operators

[9.26a] $$D_x(u_{i,n}) = (u_{i+1,n} - u_{i-1,n})/(2\Delta x)$$

[9.26b] $$D_{xx}(u_{i,n}) = (u_{i-1,n} - 2u_{i,n} + u_{i+1,n})/(\Delta x)^2$$

The Crank–Nicolson scheme for Eq. [9.25] in terms of the operators in Eq. [9.26a,b] is

[9.27]
$$D_{xx}\{[u_{i,n+1} + u_{i,n}]/2\}$$
$$= f(x_i, t_p, [u_{i,n+1} + u_{i,n}]/2, D_x\{[u_{i,n+1} + u_{i,n}]/2\})(u_t)_{i,p}$$
$$+ g(x_i, t_p, [u_{i,n+1} + u_{i,n}]/2, D_x\{[u_{i,n+1} + u_{i,n}]/2\});$$
$$p = n + 1/2$$

We note again that the stability of the C–N scheme is proven only for linear equations. No proof can be established for nonlinear equations; instead, we are relying on the scheme's behavior for linear equations to carry over to nonlinear equations.

Depending on the forms of f and g, we generally have to solve a nonlinear system of equations for u at time level $(n + 1)$. In the predictor–corrector approach, we first predict $u_{i,p}$, denoted by $[u']_{i,p}$. To do so, we replace $[u_{i,n+1} + u_{i,n}]/2$ by $[u']_{i,p}$ on the left side and by $u_{i,n}$ on the right side of the C–N scheme, and use a backward-difference approximation for the time derivative u_t. The predictor equation is then given by

Predictor:

[9.28a] $$D_{xx}[u']_{i,p} = f(x_i, t_p, u_{i,n}, D_x\{u_{i,n}\})(u_{i,p} - u_{i,n})/(\Delta t/2)$$
$$+ g(x_i, t_p, u_{i,n}, D_x\{u_{i,n}\}); \qquad p = n + 1/2$$

Note that the terms $[u']$ at level p now occur linearly and may be obtained by solving a tridiagonal system of equations.

The corrector step consists of the C–N equation with the predicted values $[u']_{i,p}$ replacing $[u_{i,n+1} + u_{i,n}]/2$ in the functions f and g. The corrector step is given by

Corrector:

[9.28b] $$D_{xx}\{[u_{i,n+1} + u_{i,n}]/2\}$$
$$= f(x_i, t_p, [u']_{i,p}, D_x\{[u']_{i,p}\})(u_{i,n+1} - u_{i,n})/\Delta t$$
$$+ g(x_i, t_p, [u']_{i,p}, D_x\{[u']_{i,p}\}); \qquad p = n + 1/2$$

The solutions we seek at level $(n + 1)$ appear linearly and may be obtained by solving another tridiagonal system.

The predictor–corrector method that we have just described is costly, because the two tridiagonal systems to be solved change at each time level. It is most useful when we are interested in accurate transient solutions that are not affected by a lag in the influence of the boundary conditions, or if an explicit scheme is just as costly because it requires a very small time step for stability.

9.2.5   Other Considerations for Parabolic Equations

The explicit scheme of §9.2.1 can easily be extended to p.d.e.s in two or three space dimensions. The basic p.d.e. in two space dimensions is

[9.29] $$u_t = \nabla^2 u = u_{xx} + u_{yy}$$

In this case, the stability condition for the explicit scheme is

[9.30] $$\Delta t[1/(\Delta x)^2 + 1/(\Delta y)^2] \leq 1/2$$

The formulation of the Crank–Nicolson scheme for Eq. [9.29] is also simple to derive. The difficulty caused by having more than one space dimension is that the matrix for u at the new time level is no longer tridiagonal. Although the matrix is block structured, excessive storage and computation time requirements arise when the domain is very large. Iterative methods such as the successive overrelaxation method of Chapter 2 are appropriate for such cases. The starting estimate for the iterative solution at each time level is the solution at the previous level.

Other methods based on the Crank–Nicolson scheme take advantage of tridiagonal solvers. Among these are the ***Peaceman–Rachford Alternating Direction Implicit method*** for two space dimensions, and the ***Douglas–Rachford and Douglas ADI methods*** for two and three space dimensions.

Our earlier prototype of Eq. [9.10a, b, d] dealt only with Dirichlet boundary conditions. We now look at the adjustments that are necessary for Neumann or mixed boundary conditions. Let us consider the general boundary condition given in Eq. [9.10 c] at the left boundary ($x = 0$). The difference equation for the explicit method of §9.2.1 at this point is

[9.31] $$u_{0,n+1} = r u_{-1,n} + (1 - 2r)u_{0,n} + r u_{1,n}$$

The value $u_{0,n}$ exists either from the initial conditions or from the solution at the previous time level; however, the value $u_{-1,n}$ does not exist. To apply the boundary condition of the form

$$a u_x + p u = g(t); \qquad x = 0, t \geq 0$$

we can use a central-difference approximation for $(u_x)_{0,n}$ to write

$$a(u_{1,n} - u_{-1,n})/(2\Delta x) + p u_{0,n} \simeq g(t_n)$$

The solution for $u_{-1,n}$ from this equation is then substituted into Eq. [9.31].

A similar technique is used to write the equation at the left boundary for other methods such as the Crank–Nicolson method. The procedure for the right boundary is also similar.

9.3 ELLIPTIC EQUATIONS

Our discussion of *elliptic equations* will be brief because the basic procedures for solving the difference equations have been treated already in Chapter 2. These procedures are the Jacobi, Gauss–Seidel, and successive overrelaxation (SOR) methods. Our focus will instead be on the formulation of the difference equation for the prototypical *Laplace equation* on a rectangular domain. The equation is

[9.32] $$\nabla^2 u = u_{xx} + u_{yy} = 0; \qquad 0 \le x \le X, 0 \le y \le Y$$

Equations such as Eq. [9.32] are often said to give steady-state solutions. This is because the solution of Eq. [9.32] can be regarded as the solution to the parabolic equation

$$u_t = \nabla^2 u$$

at infinite time if the time-invariant boundary conditions for the parabolic equation are the same as those for the elliptic equation. We can indeed obtain solutions in this fashion; however, we shall look at a more direct approach.

Let us use a rectangular grid with $[(x_i, y_j) = (i\Delta x, j\Delta y)]$ denoting grid intersections and with $u_{i,j}$ denoting $u(x_i, y_j)$. Central-difference approximations for u_{xx} and u_{yy} from Eq. [9.7] give us the expression for $u_{i,j}$ in terms of neighboring points on the grid as

[9.33] $$(u_{i-1,j} - 2u_{i,j} + u_{i+1,j})/(\Delta x)^2 + (u_{i,j-1} - 2u_{i,j} + u_{i,j+1})/(\Delta y)^2 = 0$$

For Dirichlet conditions on all boundaries, we simply solve the system of equations at all interior grid points of the domain. The spectral radius for the Jacobi iterative method for this case is

[9.34a] $$\rho_j = [(\Delta y)^2 \cos(\pi \Delta x / X) + (\Delta x)^2 \cos(\pi \Delta y / Y)]/[(\Delta x)^2 + (\Delta y)^2]$$

and the optimum acceleration parameter ω_{opt} for SOR is

[9.34b] $$\omega_{opt} = 2 / \left[1 + \sqrt{1 - (\rho_j)^2} \right]$$

A boundary with a Neumann or mixed condition may be included in the system of equations to be solved. Such boundary conditions are treated in the same way as we described for parabolic equations in §9.2.5.

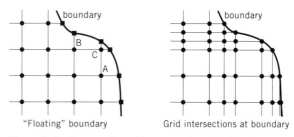

Figure 9–5   Rectangular grids near an irregular boundary.

Other situations that require special treatment occur when there is an irregular boundary. A simple approach to irregular boundaries is either to let the boundary "float" in a prescribed grid, or to modify the grid so that grid intersections occur on the boundary. These options are illustrated in Fig. 9–5.

In the case of the floating boundary with Dirichlet conditions, a new difference approximation is written for u_{xx} (as at point A of Fig. 9–5) or u_{yy} (as at point B of Fig. 9–5) or both (as at point C of Fig. 9–5). These approximations will now be first order instead of second order.

For the case in which the grid intersections are on the boundary, irregular spacings also occur. The difference approximations with such spacings are first order instead of second order. This option can be detrimental when sections of the boundary are nearly parallel to the x or y direction. The spacings in such cases can become highly irregular.

Other options for dealing with irregular boundaries are to formulate the problem in terms of finite-element approximations, or to use coordinate transformations that accommodate the shape of the boundary. Appropriate transformations may be obtained analytically or by numerical grid-generation methods. We shall discuss orthogonal transformations in Section 9.5.

The treatment of the prototype Laplace equation may be extended to problems involving the **Poisson equation**

[9.35] $$\nabla^2 u = -\sigma$$

or the biharmonic equation

[9.36] $$\nabla^4 u = c$$

The computational molecule given in Fig. 9–1 for the biharmonic operator does not possess the properties that guarantee convergence when an iterative method is used to solve the system of difference equations. A direct matrix solver may be used for very small problems. For larger problems, Eq. [9.36] is rewritten as

$$\nabla^4 u = \nabla^2 w = c$$

We solve first for w with the Laplacian operator, and then use a second step to solve for u from the equation

$$\nabla^2 u = w$$

9.4 METHODS FOR HYPERBOLIC EQUATIONS

Hyperbolic equations, like parabolic equations, are used to model propagation problems. As such, we must again be concerned about the stability of a numerical method. Another concern for hyperbolic equations are their characteristics. Consider, for example, the wave equation

[9.37] $$u_{tt} - u_{xx} = 0$$

We see from our discussion in Section 9.1 that the characteristics are given by (dx/dt) equal to (± 1). The role of the characteristics is illustrated in Fig. 9–6. The point in the (x, t) plane at the upper vertex of the triangular region on the left depends on information within the triangle. The triangular region is the ***domain of dependence***. In the figure on the right, the point at the bottom of the shaded region influences the region between the characteristics emanating from that point. The region is therefore known as the ***domain of influence***.

Our discussion of hyperbolic equations will include the Lax–Wendroff and implicit methods for the first-order hyperbolic equation in §9.4.1 and §9.4.2, methods for first-order systems in §9.4.3, and methods for the second-order equation in §9.4.4. Extensions to more than one space dimension will be discussed in §9.4.5.

9.4.1 The Lax–Wendroff Method

The prototypical first-order, hyperbolic p.d.e. is the equation

[9.38a] **P.D.E.** : $$u_t + cu_x = 0; \qquad c = \text{constant} > 0$$

We also specify an initial condition

[9.38b] **I.C.** : $$u(x, 0) = F(x)$$

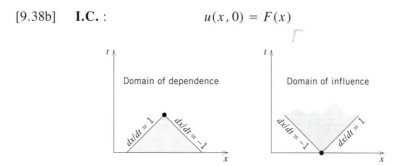

Figure 9–6 Domains of dependence and influence.

The **Lax–Wendroff method** is based on a Taylor series expansion of $u_{i,n+1}$ about the point (i, n), which is used to indicate values at (x_i, t_n). The expansion is

$$u_{i,n+1} = u_{i,n} + \{\Delta t u_t + [(\Delta t)^2/2!]u_{tt} + \cdots\}_{i,n}$$

The p.d.e. itself gives us the relations

$$u_t = -cu_x; \qquad u_{tt} = -cu_{tx} = -c(-cu_x)_x = c^2 u_{xx}$$

We then replace the time derivatives in the Taylor series expansion by space derivatives and truncate the series after the second derivative term to obtain

$$u_{i,n+1} = u_{i,n} + \{\Delta t(-cu_x) + [(\Delta t)^2/2!](c^2 u_{xx})\}_{i,n}$$

Finally, we use central-difference approximations for the space derivatives to obtain the Lax–Wendroff method

$$[9.39] \quad u_{i,n+1} = (1 - p^2)u_{i,n}$$
$$+ [(p^2 + p)/2]u_{i-1,n} + [(p^2 - p)/2]u_{i+1,n}; \qquad p = c\Delta t/\Delta x$$

The Lax–Wendroff scheme is second order in both space and time, and its stability condition is given by

$$[9.40] \qquad\qquad 0 \le p \le 1$$

The characteristic line of the p.d.e. in Eq. [9.38a] is

$$[9.41] \qquad\qquad x - ct = \text{constant}$$

We can argue intuitively that the domain of dependence for the numerical method must contain the domain of dependence for the p.d.e. The domain for the numerical method is formed by the lines of slopes $(\pm\Delta t/\Delta x)$ intersecting at $u_{i,n+1}$. The slope of the line in Eq. [9.41] is $(dt/dx = 1/c)$. For the numerical domain to contain the analytic domain, the slope $(1/c)$ must be at least as steep as the slope $(\Delta t/\Delta x)$; thus we have the condition known as the **Courant–Friedrichs–Lewy (CFL) condition**

$$[9.42] \qquad\qquad 1/c \ge \Delta t/\Delta x$$

The CFL condition coincides with the stability condition for our case study.

Other explicit schemes use a direct differencing. For u_x, the usual approximation is

$$[9.43] \qquad\qquad (u_x)_{i,n} \simeq (u_{i+1,n} - u_{i-1,n})/(2\Delta x)$$

This is combined with any of several formulas for u_t. If we use the forward-difference formula

[9.44a]
$$(u_t)_{i,n} \simeq (u_{i,n+1} - u_{i,n})/\Delta t$$

the resulting scheme is unstable. If, however, we replace $u_{i,n}$ in Eq. [9.44a] by the space-averaged value $([u_{i+1,n} + u_{i-1,n}]/2)$, we obtain

[9.44b]
$$(u_t)_{i,n} \simeq (u_{i,n+1} - [u_{i+1,n} + u_{i-1,n}]/2)/\Delta t$$

The combination of Eq. [9.43] and Eq. [9.44b] is the **Lax method**, which is first order in time.

Yet another explicit method is formed by combining [9.43] with the second-order time approximation

[9.44c]
$$(u_t)_{i,n} \simeq (u_{i,n+1} - u_{i,n-1})/(2\Delta t)$$

This is the three-level **leapfrog method**, which requires a special starting procedure. Both the leapfrog and Lax methods have the same stability condition as the Lax–Wendroff method.

9.4.2   The Wendroff Implicit Method

The **Wendroff method** for the p.d.e. of Eq. [9.38a] is formed by central differences for u_t and u_x about the point $(i + 1/2, n + 1/2)$. A value at space level $(i + 1/2)$ is taken to be the average of values at space levels i and $(i + 1)$; similarly, a value at time level $(n + 1/2)$ is taken to be the average of values at time levels n and $(n + 1)$. The resulting method is

[9.45]
$$(1 - p)u_{i,n+1} + (1 + p)u_{i+1,n+1}$$
$$= (1 + p)u_{i,n} + (1 - p)u_{i+1,n}; \qquad p = c\Delta t/\Delta x$$

This scheme is unconditionally stable, is second order in space and time, and automatically satisfies the CFL condition. There is one very important difference between the implicit Wendroff scheme and the explicit schemes of §9.4.1. The explicit schemes can be used to obtain solutions at a point whose x value is at least $(t\Delta x/\Delta t)$ purely from the initial conditions. The Wendroff scheme requires a left boundary condition to provide closure of the algebraic equations.

Let us assume that we have both an initial condition and a left boundary condition for the Wendroff scheme. Although the scheme is termed **implicit**, new values may be found one at a time without directly solving a system of equations. The sequence for obtaining these values is illustrated in the two steps of Fig. 9–7. A value may be found at any point that is immediately above a point with a known solution and immediately to the right of another point with a known solution.

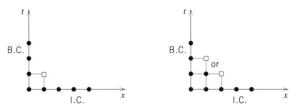

Figure 9–7   New solutions with the Wendroff method.

For negative c values (that is, a p.d.e. with a left running characteristic), the Wendroff scheme for positive time is derived by expanding about $(i - 1/2, n + 1/2)$. An initial condition and a right boundary condition are required in this case. Implicit schemes that are centered in space, such as the analog of the Crank–Nicolson scheme, would require left and right boundary conditions. For equations such as Eq. [9.38a] with only one characteristic, overdetermination or incompatibility with the problem may occur with space-centered methods.

9.4.3   First-Order Systems

The concepts associated with the scalar p.d.e. of Eq. [9.38a] may be extended to systems of the form

[9.46]
$$\mathbf{v}_t + \mathbf{C} \cdot \mathbf{v}_x = \mathbf{h}(x, t) + \mathbf{B} \cdot \mathbf{v}$$

in which $\mathbf{v}$ and $\mathbf{h}$ are m-component column vectors, and $\mathbf{B}$ and $\mathbf{C}$ are $(m \times m)$ matrices. For example, consider the scalar wave equation

[9.47]
$$u_{tt} - u_{xx} = q(x, t)$$

We may put Eq. [9.47] in the form of Eq. [9.46] by defining

$$\mathbf{v} = \begin{bmatrix} u_t \\ u_x \end{bmatrix}; \qquad \mathbf{h} = \begin{bmatrix} q \\ 0 \end{bmatrix}$$

Our system representation of the wave equation is then

$$\mathbf{v}_t + \begin{bmatrix} 0 & -1 \\ -1 & 0 \end{bmatrix} \cdot \mathbf{v}_x = \mathbf{h}$$

The solution of the system for $\mathbf{v}$ gives us u_t and u_x, which are then integrated to obtain u.

Now consider the parabolic equation

$$u_t - u_{xx} = q(x, t)$$

The vector $\mathbf{h}$ is the same as before, but $\mathbf{v}$ is now defined by

$$\mathbf{v} = \begin{bmatrix} u \\ u_x \end{bmatrix}$$

The system equation is then found to be

$$\begin{bmatrix} 1 & 0 \\ 0 & 0 \end{bmatrix} \cdot \mathbf{v}_t + \begin{bmatrix} 0 & -1 \\ -1 & 0 \end{bmatrix} \cdot \mathbf{v}_x = \mathbf{h} + \begin{bmatrix} 0 & 0 \\ 0 & -1 \end{bmatrix} \cdot \mathbf{v}$$

To put this equation in the form of Eq. [9.46], we must premultiply both sides of the equation by the inverse of the first matrix. This step introduces infinite values to our system.

In considering initial-boundary-value problems with systems, we must consider the eigenvalues of the matrix $\mathbf{C}$. We have a correspondence with a positive c in the scalar equation if $\mathbf{C}$ is a positive-definite matrix, and a correspondence with a negative c if $\mathbf{C}$ is a negative-definite matrix; otherwise, we have no counterpart in the scalar equation.

9.4.4   Direct Treatment of the Wave Equation

We can devise methods for dealing directly with the wave equation of Eq. [9.47] instead of using the system approach. A three-level explicit scheme is obtained by using central differences for u_{tt} and u_{xx} about the point (i, n). The result is

$$[9.48] \qquad \begin{aligned} u_{i,n+1} &= p^2(u_{i+1,n} + u_{i-1,n}) + 2(1 - p^2)u_{i,n} \\ &\quad - u_{i,n-1} + (\Delta t)^2 q(x_i, t_n); \qquad p = \Delta t / \Delta x \end{aligned}$$

The scheme of Eq. [9.48] is second order in space and time. Its stability and CFL conditions are coincidentally

$$[9.49] \qquad\qquad\qquad p \leq 1$$

When p is equal to 1, the domains of dependence for the difference equation and the differential equation coincide, and the local truncation error vanishes.

The implicit analog of the explicit method is formed by rewriting $(u_{xx})_{i,n}$ as

$$[9.50] \qquad\qquad D_{xx}([u_{i,n-1} + u_{i,n+1}]/2)$$

where D_{xx} is defined in Eq. [9.26b]. This scheme is unconditionally stable.

9.5   HYPERBOLIC EQUATIONS IN TWO SPACE DIMENSIONS

Equations in two space dimensions are represented by

[9.51a]
$$u_t + c_1 u_x + c_2 u_y = 0$$

[9.51b]
$$u_{tt} = u_{xx} + u_{yy}$$

The methods for the one-dimensional (in space) cases may be extended to these equations. For the first-order p.d.e. of Eq. [9.51a], the CFL condition for the Lax–Wendroff scheme with $(\Delta x = \Delta y = \Delta)$ is now

[9.52]
$$\max(|c_1|, |c_2|)(\Delta t/\Delta) \leq 1/\sqrt{8}$$

Implicit schemes require suitable boundary conditions.

For the second-order p.d.e. of Eq. [9.51b], the explicit scheme corresponding to Eq. [9.48] with $(\Delta x = \Delta y = \Delta)$ has coincident stability and CFL conditions

[9.53]
$$\Delta t/\Delta \leq 1/\sqrt{2}$$

The implicit analog of the explicit method is obtained by averaging space derivatives at the time levels n and $(n + 1)$.

9.6   ORTHOGONAL COORDINATE TRANSFORMATIONS

Much of the success of finite-difference methods depends on suitable computational grids. In this section, we present some fundamental properties associated with orthogonal coordinate systems.

Let (x_1, x_2, x_3) be Cartesian coordinate variables which are transformed to the ξ space according to

[9.54]
$$x_i = x_i(\xi_1, \xi_2, \xi_3); \qquad i = 1, 2, 3$$

Then

$$dx_i = \sum_{j=1}^{3} (\partial x_i/\partial \xi_j) d\xi_j; \qquad i = 1, 2, 3$$

We note here that we are **not** using the summation convention for tensors.

The distance ds from (ξ_1, ξ_2, ξ_3) to $(\xi_1 + d\xi_1, \xi_2 + d\xi_2, \xi_3 + d\xi_3)$ is expressed by

$$(ds)^2 = \sum_{i=1}^{3} (dx_i)^2 = \sum_{i=1}^{3} \left[\sum_{j=1}^{3} (\partial x_i/\partial \xi_j) d\xi_j \right]^2$$

which may be expanded to give

[9.55]
$$(ds)^2 = g_{11}(d\xi_1)^2 + g_{22}(d\xi_2)^2 + g_{33}(d\xi_3)^2$$
$$+ 2\{g_{12}d\xi_1 d\xi_2 + g_{23}d\xi_2 d\xi_3 + g_{31}d\xi_3 d\xi_1\}$$

The quantities g_{ij} (equal to g_{ji}) are the **metrics** of the transformation and are given by

[9.56]
$$g_{ij} = \sum_{k=1}^{3} \{(\partial x_k/\partial \xi_i)(\partial x_k/\partial \xi_j)\}$$

Now let $\mathbf{q}_i$ be a unit vector in the direction x_i and let $\mathbf{e}_i$ be a unit vector in the direction ξ_i. Then $\mathbf{e}_1$ is tangent to the curve on which $(d\xi_2 = d\xi_3 = 0)$, and we have along this vector

$$dx_1 : dx_2 : dx_3 = (\partial x_1/\partial \xi_1)d\xi_1 : (\partial x_2/\partial \xi_1)d\xi_1 : (\partial x_3/\partial \xi_1)d\xi_1$$

Similar ratios can be developed for $\mathbf{e}_2$ and $\mathbf{e}_3$, and we use the ratios to obtain

[9.57]
$$\mathbf{e}_i = \{(\partial x_1/\partial \xi_i)\mathbf{q}_1 + (\partial x_2/\partial \xi_i)\mathbf{q}_2 + (\partial x_3/\partial \xi_i)\mathbf{q}_3\}/h_i;$$
$$h_i = \sqrt{g_{ii}}$$

If we consider a vector $\mathbf{a}$ expressed in the two coordinate systems by

[9.58]
$$\mathbf{a} = a_1\mathbf{q}_1 + a_2\mathbf{q}_2 + a_3\mathbf{q}_3 = \alpha_1\mathbf{e}_1 + \alpha_2\mathbf{e}_2 + \alpha_3\mathbf{e}_3$$

we may relate the components by

[9.59]
$$\begin{bmatrix} \partial x_1/\partial \xi_1 & \partial x_1/\partial \xi_2 & \partial x_1/\partial \xi_3 \\ \partial x_2/\partial \xi_1 & \partial x_2/\partial \xi_2 & \partial x_2/\partial \xi_3 \\ \partial x_3/\partial \xi_1 & \partial x_3/\partial \xi_2 & \partial x_3/\partial \xi_3 \end{bmatrix} \cdot \begin{bmatrix} \alpha_1/h_1 \\ \alpha_2/h_2 \\ \alpha_3/h_3 \end{bmatrix} = \begin{bmatrix} a_1 \\ a_2 \\ a_3 \end{bmatrix}$$

If the ξ system is orthogonal, the dot product of any two different unit vectors is zero; that is,

$$\mathbf{e}_1 \cdot \mathbf{e}_2 = \mathbf{e}_2 \cdot \mathbf{e}_3 = \mathbf{e}_3 \cdot \mathbf{e}_1 = 0$$

Orthogonality also requires that

$$g_{ij} = 0; \quad i \neq j$$

We now consider some fundamental operations in orthogonal systems.

The **divergence** of a vector $\mathbf{a}$ represents the net flux of a quantity out of a volume of dimensions $(d\xi_1 \times d\xi_2 \times d\xi_3)$. It is written as

[9.60]   div $\mathbf{a} = \{\partial(\alpha_1 h_2 h_3)/\partial\xi_1 + \partial(\alpha_2 h_3 h_1)/\partial\xi_2 + \partial(\alpha_3 h_1 h_2)/\partial\xi_3\}/[h_1 h_2 h_3]$

The *curl* of a vector $\mathbf{a}$ indicates path independence when the tangential component of $\mathbf{a}$ is integrated around a closed path. The curl is itself a vector and is written

$$[9.61] \quad \text{curl } \mathbf{a} = [\partial(h_3\alpha_3)/\partial\xi_2 - \partial(h_2\alpha_2)/\partial\xi_3]\mathbf{e}_1/[h_2 h_3]$$
$$+ [\partial(h_1\alpha_1)/\partial\xi_3 - \partial(h_3\alpha_3)/\partial\xi_1]\mathbf{e}_2/[h_3 h_1]$$
$$+ [\partial(h_2\alpha_2)/\partial\xi_1 - \partial(h_1\alpha_1)/\partial\xi_2]\mathbf{e}_3/[h_1 h_2]$$

The *gradient* of a scalar u is the vector quantity

[9.62] $\nabla u = (\mathbf{e}_1/h_1)(\partial u/\partial\xi_1) + (\mathbf{e}_2/h_2)(\partial u/\partial\xi_2) + (\mathbf{e}_3/h_3)(\partial u/\partial\xi_3)$

The Laplacian is then obtained from

$$[9.63] \quad \nabla^2 u = \text{div}(\nabla u) = [1/(h_1 h_2 h_3)]$$
$$\times \{\partial([h_2 h_3/h_1][\partial u/\partial\xi_1])/\partial\xi_1$$
$$+ \partial([h_3 h_1/h_2][\partial u/\partial\xi_2])/\partial\xi_2 + \partial([h_1 h_2/h_3][\partial u/\partial\xi_3])/\partial\xi_3\}$$

The Cartesian forms of the operations given in Eq. [9.60] through Eq. [9.63] are obtained by setting h_i equal to 1 and x_i equal to ξ_i. Two-dimensional forms are derived from the three-dimensional forms by setting h_3 equal to 1 and derivatives with respect to ξ_3 equal to zero.

9.7   CLOSURE

We have barely touched on the vast number of issues associated with partial differential equations. We have presented basic finite-difference methods for simple types of parabolic, elliptic, and hyperbolic equations to be solved on simple domains. With this background, we can deal with problems such as those in conductive heat transfer, potential fields, torsion of shafts, voltage variations in transmission cables, vibrations, and deflections of loaded plates.

The classification of p.d.e.s is typically based on the linear second-order p.d.e. We have extended the classification by looking at how the solution behaves. Thus, propagation problems are generally classified as parabolic or hyperbolic, and steady-state problems are classified as elliptic problems.

The important considerations for parabolic equations are stability and the domain of dependence. The domain for explicit methods causes a lag in the influence of the boundary conditions. Implicit methods such as the Crank–Nicolson method do not suffer from this behavior. Implicit methods generally have better stability qualities than explicit methods, but they are more difficult to use, especially for nonlinear equations or equations in more than one space dimension. Predictor-corrector methods can alleviate the difficulty in dealing with nonlinear problems, and iterative methods may be used for solving the algebraic systems that occur in higher dimensions.

The prototype problem for elliptic equations is the Laplace equation. The methods for this equation are the iterative methods discussed earlier in Chapter 2. The same methods are applicable to the Poisson equation and to the biharmonic equation. The biharmonic problem must be split into two Poisson problems if we are to use iterative methods for its solution.

Hyperbolic equations demand the same attention to stability as parabolic equations. In addition, we must pay attention to the characteristics of the problem, and the Courant–Friedrichs–Lewy condition must be satisfied along with stability. The method used for hyperbolic equations determine what types of conditions are necessary. Explicit methods generally require only initial conditions; implicit methods require one or more boundary conditions depending on the scheme and the characteristics. Methods for the scalar first-order equation may be extended to first-order systems, which are used to represent higher-order scalar equations.

Good computational grids are a necessity for finite-difference methods. To deal with any but the simplest geometries, coordinate transformations will probably be required. We have presented the basic details for non-Cartesian coordinates in three dimensions, and we have presented some fundamental operations in terms of orthogonal systems.

9.8   EXERCISES

1. Two identical bars at different temperatures are brought together (end-to-end). The free ends are kept at their original temperatures, and the bars are insulated along their lengths. With appropriate scaling of variables, we can examine the unsteady temperature distribution *via* the following problem.

$$\text{P.D.E.: } u_t = u_{xx}; \qquad 0 \le x \le 1, t \ge 0$$

$$\text{I.C.: } \quad u(x, 0) = \begin{cases} 0; & 0 \le x < 0.5 \\ 1; & 0.5 < x \le 1.0 \end{cases}$$

$$\text{B.C.: } \quad u(0, t) = 0 \text{ and } u(1, t) = 1; \qquad t > 0$$

In this problem, all variables are nondimensional, u is the temperature, x is the space coordinate, and t is time. The initial u value at x equal to 0.5 is interpreted to be the limiting value of 0.5.

Write codes to obtain the solutions $u(x, 0.012)$ and $u(x, 0.120)$ on the rod, and $u(0.1, t)$ and $u(0.6, t)$ for t values up to 0.12. Use Δx equal to 0.1 with the following methods.

The two-level explicit method with $t = 0.004$.
The two-level explicit method with $t = 0.006$.
The Crank–Nicolson method with $t = 0.006$.

Plot your results appropriately, and discuss them.

2. The general temperature distribution problem for an insulated bar of length L, density ρ, specific heat c_p, and thermal conductivity k is governed by

$$\partial T/\partial \tau = (k/[\rho c_p]) \, \partial^2 T/\partial X^2$$

in which X is position, T is temperature, and τ is time.

Show that the equation can be reduced to the form ($u_t = u_{xx}$) by using

$$x = X/L, t = \tau/[(\rho c_p/k)L^2]; \qquad u = T/T_{\text{ref}}$$

where T_{ref} is a reference temperature.

Repeat the solutions for Problem 1 with the new set of conditions below.

I.C. :
$$u(x,0) = \begin{cases} x; & 0 \le x \le 0.5 \\ 1 - x; & 0.5 < x \le 1.0 \end{cases}$$

B.C. :
$$u(0, t) = u(1, t) = 0; \qquad t > 0$$

3. A prototypical problem for the unsteady temperature distribution between the walls of two concentric cylinders is given by

$$u_t = u_{rr} + u_r/r; \qquad 1 \le r \le 3, t \ge 0$$

in which r is a radius, u is a temperature, and t is time.

Obtain an accurate solution at time t equal to 0.5 for the conditions

I.C. :
$$u(r, 0) = 0; \qquad 1 \le r \le 3$$

B.C. :
$$u(1, t) = e^t - 1 \quad \text{and} \quad u(3, t) = 0; \qquad t > 0$$

4. Consider Burger's equation

$$u_t + u(u_x) = u_{xx}; \qquad 0 \le x \le 1$$

with boundary conditions

$$u(0, t) = u(1, t) = 0; \qquad t > 0$$

Develop the full set of matrix equations for the predictor–corrector analog of the Crank–Nicolson method for Δx values of 0.25.

5. The angle of twist for a shaft in torsion is related to the stress function ϕ, which satisfies Poisson's equation

$$\nabla^2 \phi = -2$$

on the cross section of the shaft with boundary condition ($\phi = 0$) on the perimeter of the section. Solve this problem with a grid spacing approximately equal to 0.1 when the cross section is an equilateral triangle with each side equal to 0.5 in length.

6. A plate is a polygon ABCDEA. The (x, y) coordinates of the vertices are

$$(x_a, y_a) = (0, 0); \quad (x_b, y_b) = (5, 0); \quad (x_c, y_c) = (5, 2)$$
$$(x_d, y_d) = (2, 7); \quad (x_e, y_e) = (0, 7)$$

Solve ($\nabla^2 u = 0$) for the temperature u in the plate with a grid spacing of 0.5 and with the boundary conditions

$$u = 1 \text{ on AE}; \quad u = 0 \text{ on BC and CD}; \quad u_y = 0 \text{ on AB and DE}$$

7. A 30 in. × 40 in. rectangular plate is simply supported at its edges and is under a uniform load q of 1 lb/in². The deflection u of the plate is governed by

$$\nabla^4 u = q/(10^5 \text{ lb} \cdot \text{in})$$

in which the denominator on the right-hand side is the rigidity of the plate.
    Solve the equation for u with boundary conditions ($u = 0$) on the edges. Note that the biharmonic operator may be written as

$$\nabla^4 u = \nabla^2 w; \qquad w = \nabla^2 u$$

The boundary values of w are also equal to zero.

8. Develop Lax–Wendroff schemes for the equations

$$u_t + t u_x = 0; \qquad u_t + u_x - u = 0$$

9. Find the characteristic line for the equation

$$u_t + 2x u_x = 0$$

and find a transformation $x(s)$ so that the equation can be reduced to the form

$$u_t + u_s = 0$$

10. Express the following equation as a first-order system.

$$u_{tt} + u_{xt} - u_{xx} = 0$$

11. Recast the following equation as a first-order system, and find the range(s) of the constant c for which the equation is hyperbolic.

$$(u_t + c u_x)_t + u_t + u_{xx} = 1$$

12. Solve the prototype wave equation

$$u_{tt} = u_{xx}; \qquad 0 \le x \le 1, t \ge 0$$

with the initial conditions

$$u(x, 0) = \sin(\pi x) \text{ and } u_t(x, 0) = 0; \qquad 0 \le x \le 1$$

and the boundary conditions

$$u(0, t) = u(1, t) = 0; \qquad t \ge 0$$

The solution is to be found at t equal to 0.8, and a Δx value of 0.05 is to be used.

13. A Cartesian (x, y) system and another (ξ, η) system are related by

$$x = \xi^2 - \eta^2; \qquad y = 2\xi\eta$$

(a) Show that the (ξ, η) system is orthogonal.
(b) A vector has only an x component equal to c. What are its components in the (ξ, η) system? Are there any points in the (ξ, η) system at which the components are indeterminate?
(c) Express $\nabla\phi$ in the (ξ, η) system at (ξ, η) equal to (3,4).

14. The elliptic coordinate system (ξ, η) is related to a Cartesian (x, y) system by

$$x = c \cosh \xi \cos \eta; \; y = c \sinh \xi \sin \eta$$

in which c is a constant.

(a) Show that the elliptic system is orthogonal.
(b) Express the Poisson's equation $(\nabla^2\phi = -2)$ in the elliptic system.

15. **MacCormack's predictor–corrector method** is a popular one for propagation problems. Suppose we have a p.d.e. of the form

$$u_t = \text{right-hand side}$$

The values of u at time level $(n + 1)$ are predicted as

$$(u_{n+1})_p = u_n + \Delta t [\text{right-hand side}]_n$$

The corrector step is then used to obtain a better approximation of the u values from

$$u_{n+1} = u_n + (\Delta t/2)\{[\text{right-hand side}]_n + ([\text{right-hand side}]_{n+1})_p\}$$

Space derivatives in the right-hand side are computed from central-difference approximations (typically of second order). The values of u that are used in the right-hand side are either the available solutions at time-level n or the predicted solutions at level $(n + 1)$, as appropriate.

Apply MacCormack's method to Problem 3.

10

Design and Optimization

Design is one of the principal activities of the practicing engineer. It refers to the development of an entity (an object, a system, a procedure, etc.) to perform a specified function. Design may be described as the process by which we can provide sufficient information about the entity to permit its realization. After the process is completed, we may also use the word design to mean the result or solution we obtain.

Among the goals of design is optimization of the entity. To achieve that goal, we must have some way of deciding what makes one solution better than another. Such a measure or standard of comparison is known as a ***criterion***. The process is further complicated by the existence of requirements that must be met if the design is to be acceptable. These requirements are the ***constraints*** of the problem.

Optimization with constraints within the context of design is the topic of this chapter. We begin with a formal description of the design process in Section 10.1 and develop the formulation and models for a case study problem in §10.1.1 and §10.1.2. Although the case study is aimed at engineers and is used to illustrate the optimization methods of Section 10.2, it is not so restrictive as to be irrelevant to a wider audience.

A general discussion of the case study is given in Section 10.3. In Section 10.4, we depart from the usual exercises at the end of a chapter. Instead, we pose a set of design problems whose solutions require the concepts of this chapter as well as one or more of the concepts in the previous chapters.

10.1   THE DESIGN PROCESS

The fundamental concepts of the design process are described in the following steps, which are open to relatively broad interpretation depending on the specific problem to be solved.

317

1. Define the problem.
2. Decide what output data are required to specify a solution to the problem.
3. Decide what input data are required to formulate the problem.
4. Develop models that can be used to obtain the required output from the input data.
5. Use the models to obtain a solution for a given set of input data.
6. Evaluate the solution in terms of the constraints and criteria.
7. Repeat as many parts of the process as necessary to obtain a satisfactory solution.

We now develop a problem to be used as a case study for the methods of Section 10.2. The problem description in § 10.1.1 corresponds roughly to Step 1 of the process, and the models for the problem in § 10.1.2 involve Steps 2, 3, and 4.

10.1.1   A Problem Description

Let us suppose that the need to construct a large, parabolic-trough solar concentrator has been established, and that the width s of the trough has been chosen as 40 ft. The trough is to focus the Sun's energy onto a collector pipe whose axis is at the focal point of the parabola.

The initial design problem might have involved sizing of the system and deciding on its construction. A one-piece construction of the trough might have been rejected as being too cumbersome. Suppose instead that the decision from the preliminary design is to form the trough from 1-ft-wide flat panels. Suppose further that the parabolic shape is to be achieved by attaching the panels at their midwidth points to support cables in such a way that the cables form parabolas when they are suspended. The panels are to be aligned with the tangents to the parabola at the attachment points.

A flat layout of the panel system is shown in Fig. 10–1. This layout is a top view of the ***unsuspended*** system; that is, the cable is extended in a straight line.

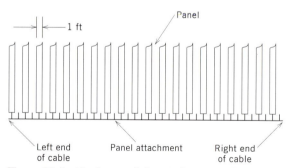

Figure 10–1   Flat layout of the panel system with alternate panels shown.

For clarity, we have shown only one cable (the other is parallel to it and on the other side of the panels), and we have removed alternate panels; however, we have retained all of the attachments. The distances between the attachment points are **not** equal in this configuration.

The parabolic shape is formed when the cables are suspended so that the horizontal distance between their left and right ends is equal to the width s of the trough. In the suspended configuration, the left end A is 20 ft above the ground and the right end B is at **most** 20 ft above the ground. The horizontal spacing between the panel attachments is constant and equal to 1 ft in this configuration, because a uniform loading of the cable is required to produce the parabolic shape. An end view of the suspended system is given in Fig. 10–2.

The geometry in Fig. 10–2 includes two coordinate systems. The (p, q) coordinate system is a physical one in which the p axis coincides with the ground and the q axis passes through the left endpoint A. If we consecutively number the panels from 1 to 40 starting from the left, uniform loading requires the p coordinate of panel j to be $(j - 0.5)$ ft. The (x, y) system is an auxiliary one with its origin at the lowest point of the parabola. This origin is located by the horizontal distance d and the height h, and the focal length of the parabola is denoted by m; the collector pipe is to be located at the focal point.

The design operating condition of the system is when the Sun is directly overhead. Because incident solar power is lost through the gaps between the panels (as seen in the magnified view in Fig. 10–2), the concentrator is not as efficient as one that consists of a one-piece trough. The available solar power is proportional to the total horizontal projection of the panel widths, and we may therefore define the collecting efficiency η of the trough as the ratio of the total horizontal projection to the span s. The objective of the design is to maximize the collecting efficiency η.

The preceding background includes a broad definition of the problem we wish to consider. We now look at additional specifications and provide a descriptive summary of the definition.

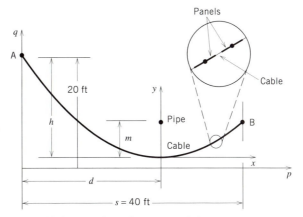

Figure 10–2   End view of the suspended system.

We have already specified the width s of the trough to be 40 ft, the height of end A in Fig. 10–2 to be 20 ft, and the height of end B to be at most 20 ft. The collector pipe must obviously be within the 40-ft span and must be above the lowest point of the parabola. We impose a further condition that it is to be no higher than 18 ft from the ground. The loading of the cable is assumed to be from the weights of the panels; the weight of the cable itself is neglected. The maximum cable tension is not to exceed 900 lb. The load w at each of the forty panel attachments is estimated to be 25 lb. The load distribution is therefore uniform at 25 lb per horizontal ft. The problem is now summarized as follows.

- **Main Objective and Criterion**: To provide sufficient detail about the location of the attachment points of the panels to permit construction of a parabolic-trough solar concentrator. The collecting efficiency η is to be maximized.

- **Constraints**: Endpoint A of a cable is to have physical coordinates (p, q) equal to (0,20) ft and is to be no lower than the other endpoint B. The focal point of the parabola is to be no higher than 18 ft from the ground and is not to fall outside the span s between the endpoints A and B. Cables must clear the ground and are limited to a maximum tension of 900 lb.

We note here that although the problem we have just summarized corresponds to Step 1 of our procedure, it represents the start of only one stage of the design and not the start of the entire design process. Rather, this stage is the result of a preliminary stage (to which we have alluded) that involved sizing and decisions about construction.

10.1.2   Modeling the Problem

We now consider the output and input data for the problem and the models that link them; that is, Steps 2, 3, and 4 of the design process. An obvious set of output data consists of the locations of the attachment points. We can define the location of the attachment point for panel j as the distance λ_j along the cable from the endpoint A. We also need the focal length m and the maximum tension $T_{\max}$ to evaluate a solution in terms of the constraints, and we need the efficiency η to implement the criterion. Other useful output data are the cable length L, the height q_b of endpoint B, and the panel locations in physical (p, q) coordinates to describe the shape of the parabolic trough.

The input data are the variables that drive the models to produce the required output; the input data that optimize the problem may also be considered as a part of the output. Two variables that are sufficient to describe the parabola are the distances d and h shown in Fig. 10–2. Other variable pairs may be chosen; however, d and h are easily understood quantities with physical significance. Furthermore, we may satisfy the remaining constraints regarding ground clearance, the elevation of endpoint B, and the horizontal location of the focal point by placing simple restrictions directly on the possible values of d and h. The other input data are the panel load w (25 lb), the span s (40 ft), and the p coordinates for the panel attachments so that the horizontal loading is uniform. Recall that these coordinates are given by

[10.1] $$p_j = j - 0.5; \qquad j = 1, 2, \ldots, 40$$

We are now ready to look at the models that allow us to obtain the output data for input values of d and h. We begin by noting that the focal point is located directly above the lowest point on the parabola, and that the minimum value of d is $s/2$ when the endpoints A and B are at the same elevation. When B is lower than A, the lowest point on the parabola is to the right of the midspan location. We also note that a zero value of h yields a horizontal cable whose focus is at infinity. Direct imposition of the conditions

[10.2] $$s/2 \leq d \leq s; \qquad 0 < h < 20 \text{ ft}$$

ensures a finite focal length and satisfies the constraints that A must be no lower than B, that the focal point must lie somewhere on the span s, and that the cable must clear the ground. These conditions do not necessarily satisfy the other constraints of the problem.

The model for the parabola in the auxiliary (x, y) coordinates of Fig. 10–2 is

[10.3] $$y = x^2/(4m)$$

The endpoint A has (x, y) coordinates equal to $(-d, h)$ and the endpoint B has (x, y) coordinates equal to $(s - d, y_b)$, with y_b to be determined as part of the design. We use these data and Eq. [10.3] to obtain the focal length m and the coordinate y_b as

[10.4] $$m = d^2/(4h); \qquad y_b = (s - d)^2/(4m)$$

The constraint on the height of the focal point may be expressed in terms of h and m as

[10.5] $$h - m \geq 2 \text{ ft}$$

The models related to the cable tension are derived from the principles of mechanics. For a uniform load w per horizontal foot, the horizontal component of tension is constant and is equal to

$$T_o = 2mw$$

The tension T at any point on the cable is tangent to the cable at an angle θ to the horizontal. T is then related to T_o by

$$T = T_o/(\cos \theta)$$

in which θ is obtained by differentiating Eq. [10.3] to get

$$\theta = \tan^{-1}[dy/dx] = \tan^{-1}[x/(2m)]$$

The maximum tension occurs when θ has the largest magnitude; that is, at the highest point A on the cable. The constraint on the maximum tension may therefore be written as

[10.6] $$T_{\max} = 2mw/\cos(\theta_a) = wd\sqrt{1 + [d/(2h)]^2} \le 900 \text{ lb}$$

The x coordinates corresponding to the physical p coordinates of Eq. [10.1] are

[10.7] $$x_j = p_j - d; \quad j = 1, 2, \ldots, 40$$

and the corresponding y coordinates from Eq. [10.3] are

[10.8] $$y_j = (x_j)^2/(4m); \quad j = 1, 2, \ldots, 40$$

The relation between the q and y coordinates gives q_b and q_j as

[10.9] $$q_b = y_b + 20 - h; \quad q_j = y_j + 20 - h, \quad j = 1, 2, \ldots, 40$$

The horizontal projection of the 1-ft width of panel j is simply the width multiplied by the cosine of θ at (x_j, y_j); it is denoted by

[10.10] $$\xi_j = \cos(\tan^{-1}[x_j/(2m)]); \quad j = 1, 2, \ldots, 40$$

The collecting efficiency η is then computed from

[10.11] $$\eta = (1/s)\sum_{j=1}^{40}(\xi_j)$$

The remaining output quantities are the arclengths λ_j from the endpoint A to the panel attachment points, and the cable length L. They may be found from the arclength integrations

$$\lambda_j = \int_{-d}^{x_j}\sqrt{1 + (dy/dx)^2}\,dx; \quad L = \int_{-d}^{s-d}\sqrt{1 + (dy/dx)^2}\,dx$$

The results of the integrations are

[10.12] $$\lambda_j = f(x_j) - f(-d); \quad L = f(s - d) - f(-d)$$

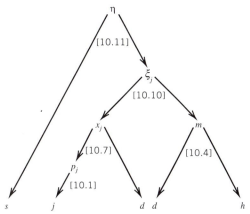

Figure 10–3   A task tree for computing the criterion function η.

in which function $f(x)$ is defined by

[10.13] $f(x) = x\zeta(x)/(4m) + m \ln[x + \zeta(x)]; \qquad \zeta(x) = \sqrt{(x^2 + 4m^2)}$

We have generated a large number of models for this problem. Some of them are intermediate models that are used to derive other models. To work our way through them to get to the required output, we use a ***top–down analysis.*** Such an analysis begins with what we want and proceeds through the appropriate models until we arrive at the input data. The analysis for computing the criterion function η is summarized by the task tree in Fig. 10–3. The quantity η at the top is the root of the tree (the tree is upside-down), and the lower extremities of the tree are the input data. The arrows show how one quantity depends on others that are lower in the tree. The computation of η is done in reverse order from the given data in what is known as ***bottom–up implementation.***

The constraints of the problem must also be verified. We use Eq. [10.5] with the values of h and m to check the focal point constraint, and we compute $T_{\max}$ from w, d, and h to check the tension constraint in Eq. [10.6]. Recall that the other values of interest were the cable length L and the arclengths λ_j, the height q_b of end B of the cables, and the physical coordinates (p, q) of the suspended system. L and λ_j are found from Eq. [10.12] and Eq. [10.13], q_b is found from Eq. [10.4] and Eq. [10.9], and the coordinates (p, q) are found from Eq. [10.1], Eq. [10.8], and Eq. [10.9].

At this point, we can perform Step 5 of the design process by writing and testing a code to produce the required solutions when the input data satisfy Eq. [10.2], Eq. [10.5], and Eq. [10.6]. We may use the code to find the input values of d and h that maximize the criterion function $\eta(d, h)$. Methods of optimizing the solution are discussed in Section 10.2.

10.2   OPTIMIZATION METHODS

Optimization of a design often involves Steps 5, 6, and 7 of the design process. It typically requires us to determine a solution for a given set of input data, and to iterate to an optimum based on evaluations of previous solutions.

We remark here that there are a variety of methods for the unconstrained global optimization problem. Some of them involve searches that require evaluation of only the criterion function. The more classical approach requires a solution of the system of equations that is obtained by setting the partial derivatives of the criterion function to zero. The root-solving methods of Chapter 3 would be useful with this approach.

The presence of constraints rules out the direct zero-derivative approach because the optimum solution of the constrained problem does not generally coincide with a stationary point of the criterion function. Instead, we often find an optimum value with nonzero derivatives at a point on one or more of the constraint boundaries.

The methods presented in §10.2.1 and §10.2.2 are search methods that require evaluation of the criterion function. Lagrange multipliers are discussed in §10.2.3 both as an alternative to search methods and as a means of explaining why the constraint boundary search of §10.2.2 is useful. Other strategies involving global optimization and gradient methods are discused in §10.2.4.

10.2.1   Exhaustive Searching

The exhaustive search is a "brute force" approach in which relatively high resolution is used to search a region for an optimum solution that satifies the constraints. In our case study, possible boundaries for a search region are given by Eq. [10.2]. A look at Eq. [10.5] quickly allows us to reduce the range of h even further to (2 ft $< h <$ 20 ft) because m is positive. A code for our case study is given in Pseudocode 10–1. It includes only the processing required for the constraints and the criterion function.

If we use a resolution of 0.5 ft for both d and h, the input values would be

$$d_{min} = 20.0, d_{max} = 40.0, n_d = 41; \qquad h_{min} = 2.5, h_{max} = 19.5, n_h = 35$$

The search would therefore be conducted on 1435 points. The results of such a search yield a maximum η value of 0.8951 at $(d_{opt}, h_{opt}) = (26.5, 14.5)$ ft. Of the 1435 points searched, only 220 did not violate at least one constraint. A contour map of the η values on a part of the search region is shown in Fig. 10–4. A search point (d, h) is actually represented by an area bounded by $(d \pm 0.25, h \pm 0.25)$.

It is clear that the exhaustive search is inefficient. Improved efficiency may be achieved by a variation in which successive searches are performed starting with a low resolution grid and ending on one with high resolution. The resolution of the initial search should still be good enough to retain the behavioral characteristics of the criterion function. Table 10–1 shows the details and results of three successive searches to obtain the same results as before in a more efficient manner.

Pseudocode 10–1   Exhaustive search for an optimum solution

Initialize: s, w

Initialize: d_{min} \ minimum d value
$\qquad\quad d_{max}$ \ maximum d value
$\qquad\quad n_d$ \ number of d values from d_{min} to d_{max}

Initialize: h_{min} \ minimum h value
$\qquad\quad h_{max}$ \ maximum h value
$\qquad\quad n_h$ \ number of h values from h_{min} to h_{max}

$\quad\eta_{max} \leftarrow 0$ \ initial reference
$\quad\Delta d \leftarrow (d_{max} - d_{min}/(n_d - 1)$
$\quad\Delta h \leftarrow (h_{max} - h_{min})/(n_h - 1)$
For $[k_d = 1, 2, \ldots, n_d]$
$\quad d \leftarrow d_{min} + \Delta d (k_d - 1)$

$\quad$**For** $[k_h = 1, 2, \ldots, n_h]$
$\qquad h \leftarrow h_{min} + \Delta h (k_h - 1)$
$\qquad m \leftarrow$ from Eq. [10.4]

$\qquad$**If** [**not** $([h - m] < 2)$] **then**
$\qquad\quad T_{max} \leftarrow$ from Eq. [10.6]

$\qquad\quad$**If** [**not** $(T_{max} > 900)$] **then**

$\qquad\qquad \eta \leftarrow 0$
$\qquad\qquad$**For** $[j = 1, 2, \ldots, 40]$
$\qquad\qquad\quad p_j \leftarrow$ From Eq. [10.1]
$\qquad\qquad\quad x_j \leftarrow$ From Eq. [10.7]
$\qquad\qquad\quad \xi_j \leftarrow$ From Eq. [10.10]
$\qquad\qquad\quad \eta \leftarrow \eta + \xi_j$
$\qquad\qquad$**End For** $\qquad$ \ with j counter
$\qquad\qquad \eta \leftarrow \eta / s$

$\qquad\qquad$**If** $[\eta > \eta_{max}]$ **then**
$\qquad\qquad\quad \eta_{max} \leftarrow \eta \qquad$ \ set new maximum for η
$\qquad\qquad\quad d_{opt} \leftarrow d \qquad$ \ set corresponding d value
$\qquad\qquad\quad h_{opt} \leftarrow h \qquad$ \ set corresponding h value
$\qquad\qquad$**End If** $\qquad$ \ for resetting η_{max}
$\qquad\quad$**End If** $\qquad$ \ for tension constraint
$\qquad$**End If** $\qquad$ \ for focal point constraint
$\quad$**End For** $\qquad$ \ with k_h counter
End For $\qquad$ \ with k_d counter
Write: $d_{opt}, h_{opt}, \eta_{max}$

Table 10–1 Results of the Optimization with Successive Searches

Search number:	1	2	3
d_{min} [ft]:	20	20	23
d_{max} [ft]:	40	28	27
n_d:	11	9	9
h_{min} [ft]:	4	10	12
h_{max} [ft]:	18	18	16
n_h:	8	9	9
resolution [ft]:	2.0	1.0	0.5
η_{max}:	0.8793	0.8892	0.8951
d_{opt} [ft]:	24.0	25.0	26.5
h_{opt} [ft]:	14.0	14.0	14.5
valid points:	13	44	46
total points:	88	81	81

We see in Table 10–1 that a total of 250 points were examined in the three searches, and that the total number of valid points was 103 (compared with 1435 examined and 220 valid in the original search). The rule of thumb that was used for the second and third searches was to set the limits of d and h to two grid spaces on either side of the optimum values from the previous search. The adequacy of this rule is partly due to the fact that multiple local maxima for η do not exist in the region of valid d and h values. The existence of a single maximum is verified by the contour map of Fig. 10–4.

Given the behavior of η, the rule of thumb may appear to be too conservative; that is, it produces an excessively large region for the next search. A brief examination of the the results will explain why the rule of thumb is necessary and will set the stage for the discussion of Section 10.2.2.

We see from Fig. 10–4 that the maximum η value is limited by the constraints. Such behavior is typical in an engineering application where a constraint has a real significance. Weaker constraints that do not limit the optimum also exist, however. These are still useful, because they limit the region on which we must search for an optimum solution.

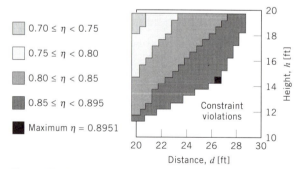

Figure 10–4 Contour map of the criterion function η on a (d, h) region.

The optimum solution found by any of the searches will tend to be at a search point near the constraint boundaries as resolution is increased. Because these boundaries are not necessarily aligned with the lines of the search grid, we cannot rely on the true optimum to be within one grid spacing of the current optimum. We must therefore be careful not to reduce the search region too quickly and exclude the location of the true optimum. The need for such care is easily seen in the results of Table 10–1. The optimum d from the second search was 25 ft with a resolution of 1 ft; the optimum d from the third search was more than 1 ft away at 26.5 ft.

The results of the searches point out properties that can be used to advantage in constrained optimization problems; namely, in a problem whose optimum solution is limited by the constraints, the search for the optimum can be restricted to the constraint boundaries. We shall look at this approach in § 10.2.2.

10.2.2   Searching on Constraint Boundaries

The problem to be solved in our case study is the maximization of the criterion function $\eta(d, h)$ subject to the constraints of Eq. [10.2], Eq. [10.5], and Eq. [10.6]. We assume that the solution we seek will lie somewhere on the constraint boundaries. The preliminary step in solving the problem is the determination of those boundaries.

The boundaries corresponding to Eq. [10.2] are the lines $d = 20$, $d = 40$, $h = \epsilon$, and $h = 20 - \epsilon$, with ϵ as an arbitrarily small, positive quantity. The boundaries corresponding to Eq. [10.5] and Eq. [10.6] are represented by equalities in the constraints and may be expressed in functional form by

[10.14] $$z_1(d, h) = h - d^2/(4h) - 2 = 0$$

[10.15] $$z_2(d, h) = (wd)^2(1 + [d/(2h)]^2) - 900^2 = 0$$

It is a simple matter to obtain d in terms of h from Eq. [10.14] and h in terms of d from Eq. [10.15]. The resulting constraint boundaries are

[10.16] $$d = 2\sqrt{h^2 - 2h}$$

[10.17] $$h = (wd^2/2)/\sqrt{900^2 - (wd)^2}$$

The boundaries from Eq. [10.16] and Eq. [10.17] are plotted on the map of Fig. 10–5. Solutions that do not violate the constraints are possible in the region PQRS of the map. Let us use subscripts p, q, r, and s to refer to the points P, Q, R, and S; a search along the boundaries may be performed conveniently in four segments as follows.

1. From d_p to d_q with h equal to $20 - \epsilon$.
2. From d_q to d_r with h from Eq. [10.17]
3. From h_r to h_s with d from Eq. [10.16]
4. From h_s to h_p with d equal to 20

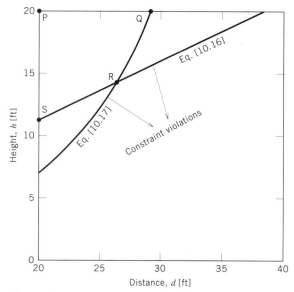

Figure 10–5   Map of the constraint boundaries.

We must first determine the intersection coordinates d_p, d_q, d_r, h_r, h_s, and h_p. If we choose ϵ as 0.001 ft, we immediately have

$$d_p = 20.000 \text{ ft}; \qquad h_p = 19.999 \text{ ft}$$

The value of d_q is found from the quadratic equation in d^2 that results from Eq. [10.17] when h is 19.999 ft, and the value of h_s is found from the quadratic equation in h that results from Eq. [10.16] when d is 20 ft. The results are

$$d_q = 29.108 \text{ ft}; \qquad h_s = 11.050 \text{ ft}$$

Finally the values of d_r and h_r are those that simultaneously satisfy Eq. [10.16] and [10.17]. Substitution of the expression for d in Eq. [10.16] into Eq. [10.17] and subsequent algebraic manipulation yield another quadratic equation in h. This equation is solved for h_r, which is then used in Eq. [10.16] to obtain d_r. The results are

$$d_r = 26.399 \text{ ft}; \qquad h_r = 14.238 \text{ ft}$$

The last digit in each of the six coordinates is chosen so that inexact representations do not cause violations of the constraints. The effects of finite precision should also be kept in mind as we search along the constraint boundaries. In practice, the h value for the QR segment of the search would be increased slightly from the value given by Eq. [10.16], and the d value for the RS segment would be decreased slightly from the value given by Eq. [10.17]. The effect of these modifications is to bring the search point far enough within the constraint boundaries so that constraint checks do not indicate a false violation because of small precision errors.

The technique of successive searches used for a region in §10.2.1 may be applied to the constraint boundaries. The result of a search along the boundaries shows that the maximum efficiency η is 0.89725 at the intersection R where d is equal to 26.399 ft and h is equal to 14.238 ft. The computer output for the final design is given in Table 10–2.

The occurrence of the optimum solution at the intersection of two constraint boundaries means that both of the constraints play a role in limiting the collecting efficiency. If we had been certain *a priori* that this was the case, we could have looked directly at the intersection point R. Unfortunately, we cannot always assume that an optimum will occur at boundary intersections. For example, if we modify the problem to allow a maximum tension of 1200 lb instead of the previous 900 lb, the boundary similar to the QR segment of Fig. 10–5 would be moved to the right. We would then find an optimum on the left of the intersection point at (d, h) approximately equal to (28, 15) ft; that is, only the focal point constraint would limit the efficiency.

10.2.3   Lagrange Multipliers

Another approach that we shall consider for the problem of optimization with constraints is the ***Lagrange multiplier method***, whose theory is described in any text on advanced calculus (Ref. 18). We shall again use the case study developed in Section 10.1 to discuss the interpretation of the method.

The general problem for the Lagrange multiplier method is to optimize a function of n variables subject to m constraints with m no greater than n. Let the criterion function and the constraints be expressed as

Function : $f(x_1, x_2, \ldots, x_n)$

Constraints : $z_j(x_1, x_2, \ldots, x_n) = 0;$ $j = 1, 2, \ldots, m \leq n$

The procedure for optimizing the criterion function begins with forming a new function G from the criterion and constraint functions according to

$$G(x_1, x_2, \ldots, x_n, \lambda_1, \lambda_2, \ldots, \lambda_m) = f(x_1, x_2, \ldots, x_n) + \sum_{j=1}^{m} [\lambda_j \, z_j(x_1, x_2, \ldots, x_n)]$$

in which the as yet undetermined constants λ_j (not to be confused with the arclengths λ_j of Eq. [10.13]) are the Lagrange multipliers. The system of equations

Table 10–2   Computer Output for the Final Design of the Solar Concentrator

```
          Input value for the distance d [ft]:   26.3990
          Input value for the height h [ft]:     14.2380

              Height of focal point [ft]:        17.9987
              Maximum cable tension [lb]:        899.9512

                      Cable length [ft]:         45.1396
                    Height of endpoint B [ft]:    9.5413


   Panel    p [ft]      q [ft]     lambda [ft]

     1       0.5        19.466       0.732
     2       1.5        18.428       2.173
     3       2.5        17.431       3.585
     4       3.5        16.475       4.969
     5       4.5        15.560       6.324
     6       5.5        14.685       7.653
     7       6.5        13.852       8.954
     8       7.5        13.059      10.231
     9       8.5        12.307      11.482
    10       9.5        11.596      12.709
    11      10.5        10.926      13.912
    12      11.5        10.297      15.094
    13      12.5         9.709      16.254
    14      13.5         9.161      17.394
    15      14.5         8.655      18.515
    16      15.5         8.189      19.619
    17      16.5         7.764      20.705
    18      17.5         7.380      21.776
    19      18.5         7.037      22.834
    20      19.5         6.734      23.879
    21      20.5         6.473      24.912
    22      21.5         6.252      25.936
    23      22.5         6.073      26.952
    24      23.5         5.934      27.962
    25      24.5         5.836      28.967
    26      25.5         5.779      29.969
    27      26.5         5.762      30.969
    28      27.5         5.787      31.969
    29      28.5         5.852      32.971
    30      29.5         5.958      33.977
    31      30.5         6.106      34.988
    32      31.5         6.294      36.006
    33      32.5         6.522      37.031
    34      33.5         6.792      38.067
    35      34.5         7.103      39.114
    36      35.5         7.454      40.174
    37      36.5         7.847      41.249
    38      37.5         8.280      42.339
    39      38.5         8.754      43.445
    40      39.5         9.269      44.570
   Collecting Efficiency, ETA: 0.8972   <= = = = =
```

$$\partial G/\partial x_i = 0; \qquad i = 1, 2, \ldots, n$$
$$\partial G/\partial \lambda_j = 0; \qquad j = 1, 2, \ldots, m$$

is then to be solved for the x_i and λ_j values. There may be more than one set of solutions; nevertheless, the x_i values that optimize $f(x_1, x_2, \ldots, x_n)$ are among the possibilities.

We note that the partial derivatives of G with respect to λ_j are simply the functions $z_j(x_1, x_2, \ldots, x_n)$ that specify the constraints. Therefore, if m is equal to n, the optimum solution will occur at the intersection of all of the constraints. In this case, the Lagrange multiplier method is equivalent to finding x_i from the reduced system of equations

$$z_j(x_1, x_2, \ldots, x_n) = 0; \qquad j = 1, 2, \ldots, m = n$$

Let us consider the application of the Lagrange multiplier method to the case study. Because the criterion function $\eta(d, h)$ depends on two variables, we can deal with no more than two constraints at a time. The case study has four significant constraint boundaries as seen in Fig. 10–5. We must then select the two boundaries that we expect to limit the solution. If we suppose that the two boundaries are those for the focal length and the maximum tension given by Eq. [10.15] and [10.16], the system to be solved consists of

$$\partial G/\partial d = 0 = \partial G/\partial h$$
$$z_1(d, h) = 0 = z_2(d, h)$$

with G defined as

$$G(d, h, \lambda_1, \lambda_2) = \eta(d, h) + \lambda_1 z_1(d, h) + \lambda_2 z_2(d, h)$$

We see immediately that the solution is at the intersection of the two constraint boundaries (as we had found earlier).

Suppose now that we allow the maximum tension to be 1200 lb. If we use the same formulation of the Lagrange multiplier problem, we would again obtain a solution at the (new) intersection of the constraint boundaries, but this solution would not correspond to the maximum value of η that is permitted in the design problem!

It is important to distinguish between the design problem formulation and the Lagrange multiplier formulation. The modified design problem is to maximize η so that T_{max} does not **exceed** 1200 lb and so that $(h - m)$ is not **less** than 2 ft. The Lagrange multiplier problem maximizes η so that T_{max} is **equal** to 1200 lb and $(h - m)$ is **equal** to 2 ft. Therefore, when we choose to use Lagrange multipliers, we are already presuming that the optimum solution is actually **on the boundaries** of the chosen design constraints.

The modified problem with 1200 lb as the upper limit on T_{max} has an optimum solution on the boundary given by Eq. [10.16]. The Lagrange multiplier problem is then to solve the 3-equation system

[10.18] $$\partial G / \partial d = 0; \quad \partial G / \partial h = 0; \quad z_1(d, h) = 0$$

with

[10.19] $$G(d, h, \lambda_1) = \eta(d, h) + \lambda_1 z_1(d, h)$$

The solution of Eq. [10.18] may be accomplished by the extended Newton–Raphson method of Chapter 3. That method will require evaluation of both first and second partial derivatives of η and z_1 with respect to d and h. Derivatives of z_1 can easily be expressed analytically, but those for η would have to be found numerically from difference formulas as discussed in Chapter 6.

The alternative to a direct solution of Eq. [10.18] is to search along the constraint boundary as described in §10.2.2. Because the definition in Eq. [10.19] already presumes that the optimum is on the constraint boundary described by $[z_1(d, h) = 0]$, we can restrict our search to that boundary. The choice between a direct solution and a boundary search is dictated by ease of implementation, computational effort required, and the desired level of accuracy.

10.2.4   Other Strategies

Searching for an optimum on a constraint boundary is more efficient than searching in the region bounded by the constraints. However, we do not know *a priori* if the optimum will occur on the boundaries. One strategy for dealing with this situation is to look first for a solution on the boundary, and then to attempt a global optimization. If global optimization takes us outside the constraint boundaries, the boundary solution we found earlier is the one we want.

As we have stated earlier, a global optimum occurs at a stationary point where the partial derivatives of the criterion function with respect to the problem variables are zero. To solve such a problem by the Newton–Raphson method involves computations of the second partial derivatives.

For complicated problems, it is best to use numerical differentiation to compute the derivatives of the criterion function. Three function values are required for every second derivative if we accept second-order accuracy. Nevertheless, a quickly converging method may be less costly than an exhaustive search.

Another approach to global optimization is the **steepest gradient method**. Let the criterion function be

Function: $$f(x_1, x_2, \ldots, x_n)$$

and let the unit vector in the direction x_i be $\mathbf{e}_i$. The **gradient** of f and its magnitude are

$$\nabla f = \sum_{i=1}^{n} (\partial f / \partial x_i) \mathbf{e}_i; \quad |\nabla f| = \sqrt{(\partial f / \partial x_1)^2 + (\partial f / \partial x_2)^2 + \cdots + (\partial f / \partial x_n)^2}$$

For maximization problems, we move from a current location in the direction of steepest ascent; that is, we move by an amount $(c\nabla f / |\nabla f|)$, where c is a positive value. For minimization problems, we move in the opposite direction. The difficulty with gradient methods is in choosing c. Too small a value will require many steps to reach the solution, and too large a value will result in (oscillating) overshoot or divergence.

The steepest gradient method usually converges quickly at first, but then slows down as the target is approached. The Newton–Raphson method converges rapidly once we are in the vicinity of the solution. We might therefore combine the two methods, starting with the gradient method and switching to the Newton–Raphson method later. In any case, we abandon the global optimization attempt as soon as it is clear that it will take us outside the constraint boundaries.

Other optimization methods involve only a global optimization scheme. In these cases, the criterion function is modified by a **penalty function** whenever the constraints are violated. For example, we add a negative penalty to the criterion function for the maximization problem. The design of penalty functions that exhibit good behavior under global optimization methods is not trivial and is not treated here.

10.3   GENERAL DISCUSSION

The design problem used as our case study is relatively simple. All models were analytically expressed, and numerical methods were required only in the search for an optimum solution. More complicated problems would generally require the use of numerical or even experimental methods to obtain some of the required data. An inefficient optimization process can be very costly in those cases.

The case study exercise illustrates the role of the constraints in a typical design problem. A good understanding of the design problem is essential in identifying the constraints that are likely to limit the criterion function; the optimization effort can then be minimized by focusing the search for an optimum solution on the boundaries of those constraints.

An example of identifying the important constraints is available from the case study. We know that an ideal collecting efficiency η equal to 1 occurs only when the concentrator panels cover the entire span of 40 ft; that is, when the cable is attached to two endpoints at the same elevation and has no sag. This configuration would obviously produce a focal point at infinity and an infinite cable tension. We must accept some loss in collecting efficiency to achieve realistic values of the focal height and cable tension. It is then easy to see that the constraints on the focal point and the cable tension are the ones that are potentially significant in limiting the criterion function.

Although the most significant constraints can be identified, we may not be able to conclude that they simultaneously limit the criterion function unless we have some prior experience with the problem or evidence from preliminary investigations. If we do have such experience or evidence, we can restrict our search to the intersections of the constraint boundaries; otherwise, we must search on the entire boundaries.

A useful approach to efficient searching is the use of successive searches with increasing resolution. The results shown in Table 10–1 demonstrate the effectiveness of this strategy. However, as we have stated earlier, the resolution for the initial search must be good enough to capture the behavioral characteristics of the criterion function.

One alternative to the search approach is the Lagrange multiplier method. The formulation of the system of equations for that method contains implicit assumptions about the roles of the constraints. The constraints of the Lagrange multiplier problem are not necessarily the constraints of the design problem; rather, they are the design problem's constraint boundaries on which the optimum solution is expected to lie. The number of constraints that can be included in any Lagrange multiplier formulation is also limited by the number of independent variables of the criterion function.

After an optimum solution is found on the constraint boundaries, an attempt at global optimization may be carried out by a method such as the Newton–Raphson or steepest gradient method. The global attempt is abandoned if it takes us outside the constraints.

In closing, we state once again that the case study we have considered is only one part of the overall design problem. The iteration in Step 7 of the design process applies to the entire design and not only to the optimization part of it. In general, the solutions obtained by choosing a particular approach may fall short of expectations, or the process used to find a solution may show characteristics that lead to variations in approach or modifications of the constraints and criteria.

10.4 DESIGN PROJECTS

We now provide a set of design problems that involve optimization with constraints and at least one of the numerical methods that are discussed in the text. These problems replace the usual exercises at the end of a chapter and are drawn from various engineering disciplines; however, they are not so specific as to require strong expertise in a particular discipline. The major intent of these problems is to lend relevance to the application of computational methods to engineering design problems. The subjects of the designs are therefore chosen to be familiar to students in any discipline, and the necessary background information for the mathematical formulation of the problems is provided. Some physical models are also given in Appendix D.

10.4.1   Problem 1: Design of an Airship Gas Envelope

Airships, also known as **blimps** or **dirigibles**, are lighter-than-air vehicles. The primary mechanism for lift is buoyancy, which is obtained by filling a gas envelope with helium. Airships are used mainly for advertising and for added television coverage of sporting events. Other possible uses are as platforms for border surveillance, for damage assessment following natural disasters, and for early warning and detection systems.

The common shape for the gas envelope of an airship is similar to an ellipsoid. We shall consider a body of revolution whose radius r varies with the coordinate z on the axis of revolution according to

[P1.1]
$$r^2 = b^2(1 - 4z^2)(0.5 - z)^c; \qquad -0.5 \le z \le 0.5$$

The quantities r and z are lengths that are nondimensionalized with respect to the axial length L of the envelope. The quantity b is a fineness parameter that is related to the maximum radius r_{max} of the envelope. The quantity c is a shape parameter that is related to the location z_m at which the radius is maximum. When c is equal to zero, the envelope is an ellipsoid with z_m equal to zero and r_{max} equal to b. Positive values of z_m produce negative values of c.

The shape of the envelope is to be specified *via* the maximum nondimensional radius r_{max} and its nondimensional location z_m. The parameter c is obtained from z_m by setting the derivative $d[r^2]/dz$ to zero when z is equal to z_m, and is then used to compute the parameter b from Eq. [P1.1]. An envelope is illustrated in Fig. P1–1.

The envelope is to have a volume of 2000 m³. The nondimensional volume V is found from

[P1.2]
$$V = \pi \int_{-0.5}^{0.5} r^2 \, dz$$

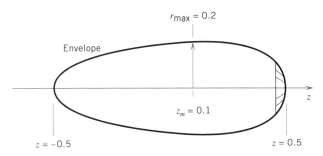

Figure P1–1   Gas envelope with ($r_{max} = 0.2$) and ($z_m = 0.1$).

The integral in Eq. [P1.2] may be obtained analytically by performing two integrations by parts. The length L of the envelope (in meters) is then given by

[P1.3]
$$L^3 = 2000/V$$

The nondimensional surface area S of the envelope is found from

[P1.4]
$$S = 2\pi \int_{-0.5}^{0.5} r \sqrt{1 + (dr/dz)^2} \, dz$$

This integral is best evaluated by one of the numerical integration methods of Chapter 6. Problems 18 and 19 of Chapter 6 are especially useful. It is suggested that the transformation

$$z = -0.5 \cos \theta$$

be used to change the variable of integration to θ between the limits 0 and π. This change clusters the control points for the integration at locations where the integrand varies rapidly. Regardless of the variable of integration, pay special attention to the integrand values at the limits of the integration.

The "cost" of the design is proportional to the actual surface area (SL^2). We wish to maximize the buoyancy, which is proportional to the actual volume (VL^3). We also wish to minimize the drag on the airship. This drag is proportional to $(C_d[r_{max}L]^2)$, where C_d is a nondimensional drag coefficient. An empirical relation for the behavior of C_d is

[P1.5] $C_d = 0.1136 - f\{0.04858 - f(0.01170 - 0.0008167f)\}$; $2 \le f \le 5$

in which f is the envelope's fineness ratio

[P1.6]
$$f = 1/(2r_{max})$$

An appropriate criterion function for the design of the gas envelope is one that embodies the objectives of maximizing buoyancy and minimizing cost and drag. We therefore specify a criterion function ϕ by

[P1.7]
$$\phi = LC_d(r_{max})^2 S/V$$

One of the constraints for the problem is that the fineness ratio f in Eq. [P1.6] must be in the range specified in the empirical relation of Eq. [P1.5]. This constraint is imposed for the practical reason that the drag coefficient is available only for this range. The equivalent constraint in terms of r_{max} is

[P1.8]
$$0.1 \le r_{max} \le 0.25$$

A second, *ad hoc* constraint is imposed for esthetic as well as physical reasons. The drag coefficient given by Eq. [P1.5] is for conventional "fuselage-like" shapes. To avoid excessive bluntness at the "nose" of the gas envelope, the axial location z_m of the maximum radius is restricted to the range

[P1.9] $$2r_{max} - 0.5 \le z_m \le 0.5 - 2r_{max}$$

Note that the constraint on z_m allows either end ($z = -0.5$) or ($z = 0.5$) to be the nose of the envelope, and allows only a zero value of z_m when r_{max} is at the upper limit of the range in Eq. [P1.8].

In summary, the problem is to design the gas envelope of an airship according to the shape given in Eq. [P1.1]. The actual volume of the envelope is to be 2000 m^3, the criterion function ϕ of Eq. [P1.7] is to be minimized, and the constraints of Eq. [P1.8] and [P1.9] are to be satisfied. Values to be reported for the final design are the nondimensional quantities r_{max} and z_m, the criterion function ϕ, the fineness and shape parameters b and c, the nondimensional volume V and surface area S, the length L of the envelope (in meters), the fineness ratio f, the drag coefficient C_d, and a table of nondimensional r and z. An illustration of the envelope, similar to the one in Fig. P1–1, is also to be provided.

The following properties of an ellipsoid with a unit major axis and a maximum radius r_{max} may be useful as a preliminary check of the results for V and S. The corresponding values for an ellipsoid (corresponding to a zero value of z_m) are

$$V_e = 2\pi(r_{max})^2/3$$

$$S_e = \pi(r_{max}/\beta)\left[\sqrt{\beta^2 - 0.25} + 2\beta^2 \sin^{-1}(0.5/\beta)\right]; \qquad 1/\beta = 2\sqrt{1 - (2r_{max})^2}$$

An open-ended version of the problem is to design the envelope without the shape function of Eq. [P1.1]. The envelope is still required to be a body of revolution with infinite slopes (dr/dz) at the nose and tail, and one location z_m at which (dr/dz) is zero. The criterion function of Eq. [P1.7] and the constraints on r_{max} and z_m in Eq. [P1.8] and Eq. [P1.9] remain in effect.

A new constraint to define acceptable body shapes must now be introduced to replace the one that was inherent in the shape function of Eq. [P1.1]. This constraint requires the envelope to be nonconcave and is given by

[P1.10] $$d^2r/dz^2 \le 0; \qquad -0.5 \le z \le 0.5$$

10.4.2 Problem 2: Design of a Plane Truss

Trusses are lightweight structures for supporting heavy loads. Examples of trusses are seen on bridges and on roof structures. A discussion of plane trusses is provided in §2.2.5 and Appendix D.

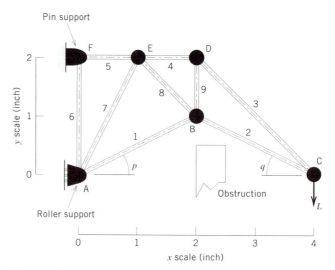

Figure P2–1 Illustration of the truss.

The truss for our design problem consists of six joints and nine members as shown in Fig. P2–1. The joints are labeled A through F, and members are numbered 1 through 9. All lengths are in inches, and the coordinates (x, y) of all joints except B are fixed as follows.

[P2.1] $\quad (x_a, y_a) = (0, 0); \quad (x_c, y_c) = (4, 0);$
$\qquad (x_d, y_d) = (2, 2); \quad (x_e, y_e) = (1, 2); \quad (x_f, y_f) = (0, 2)$

The external load L is in the negative y direction at joint C. The truss has a pin support at joint F; the external force at that joint thus has two components. The support at joint A is a roller support; the external force at that joint has only an x component.

The location of joint B must be such that members 1 and 2 clear an obstruction. Clearance is satisfied by the following conditions on the angles p and q.

[P2.2] $\qquad\qquad\qquad p \geq 16°; \qquad q \geq 21°$

In addition, joint B is constrained to lie in the region formed by members 3, 4, and 7, and it is to be no closer than 0.5 in. from the centerline of any of those members. The objective of the design is to maximize the load per weight of the truss. If $L_{\max}$ is the maximum load that the truss can support without failure, and W is the weight of the truss, the criterion function to be maximized is

[P2.3] $\qquad\qquad\qquad\qquad \phi = L_{\max}/W$

The truss is to be constructed of 1/16-in.-diameter steel rods. Let r_m be the length of member m; then the weight W in pounds is given by

[P2.4]
$$W = 0.00087 \sum_{m=1}^{9} r_m$$

The lengths r_m may be obtained by the distance formula (based on Pythagoras' theorem) for a line between two known points.

The determination of the maximum load L_{max} proceeds as follows. We first compute the member forces f_m for a load L equal to 1 lb, with tensions as positive forces and compressions as negative forces in the formulation of the problem. We then compute for each member the fraction Q_m of the maximum safe load that each force f_m represents. The values of Q_m for the nine members are obtained from

[P2.5]
$$Q_m = \begin{cases} f_m/250; \ f_m \geq 0 \\ -(f_m/450)(r_m)^2; \ \ f_m < 0 \end{cases}$$

The value for positive f_m is for failure in tension, and the value for negative f_m is for failure in compression due to buckling. The maximum external load L_{max} (in pounds) can now be computed from

[P2.6]
$$L_{max} = 1/\{max(Q_m)\}$$

An open-ended version of this problem retains the locations of joints A, C, and F. The constraints of Eq. [P2.2] are also retained. The new design problem is to locate the remaining joints B, D, and E so as to maximize (L_{max}/W). The basic truss geometry is to be preserved, and no length r_m is to be less than 0.5 inch.

10.4.3 Problem 3: Design of a Four-Bar Linkage

A *four-bar linkage* is shown in Fig. P3–1. It consists of an input crank AB of length p, an output crank OC of length q, and a connecting bar BC of length r. The fourth

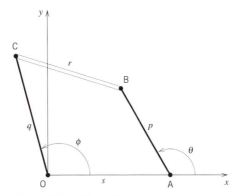

Figure P3–1 Illustration of a four-bar linkage.

bar of length s does not actually exist; rather it is the fixed foundation for the pivot points O and A.

Four-bar linkages are typically used to convert the rotary motion of the input crank to an oscillatory motion of the output crank. The output crank may then drive mechanisms such as windshield wipers, agitators for washing machines, and the oscillating arm of lawn sprinklers.

The pivot O of the output crank is the origin of an (x, y) coordinate system as shown in Fig. P3–1. The angles θ and ϕ of the input and output cranks are measured counterclockwise from the positive x axis.

The maximum output angle ϕ_{max} occurs when θ is equal to θ^+, and the minimum output angle ϕ_{min} occurs when θ is equal to θ^-. The output angles are constrained to lie in the range

[P3.1] $$0 \leq \phi_{min} < \phi_{max} \leq \pi \text{ radian}$$

The angles ϕ_{max}, θ^+, ϕ_{min}, and θ^- may be found with the help of Fig. P3–2. Maximum output occurs when AB and BC are fully extended to form a straight line; minimum output occurs when AB and BC overlap.

The angles for minimum and maximum output may be obtained by applying the cosine law to the triangle OAC in each of the configurations. The more general problem of computing ϕ for an arbitrary value of θ is solved with the aid of Fig. P3–3. Two cases are shown — one with point B above the x axis, and one with point B below the x axis.

Using polar coordinates, we obtain the coordinates (x_b, y_b) of point B as

[P3.2] $$x_b = p \cos \theta + s; \qquad y_b = p \sin \theta$$

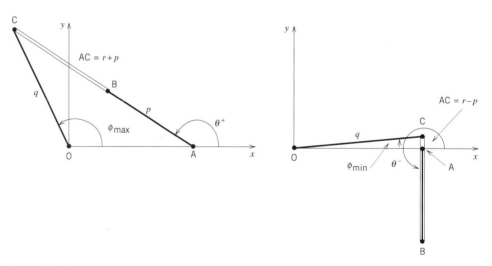

Figure P3–2 Geometries for minimum and maximum output angles.

The distance h from O to B is then found from the distance formula. For either configuration, the cosine law may be applied to triangle OBC to obtain the angle γ. The angle ϕ is then found from

[P3.3]
$$\phi = \gamma + \tan^{-1}(y_b/x_b)$$

The objective of the design is to choose the lengths p, q, and r with s equal to 1 so that the behavior of the output angle ϕ with θ is as close to sinusoidal as possible. To define sinusoidal for a particular set of lengths, we use the concept of least squares from Chapter 4.

Let ϕ_i be the output angles at input angles

[P3.4]
$$\theta_i = i(\pi/12); \qquad i = 0, 1, \ldots, 23$$

The sinusoidal model for ϕ is expressed by

[P3.5]
$$\phi_{\text{model}} = \alpha + A \sin(\theta - \beta)$$

where

[P3.6]
$$\alpha = (\phi_{\max} + \phi_{\min})/2; \qquad A = (\phi_{\max} - \phi_{\min})/2$$

The phase angle β is chosen to minimize the root mean square deviation d_{rms}, which is given by

[P3.7]
$$(d_{\text{rms}})^2 = (1/24)\left[\sum_{i=0}^{23}(\phi_i - [\phi_{\text{model}}]_i)^2\right]$$

An approximate starting value for β in the minimization of d_{rms} is $[(\theta^+ + \theta^-)/2 - \pi]$.

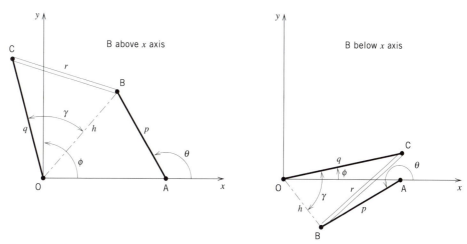

Figure P3–3 Configurations for arbitrary θ values.

The root mean square deviation d_{rms} is the criterion function for the problem. We have already given a constraint in Eq. [P3.1], and we have specified the length s of the fixed bar to be equal to 1. We now give additional constraints for the problem.

The additional constraints include a required range for the amplitude A; namely,

[P3.8] $0.95 \text{ radian} \leq A \leq 1.00 \text{ radian}$

Other constraints involve the lengths p, q, and r as follows.

[P3.9] $0.2 < p < 1$

[P3.10] $p < q < 2$

[P3.11] $0.2 < r < 2$

These last three constraints are minimal constraints, which help to size the problem. They may need to be stronger to meet the other constraints of the problem and to satisfy the condition that the link must be geometrically possible for all input crank angles θ.

To determine if the linkage is possible, consider the procedure for computing the output angle ϕ. This procedure involves the use of the cosine law to find the angle γ; that is, γ is the inverse cosine of some argument involving $p, q, r, s,$ and θ. Since the argument of the inverse cosine cannot exceed 1 in magnitude, the link is possible if the minimum and maximum values of the function are at most 1 in magnitude.

10.4.4   Problem 4: Design of a Rack and Pinion Steering Linkage

Steering of a car is usually accomplished by turning the front wheels. If the wheels have the same steering angle, the circular arcs that the individual wheels try to follow have different centers. Although the centers are different, the car will seek to turn about a common center. The tires will therefore have to undergo some deformation or sliding (also known as scrubbing) to accommodate the motion. Both of these are undesirable because they adversely affect the lateral handling of the car, and they impose severe stresses on the materials and mechanisms of the wheel assemblies.

To avoid deformation and scrubbing, all four wheels should roll in circular paths about a common center of rotation. To achieve this, the inner wheel should be turned through a greater angle than the outer wheel. The *ideal* steering angle $(\theta_I)_{\text{ideal}}$ of the inner wheel for a given steering angle θ_o of the outer wheel is illustrated in Fig. P4–1.

The steering angles are related through the right-angled triangles CAA$'$ and CBB$'$ in Fig. P4–1. The wheels pivot about points A and B, each of which is at a distance d from the centerline of the car. L is the car's wheelbase and represents a common height for the two triangles; the bases are b_i for CAA$'$ and b_o for CBB$'$.

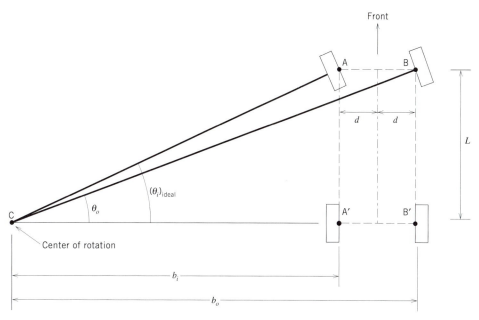

Figure P4–1 Front-wheel steering of a car.

Simple geometry gives us

$$\tan \theta_o = L/b_o; \quad \tan(\theta_i)_{\text{ideal}} = L/b_i; \quad b_i = b_o - 2d$$

L is given as 2.40 m, and d is given as 0.65 m. When θ_o is also specified, we obtain

[P4.1] $$\tan(\theta_i)_{\text{ideal}} = L/[(L/\tan \theta_o) - 2d]$$

We shall use Eq. [P4.1] to help us design a *rack and pinion* steering system, which is shown in Fig. P4–2 and is described as follows.

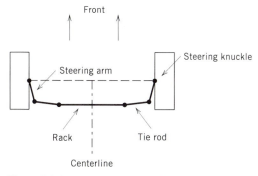

Figure P4–2 Illustration of a rack and pinion linkage.

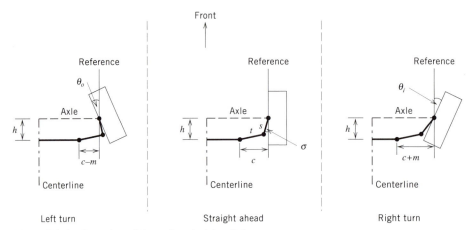

Figure P4–3 Operation of the rack and pinion linkage.

A toothed rack is driven by a small gear known as a *pinion*. The pinion converts the rotary steering wheel input to linear motion of the rack. The operation of the linkage is illustrated in Fig. P4-3.

The rack is at a fixed distance h from the axle line of the front wheels. When the rack is in a neutral position, the end joint is at a distance c from a reference line through the steering knuckle. A rack motion m produces *actual* steering angles θ_o and θ_i (depending on the direction) by pivoting the entire wheel and steering arm assembly *via* tie rods of length t; the tie rods are necessary because the circular paths of the arms and the linear path of the rack are not directly compatible. Each steering arm has a length s and an inboard angle σ.

The lengths c and h are fixed at 0.30 m and 0.22 m, respectively, for this design. We now consider how to obtain the length t of the tie rod for given inputs of s and σ, and how to obtain the required motion m and the inner steering angle θ_i for a given value of θ_o. Let us consider an (x, y) coordinate system with origin at the right steering knuckle, x to the right and coincident with the axle line, and y to the front and coincident with the reference line.

We see from the straight-ahead geometry that the ends of the tie rod are at (x, y) equal to $(-c, -h)$ and at (x, y) equal to $(-s \sin \sigma, -s \cos \sigma)$. We therefore obtain t from

[P4.2] $$t^2 = (c - s \sin \sigma)^2 + (h - c \cos \sigma)^2$$

The rack motion m required to produce a steering angle θ_o in the left-turn geometry requires more manipulation. We can show that m is related to θ_o by

[P4.3] $$m = c - s \sin(\sigma - \theta_o) - \sqrt{t^2 - [h - s \cos(\sigma - \theta_o)]^2}$$

The steering angle θ_i produced by the same rack motion is found from the right-turn geometry. It is given by

[P4.4] $\alpha + \sigma + \theta_i = \sin^{-1}\{([c + m]^2 + h^2 + s^2 - t^2)/(2s\,p)\};$

$p^2 = h^2 + (c + m)^2; \qquad \tan \alpha = h/(c + m)$

The objective of the design is to specify s and σ so that the root-mean-square deviation of θ_i from $(\theta_i)_{\text{ideal}}$ is minimized for θ_o ranging from 0 to 30°. The root mean square deviation d_{rms} is the criterion function and is given by

[P4.5] $$(d_{\text{rms}})^2 = (1/30)\int_0^{30} [\theta_i - (\theta_i)_{\text{ideal}}]^2 \, d\theta_o$$

Constraints on the geometry are imposed as follows. The length s of the steering arm is required to satisfy

[P4.6] $$s \leq 0.22 \text{ m}$$

To make sure that the joints clear the wheel, the angle σ must satisfy

[P4.7] $$s \sin \sigma \geq 0.02 \text{ m}$$

In Eq. [P4.6] and [P4.7], m denotes meter and not the rack motion. A final constraint guarantees that the linkage is geometrically possible and will not be overextended. It is expressed as

[P4.8] $$\sqrt{s^2 - 0.0004} + \sqrt{t^2 - 0.0004} \geq \sqrt{h^2 + (c + m_{30})^2}$$

where m_{30} is the motion of the rack that is required for an outer steering angle θ_o of 30°.

A somewhat more difficult version of this problem is obtained by including the length c as one of the design parameters. An additional constraint must be imposed to ensure that the rack is long enough to produce the required steering angles. This constraint is expressed by

[P4.9] $$d - c \geq m + 0.02$$

10.4.5 Problem 5: Design of a Water Park Slide

Among the rides that we might see at a water amusement park is a slide. The rider sits on a slider to descend and is launched horizontally onto the surface of a pool from the end of the slide.

A Cartesian (x, y) system with x horizontal and y vertical is used to describe the equation of the slide. The top of the slide is at (x, y) equal to $(0, h)$, and the bottom is at (x, y) equal to $(L, 0)$. The rider is to leave the top with an initial horizontal velocity v_0 equal to 1.5 m/s, and is to exit horizontally at the bottom; therefore, the equation $y(x)$ representing the shape of the slide must satisfy the relations

[P5.1] $$y(0) = h; \quad y(L) = 0; \quad y'(0) = y'(L) = 0$$

in which y' denotes (dy/dx).

The equation that we shall use for the slide is given by

[P5.2] $$y/h = a\xi^4 + b\xi^3 + c\xi^2 + 1; \qquad \xi = x/L$$

The slide has a negative curvature on the left (that is, y'' equal to d^2y/dx^2 is negative), and it has a positive curvature on the right. At some neutral point, the curvature is zero. We shall choose the location of the neutral point through a parameter λ so that

[P5.3] $$y''(x) = 0 \text{ at } x = \lambda L \text{ (or at } \xi = \lambda)$$

and use λ as a parameter to vary the shape of the slide. The coefficients of Eq. [P5.2] that satisfy Eq. [P5.1] may be found in terms of λ.

The choice of λ is not arbitrary. To ensure that the ramp has a negative slope at all points except at the top and bottom, the second derivative must be negative when ξ is equal to 0, and it must be positive when ξ is equal to 1. The constraint that satisfies the requirement of negative slope is

[P5.4] $$(1/3) < \lambda < (2/3)$$

A complete description of the slide is obtained when h, λ, and L are specified. We shall set h equal to 5 m and treat λ and L as independent variable inputs.

Let us now consider the forces acting on a body of mass M as it descends the slide. The normal forces are perpendicular to the slide and consist of a weight component of the body and the reaction of the slide on the body. Tangential forces act along the tangent to the slide and consist of a weight component of the body and the retarding frictional force. A schematic of the problem showing the free-body diagram and the accelerations is provided in Fig. P5–1.

The weight components are the normal component ($Mg \cos \alpha$) and the tangential component ($Mg \sin \alpha$), in which α is the negative angle

[P5.5] $$\alpha = \arctan(y')$$

The normal force exerted on the body by the slide is denoted by N, and the tangential friction force is μN, with μ equal to 0.1 denoting the dynamic coefficient of friction.

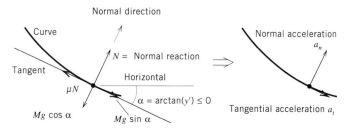

Figure P5–1   Schematic for the slide problem.

From Newton's Second Law, we obtain the normal equation of motion as

$$N - Mg \cos \alpha = Ma_n$$

Rearrangement and division by Mg yields

[P5.6] $$\Gamma = N/(Mg) = \cos \alpha + a_n/g$$

The quantity Γ is the number of "gees" experienced by the body. We shall restrict this value under *frictionless* conditions to be no more than 1.5.

The normal component of acceleration is

[P5.7] $$a_n = v^2/\rho$$

where v is the speed of the rider and ρ is the radius of curvature given by

[P5.8] $$\rho = [1 + (y')^2]^{1.5}/y''$$

The speed v with no friction is obtained from the energy conservation principle. It is found from

[P5.9] $$v^2 = (v_o)^2 + 2g(h - y); \qquad g = 9.81 \text{ m/s}^2; \qquad \mu = 0$$

Thus the constraint on Γ is

[P5.10] $$\Gamma = \cos \alpha + [(v_o)^2 + 2g(h - y)]/[\rho g] \le 1.5$$

We now consider the tangential motion to determine the speed v when friction affects the motion. The tangential equation of motion is

$$a_t = d[v^2/2]/ds = g \sin \alpha - \mu N/M$$

where s is the arclength of the slide, and

$$(ds)^2 = (dx)^2 + (dy)^2$$

With N obtained from Eq. [P5.6], the ordinary differential equation and initial condition for the motion of the rider are

[P5.11] $d(v^2)/dx = 2[g(\sin \alpha - \mu \cos \alpha) - \mu v^2/\rho]\sqrt{1 + (y')^2};$
           $v^2 = (v_0)^2$ at $x = 0$

The objective of the design is to define the slide that maximizes the exit speed v_e when the rider reaches the end of the slide. For the rider to reach the end of the slide, the solution for (v^2) cannot be negative for any x value from 0 to L.

An open-ended version of this problem is obtained by not restricting the shape of the slide to the function given in Eq. [P5.2]. References to the shape parameter λ are now meaningless, but negative slopes in the range $(0 < x < L)$ are still required. With the shape restriction removed, we must impose another constraint to ensure that the rider does not leave the slide. This constraint is

[P5.12] $\Gamma = \cos \alpha + [(v_0)^2 + 2g(h - y)]/[\rho g] > 0$

10.4.6   Problem 6: Design of a Ventilation System

This problem deals with a highly simplified **ventilation system** for a rectangular region with two openings. The geometry of the region is shown in Fig. P6–1.

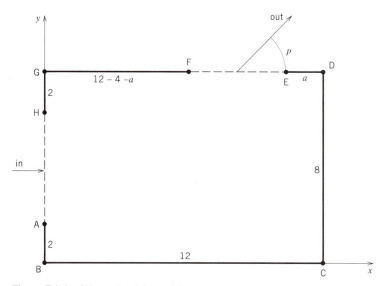

Figure P6–1   Schematic of the problem.

Dimensions in arbitrary length units are shown for each solid boundary. An exhaust fan pulls air out of the opening EF, and air enters through the other opening AH at an average nondimensional speed of 1. It is assumed that the inflow and outflow are in fixed directions and that the outflow is directed at an angle p to the line ED.

The model for the system is in terms of a stream function ψ, which is related to the x component u and the y component v of the velocity by

[P6.1]
$$u = \psi_y; \qquad v = -\psi_x$$

The governing equation for the flow in the region is

[P6.2]
$$\nabla^2 \psi = 0$$

The boundary conditions for the problem are as follows.

[P6.3a]
$$\psi = 0 \text{ on ABCDE}$$

[P6.3b]
$$\psi_y = -\psi_x \cot p \text{ between E and F}$$

[P6.3c]
$$\psi = \text{constant} = 4 \text{ on FGH}$$

[P6.3d]
$$\psi_x = 0 \text{ between H and A}$$

The problem is to be solved on a grid with x and y spacings equal to 1. The speed U of the flow at all points not on the boundary is to be found from

[P6.4]
$$U^2 = u^2 + v^2$$

The objective of the problem is to choose the angle p and the distance a so that the population standard deviation of U at the interior points of the field is minimized. If possible, provide plots of the streamlines (ψ = constant) to show the flow pattern.

The constraints for the problem are simple. The distance a must be at least 2 and no more than 6. The angle p must fall in the range ($45° \leq p \leq 135°$).

10.4.7 Problem 7: Design of a Software Application

This problem is to design a *software package* to compute the area of closed, plane shape. The boundary of the shape is to be specified by a set of (x, y) coordinates in an arbitrary sequence as shown in Fig. P7–1. Allowable shapes are those whose boundaries can be expressed in polar coordinates (r, θ) by a smooth, single-valued function $r(\theta)$. The origin of the polar system is to be the mean position of the boundary data. The points describing the shape are to be sorted in ascending polar angles.

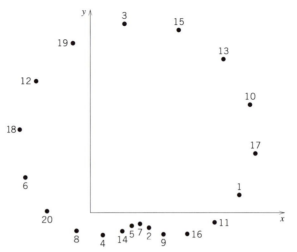

Figure P7–1 Arbitrarily sequenced data for the boundary of a closed shape.

The area A is to be found by an accurate area integration in polar coordinates from

[P7.1]
$$A = \int_{\theta_1}^{\theta_1+2\pi} (r^2/2)d\theta$$

The integration scheme may use more data points than are given. The additional data are to be found by a curve-fitting process. The criteria for packages such as this are usually computing time and accuracy. Since we have no good way to measure time across the possible platforms, we will look only at accuracy. The test data are 20 (x, y) pairs of coordinates for a shape whose correct area is 1.32π. They are to be obtained from

[P7.2]
$$x = 0.5 + \cos\sigma + 0.4\sin(2\sigma)$$
$$y = 0.5 + \sin\sigma - 0.4\cos(2\sigma)$$

The data are to be entered for σ_i equal to $[0.08 + (i-1)\pi/10]$ with i values in the sequence

$$1, 17, 6, 13, 15, 10, 16, 12, 18, 3, 20, 8, 4, 14, 5, 19, 2, 9, 7, 11$$

The objective of the design is to minimize the error in the computed area.

10.4.8   Problem 8: Design of a Rocket Launch Configuration

A small rocket has an initial total mass m_0 of 300 kg, which includes 180 kg of propellant. The rocket is to be launched from sea level at an initial angle γ_0 to the horizontal. After launch, the axis of the rocket is aligned with the flight path so that both the thrust T and the drag D are tangent to the flight path. The objective of the design is to configure the launch parameters so that the rocket has maximum (horizontal) range.

The equations of motion for the rocket are given in terms of a Cartesian (x, y) system, with x [m] denoting the horizontal (positive in the flight direction) and y [m] denoting the vertical (positive upward). A free-body diagram is given in Fig. P8–1.

Let u and v be the x and y components, respectively, of the rocket's velocity, and let t [s] denote time. Then we use the definition of velocity, Newton's Second Law, and the free-body diagram to write the governing system of ordinary differential equations as

[P8.1a] $dx/dt = u; \quad dy/dt = v$

[P8.1b] $du/dt = [(T - D)\cos\gamma]/m$

[P8.1c] $dv/dt = [(T - D)\sin\gamma]/m - g; \qquad g = 9.81 \text{ m/s}^2$

The flight path angle γ in Eq. [P8.1b, c] is given by

[P8.2] $\gamma = \tan^{-1}(v/u); \qquad -\pi/2 \text{ radian } < \gamma < \pi/2 \text{ radian}$

To consider the other quantities (except the drag D) in those equations, we must first look at the consumption rate of the propellant.

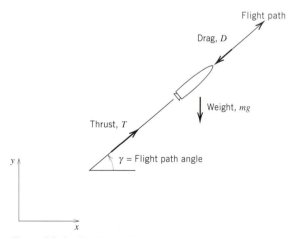

Figure P8–1   Free-body diagram for the rocket.

We shall assume that the propellant consumption rate q [kg/s] can be varied linearly with time from the instant of launch ($t = 0$) to the burnout time t_b (at which the propellant is completely consumed). The model for the consumption rate is

[P8.3] $$q = q_0 + (q_b - q_0)(t/t_b); \quad 0 \le t \le t_b$$

where q_0 [kg/s] is the initial rate at launch and q_b [kg/s] is the final rate at burnout. The consumption rates are constrained according to

[P8.4] $$7.5 \text{ kg/s} \le q_0, q_b \le 12.5 \text{ kg/s}$$

The mass of propellant that is consumed at time τ less than or equal to t_b is

[P8.5] $$m_p(\tau) = \int_0^\tau q \, dt; \quad 0 \le \tau \le t_b$$

The burnout time t_b may therefore be found from

[P8.6] $$m_p(t_b) = 180 \text{ kg} = \text{total propellant mass}$$

and the instantaneous mass m [kg] of the rocket is given by

[P8.7] $$m = \begin{cases} 300 \text{ kg} - m_p(t); & 0 \le t \le t_b \\ 120 \text{ kg}; & t > t_b \end{cases}$$

The thrust T is equal to $q v_e$, where v_e is the exhaust speed of the propellant relative to the rocket and is equal to 500 m/s, as long as the propellant lasts. Thus

[P8.8] $$T = \begin{cases} q(500 \text{ m/s}); & 0 \le t \le t_b \\ 0; & t > t_b \end{cases}$$

The drag D [N] is given by

[P8.9] $$D = (0.1 \text{ N} \cdot \text{s}^2/\text{m}^2)(u^2 + v^2)(\rho/\rho_s)/\beta$$

The term (ρ/ρ_s) is a correction for the atmospheric density ρ at altitude y relative to the density ρ_s at sea level. A model for the density ratio is

[P8.10] $$\rho/\rho_s = (1 - [2.25 \times 10^{-5} \text{ m}^{-1}]y)^{4.25}$$

The quantity β is the Prandtl–Glauert compressibility correction as the Mach number M of the rocket becomes appreciable. The expression for β is

[P8.11]
$$\beta = \sqrt{1 - M^2}; \quad M^2 = (u^2 + v^2)/c^2$$

in which c is the speed of sound in the atmosphere. For the altitudes that the rocket will attain, c may be taken as a constant equal to 340 m/s.

The range R [m] is the horizontal distance that the rocket flies before returning to sea level ($y = 0$). The objective of the design is to choose the initial flight path angle γ_0 and the propellant consumption rates q_0 and q_b so that the range is maximized. A table and a plot for the flight path of the rocket should also be provided; these should indicate when burnout occurs.

In solving the problem, care should be taken to obtain an accurate solution in the early phase of the flight, because the flight path angle may change rapidly during that period. The discontinuity in the thrust T and the discontinuous rate of change of the mass m at burnout must also be treated carefully.

Appendix A

Introduction to the Great Mathematicians

Many of the methods and equations in the text are associated with the names of famous mathematicians and scientists. Here, we provide biographical sketches of the more notable pretwentieth century figures of the modern mathematical era (Ref. 19, 20). Only those whose names appear in the text are presented.

As will be seen in the sketches, even the most well recognized pure mathematicians worked on applied problems; indeed, some of their advances were made on the way to solving such problems. To appreciate their work, we must remember that they did not have the tools we take for granted — they developed them! To help with their places in history, Fig. A–1 shows the lifespans of those that are discussed.

John Couch Adams   [1819–1892]

Adams was born in Cornwall and educated at Cambridge University. He was later appointed Lowndean Professor and Director of the Observatory at Cambridge. In 1845, he calculated the position of a planet beyond Uranus that could account for perturbations in the orbit of Uranus. His requests for help in looking for the planet, Neptune, met with little response among English astronomers. An independent set of calculations was completed in 1846 by Leverrier, whose suggestions to the German astronomer Johann Galle led to Neptune's discovery.

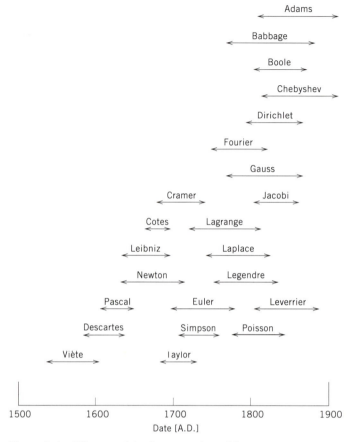

Figure A–1   Lifespans of the famous mathematicians.

Adams published a memoir on the mean motion of the Moon in 1855 and computed the orbit of the Leonids in 1867. The Leonids are meteor showers that appear to originate in the constellation Leo. They were especially prominent every 33 years from 902 to 1866.

Charles Babbage   [1792–1871]

Babbage's design of the Analytical Engine is considered to be the forerunner of the modern computer. Lack of technology and money prevented Babbage from realizing his design; however, a model built from his plans at a later date worked as Babbage had predicted. Babbage's ideas on the Analytic Engine would have been lost if Ada Lovelace had not clearly described them along with her own ideas.

Although computer scientists associate Babbage with the computer, he was better known as a prominent mathematician of his time, and he held the position of

Lucasian Professor at Cambridge. His important contributions were on the calculus of functions.

Along with George Peacock and John Herschel, Babbage formed the Analytical Society to promote analytical methods and the use of Leibniz's differential notation (the geometrically suggestive dy/dx form that we know today). English mathematicians used Newton's fluxion notation almost exclusively until then, partly because it was widely held among them that Leibniz appropriated Newton's ideas about calculus and claimed them as his own. The fluxion notation (the use of a dot above the variable x to indicate the derivative, still used in many texts today) was abstract enough to hamper developments in calculus. Babbage's objective was to replace this "dot-age" with "d-ism" at Cambridge.

George Boole [1815–1864]

Boole was born in Lincoln, in eastern England. In addition to his mathematical prowess, he studied classics on his own. Boole's work on linear transformations led to some aspects of the theory of invariants. He also performed research on differential equations and the calculus of finite differences. Boole is best remembered as one of the creators of mathematical logic, which is one of the foundations of modern computer technology.

Pafnuti Lvovich Chebyshev [1821–1894]

Chebyshev was born in Borovks, was educated at the University of Moscow, and was a Professor of Mathematics at St. Petersburg. His mathematical work dealt with the convergence of the Taylor series, prime numbers, probability, quadratic forms, and integrals. He also worked on applied problems of linkages and gears.

Roger Cotes [1682–1716]

Cotes was educated at Cambridge and later was the university's Plumian Professor of Astronomy. He was well regarded by his contemporaries, including Newton. Much of Cotes's time was spent in editing a second edition of Newton's *Principia Mathematica*. His other work included hydrostatics, treatments of rational algebraic expressions, the earliest attempt to form a theory of errors, applications of the method of differences, and problems in particle dynamics.

Gabriel Cramer [1704–1752]

Cramer was born in Geneva and later became a Professor at Geneva. His best known work is on algebraic curves, which includes a demonstration that an n-th degree curve is determined by $[n(n+3)/2]$ points. He amplified the rules for the solutions of systems of linear algebraic equations, and he formally introduced the y axis.

René Descartes   [1596–1650]

Descartes was born in Tours, France and treated mathematics as a hobby while in the army as a young man. His primary contributions to mathematics are in analytical geometry and the theory of vortices. His work laid the foundation for analytical geometry and focused on the two-dimensional rectangular coordinate system; however, it is clear that he was well aware of three-dimensional representations of a point in space. He formulated the rule of signs for the positive and negative roots of polynomials. Newton later attempted to formulate a similar rule for the complex roots. Descartes is responsible for the custom of using early letters of the alphabet for known quantities and those near the end of the alphabet for unknown quantities. He also invented the notation for expressing powers.

Descartes attempted at one time to give a physical theory of the universe. He abandoned it when he realized that it would result in conflict with the Church. In any case, eight of the ten laws of nature he proposed were incorrect. The first two, however, were almost identical to Newton's.

As a child, Descartes was allowed to stay in bed until late in the morning because of his frail health. He continued this practice later in life, and stated the opinion, obviously shared by many, that good mathematics and good health were possible only if one did not wake up too early in the morning.

Peter Gustav Lejeune Dirichlet   [1805–1859]

Dirichlet was the student of Gauss and the son-in-law of Jacobi. He succeeded Gauss as Professor of Higher Mathematics at Göttingen. He devoted much time to expositions of works by Gauss and Jacobi. His own work established Fourier's Theorem (on heat conduction) and dealt with the theory of numbers, the theory of the potential, fifth-degree equations, and definite integrals.

Leonhard Euler   [1707–1783]

Euler was born in Switzerland, studied under Johann Bernoulli at Basel, and completed his Master's degree at age 16. He formed a lifelong friendship with Bernoulli's sons Daniel and Nicholas. When they went to Russia at the invitation of Catherine I, Empress of Russia, they obtained a place for Euler at the Academy of Sciences in St. Petersburg. Euler eventually became Professor of Mathematics in 1733 when the chair was vacated by Daniel Bernoulli.

In 1741, Euler joined the Berlin Academy of Sciences at the strong request of Frederick the Great. He returned to St. Petersburg 25 years later (and was succeeded at Berlin by Lagrange). Euler was responsible for establishing Newtonian thought in Russia and Prussia.

Euler was blind in one eye by the time he was in his late 20s; within a few years of returning to Russia from Berlin, he was almost totally blind. Despite this and other misfortunes (including a fire that destroyed many of his papers), Euler was one of the most competent and prolific mathematicians of any time.

Among Euler's contributions to mathematics were extensive revisions of almost all of the branches of mathematics. He gave a full analytic treatment of algebra, the theory of equations, trigonometry, and analytical geometry. He treated series expansions of functions and stated the rule that only convergent infinite series could be used safely. He dealt with three-dimensional surfaces, calculus and calculus of variations, number theory, and imaginary numbers among other subjects. He introduced the current notations for the trigonometric functions (at about the same time as Simpson) and showed the relation between the trigonometric and exponential functions in the equation that bears his name — ($\exp(i\theta) = \cos\theta + i\sin\theta$). Another Euler equation ($v + f - e = 2$) relates the number of vertices v, the number of faces f, and the number of edges e of a polyhedron. The Beta and Gamma functions were invented by Euler.

Outside of pure mathematics, Euler made significant contributions to astronomy, mechanics, optics, and acoustics. Yet another Euler equation is the inviscid equation of motion in fluid dynamics. Even current forms of Bernoulli's hydrostatic equation, Lagrange's description of fluids, and Lagrange's calculus of variations have been given an Eulerian flavor. In astronomy, Euler tackled the three-body problem of celestial mechanics. Euler's results enabled Johann Mayer to construct lunar tables, which earned his widow £5000 from the English Parliament; £300 was also sent to Euler as an honorarium.

In short, almost every traditional subject in physics and mathematics that the modern engineering student is likely to encounter has Euler's imprint. This extends even to the symbol π, the exponential symbol e, the functional notation $f(x)$, the imaginary number i, and the summation symbol Σ.

To close the introduction to Euler, the particularly extraordinary Euler magic square (Ref. 21) is shown in Fig. A–2. In magic squares, the integers from 1 to n^2 fill the ($n \times n$) cells of a matrix in such a way that all row sums, column sums, and diagonal sums are identical. Most people are familiar with the (3×3) square. Euler's square is an (8×8) matrix in which the row sums and column sums (but not the diagonal sums) are identical. The interesting features are that the sum for half a row or column is half of the full sum, and that the numbers represent consecutive moves that a knight makes on a chessboard to hit every square once.

1	48	31	50	33	16	63	18
30	51	46	3	62	19	14	35
47	2	49	32	15	34	17	64
52	29	4	45	20	61	36	13
5	44	25	56	9	40	21	60
28	53	8	41	24	57	12	37
43	6	55	26	39	10	59	22
54	27	42	7	58	23	38	11

Figure A–2 The Euler magic square.

(Baron) Jean Baptiste Joseph Fourier   [1768–1830]

Fourier was among the prominent French physicists who also had superb abilities in mathematics. He is famous for his experiments on heat conduction which, along with ideas drawn from Newton's Law of Cooling, gave rise to Fourier's Theorem. Others had proposed similar ideas — Lagrange had given specific cases and Budan had stated the same theorem without satisfactory proof. Fourier's work on the analytical theory of heat contained the Fourier sine series, which is widely used in modern analysis.

Fourier had accompanied Napoleon's eastern expedition to Egypt and served as Governor of Lower Egypt from 1798 until the French surrendered to British forces in 1801. He was created a Baron in 1808 by Napoleon.

Karl Friedrich Gauss   [1777–1855]

Gauss was born in Braunschweig, Germany. Gauss, Lagrange, and Laplace are widely considered to be giants of analysis. Gauss's interests were so far ranging that they opened avenues of investigation for many others. His notable mathematical work included the theory of numbers, various branches of algebra, and the theory of determinants; the last formed the basis for Jacobi's work in that area. He had also obtained certain results on the theory of functions that were later found by Abel and Jacobi; however, these were not published. Gauss also developed the method of least squares and the fundamental laws of probability distributions.

The reluctance to publish was perhaps also related to Gauss's style. His oral presentations contained much of the analysis that was obscured in his published work, but he was unwilling to allow his students to take notes. In his published work, he removed all of his analytical steps and replaced them with extremely brief, though rigorous, proofs. As a result, his published work was often difficult to follow.

Gauss's interests included astronomy (he calculated the orbital elements of the asteroid Ceres following its discovery by Piazzi). His analysis resulted in an appointment as Director of the Göttingen observatory and as Professor of Astronomy. Although he retained these positions until he died, Gauss moved on to other subjects.

Among the other subjects were geodesy, optics, and electricity and magnetism. His work on the last is commemorated by the Gauss as the unit of magnetic flux density. Gauss and Weber invented the declination instrument and the magnetometer, and they built an iron-free magnetic observatory at Göttingen. Among their researches, they demonstrated the feasibility of telegraphic communications.

Carl Gustav Jacob Jacobi   [1804–1851]

Jacobi was born in Potsdam and earned his doctorate at the University of Berlin at age 21. He held a professorship in mathematics at Könisberg from 1827 to 1842. His major work was on elliptic functions, which he conducted at the same time as (but independently of) Abel. Jacobi's and Abel's theories in this area superseded Legendre's work in the same field.

Jacobi's other interests included the theory of numbers, ordinary and partial differential equations, calculus of variations, and problems in dynamics. Especially important are his work on determinants and the introduction of the Jacobian.

Joseph Louis Lagrange   [1736–1813]

Lagrange, born in Turin, was one of the greatest of the eighteenth-century mathematicians. He did not show any taste for mathematics until he was 17. Then, self-taught, he became a lecturer at 18 after only a year's study. At 19, he wrote to Euler with the solution of an isoperimetrical problem that had been discussed for over fifty years. The method used by Lagrange contained the principles of calculus of variations. Euler, recognizing the superiority of Lagrange's approach, withheld his own paper on the problem. Lagrange was thus allowed to complete his work and to receive the credit for the invention of a new form of calculus.

Lagrange's later works included corrections or improvements to works by such eminent mathematicians as Newton, Euler, Taylor, and D'Alembert. He gave the complete solution for the transverse vibration of a string and discussed echoes, the phenomenon of beats, and compound sounds. Other major work contained solutions of several problems in dynamics by the calculus of variations.

Lagrange's style was to seek general solutions to problems; even so, his work was easy to follow because of the meticulous care he used to explain his procedures.

Pierre Simon (Marquis de) Laplace   [1749–1827]

Laplace was born in Normandy. He began his professional life on the basis of a recommendation from D'Alembert, who was impressed by a paper on mechanics. Among his early contributions were proofs of the stability of planetary motions and work on integral calculus, finite differences, and differential equations. In the 1780s, he determined the attraction of a spheroid on an exterior particle; in so doing, he introduced spherical harmonics (or Laplace coefficients) and developed the concept of the potential. Similar coefficients for two-dimensional space had been presented earlier by Legendre, and the idea of the potential was taken from Lagrange's earlier works.

Because of his personality, Laplace was not well liked. He gave either little or no acknowledgment of results that he had appropriated from others. He did not care if proofs of his work were nonexistent or presented incorrectly; he was satisfied that his results were correct. Despite his pettiness, Laplace was a very capable mathematician. He developed the Laplacian equation for potentials, and did extensive work on celestial mechanics. In his volumes on celestial mechanics, Laplace put forth the nebular hypothesis; that is, that the solar system evolved from a rotating gaseous nebula.

Laplace also presented the formal proofs for the method of least squares, which had been given empirically by Gauss and Legendre. These proofs contain examples of Laplace's approach — his results were correct, but the analysis was so scanty and had so many errors that many people questioned if he had actually done the work he presented.

Other contributions were on determinants (at the same time as Vandermonde), on quadratic factors for equations of even degree, on definite integrals as solutions to linear differential equations, and on solutions to the linear partial differential equation. The theory of capillary attraction is also due to Laplace.

Aside from scientific recognition, Laplace sought social prominence. He was given the post of Minister of the Interior by Napoleon, who sought support from the scientific community, but he was removed in less than two months because of incompetence. Later, when it was clear that Napoleon's empire was crumbling, Laplace offered his services to the Bourbons and was granted the title of Marquis.

Adrian Marie Legendre   [1752–1833]

Legendre was born in Toulouse and educated in Paris. He had the misfortune of having lived at the same time as Laplace. In addition to a professorial appointment, Legendre held various public service and minor governmental positions. Any ambitions he may have had for greater recognition were stifled by Laplace's influence and hostility.

Legendre's major contributions were in geometry, the theory of numbers, various topics in integral calculus, and elliptic functions. Among these are specific instances of spherical harmonics and work on the method of least squares. In both cases, he was upstaged by Laplace who developed the full form of spherical harmonics and gave formal proofs for the method of least squares. His treatment of elliptic integrals also gave way to later superior methods by Abel and Jacobi.

Urbain Jean Joseph Leverrier   [1811–1877]

Leverrier was born in St. Lô, was educated at the Polytechnic School in Paris, and was later appointed as a lecturer there. He, independently of and later than Adams, calculated the orbit of Neptune. It was his suggestion to Johann Galle that actually led to Neptune's discovery within 1 degree of the predicted location. Leverrier's main work was in revising tables of planetary motion.

(Baron) Gottfried Wilhelm Leibniz   [1646–1716]

Leibniz was born in Leipzig. His mastery of topics ranged over mathematics, classical languages, philosophy, theology, and law. His early mathematical contributions included work on combinations and an improvement of Pascal's calculating machine.

His more important mathematical contribution was in the development of calculus. Despite controversy about the source of Leibniz's ideas (some thought that he had access to Newton's work), it is clear that his differential (dy/dx) notation was instrumental in the development of calculus. Other notational conveniences that were introduced by Leibniz include the dot as a symbol for multiplication, the equal sign, the integral sign, and the decimal point. He is also credited with the development of the binary number system.

Leibniz is also a major figure in the history of philosophy. He held that beings called *monads* were the ultimate elements of the universe, and inferred the existence of God from the harmony that existed among the monads. Euler was one who strongly opposed this philosophy.

Leibniz dabbled in dynamics, but it is clear that his knowledge in that area was limited. He also urged Peter the Great to establish the Academy of Sciences at St. Petersburg.

(Sir) Isaac Newton   [1642–1727]

Newton was born in Lincolnshire and was educated at Cambridge. He later held the Lucasian Chair at Cambridge (the same one later held by Babbage). Newton holds a prominent place in science and mathematics for his concept of infinitesimal calculus, his Law of Gravitation, his Laws of Motion, and his work on optics. The last included inventions of a refracting telescope, a reflecting microscope, and the sextant.

Newton's work on calculus used the fluxion notation, which was very difficult to master. The controversy with Leibniz caused many English scholars to persist in using this notation and resulted in the hindrance of mathematical developments until Babbage and his colleagues broke free of that prejudice.

Newton's genius was so widely recognized that he was always consulted or challenged. For example, he acted as editor for other works, and was consulted by Leibniz on infinite series, by Halley on gravitation, and by Hooke on the Earth's diurnal motion. Newton had another controversial relation with Hooke and others regarding the theory of colors. He was challenged by Johann Bernoulli to solve the brachistochrone problem (the curve, now known to be the *cycloid*, which allows quickest descent from one point to another under gravity) and another locus problem. Newton accomplished in a day what had taken Leibniz six months to solve. Another challenge resulted in Newton's laying down the principles of trajectories in a matter of hours.

It also seemed that Newton took the least obvious route in demonstrating some of his hypotheses. For example, he sought to verify his early hypothesis on gravitation by considering the orbit of the Moon. Incorrect estimates of distances caused the first attempt at verification to fail. He later repeated the calculations successfully with more accurate estimates obtained in the course of Hooke's consultation. Another example is his development of the series expansion for the inverse sine function, from which he then deduced the expansion for the sine.

The genius in Newton is exemplified by the praise he received from Lagrange and even from Laplace. Above all, there is the tribute paid to him by Gauss, another of the truly great minds. Gauss used words like *mangus* or *clarus* to describe other great mathematicians and philosophers, but he reserved the word *summus* (the best) only for Newton. Newton was knighted in 1705.

Blaise Pascal   [1623–1662]

Pascal was born in Clermont. Because of his health, his father restricted his studies to languages and prohibited the study of mathematics so that he would not be overworked. Pascal's curiosity soon led him to disregard his father's injunction, and he undertook the study of geometry. Pascal wrote a paper on conic sections at age 16 and built his celebrated adding machine at age 18. In later years, he went back and forth between mathematics and religious philosophy. Among his mathematical works are those related to the physics of gases and liquids, creation of the theory of probability (along with Fermat), and the creation of Pascal's triangle. He also devoted time to the study of cycloids, in which he effectively found the definite integrals of some trigonometric functions by summation techniques.

Siméon Denis Poisson   [1781–1840]

Poisson was born in Pithiviers and was educated by his father to be a physician. His aversion to that profession turned into permanent abandonment when one of the first patients he treated by himself died (though not through any fault of Poisson). He turned to mathematics and became a prolific contributor on the applications of mathematics to problems in physics. The fields with which he dealt included probability, mechanics, capillary action, heat, electrostatics, and magnetism. These last two spawned new branches of mathematical physics.

A major mathematical accomplishment was on the application of Fourier series to the solutions of physical problems. Another, by which he is best remembered, is the correction of Laplace's equation to produce the Poisson equation for the potential.

Thomas Simpson   [1710–1761]

Simpson was born in Leicestershire. He exerted considerable time and effort to teach mathematics to himself. His natural aptitude for the subject led him to produce a large number of works on a variety of subjects. These ranged from his applications of fluxions to several problems in astronomy and physics, and to publications on algebra, geometry, and trigonometry. He introduced the current abbreviations for trigonometric functions.

Brook Taylor   [1685–1731]

Taylor was born in Edmonton and was educated at Cambridge. He is best remembered for the Taylor series expansion for the function of a single variable, but he also studied projectile motion, vibrations, capillarity, the path of light in a heterogeneous medium, and the form of the catenary curve. Taylor abandoned the study of mathematics in 1719.

François Viète (Franciscus Vieta)   [1540–1603]

Viète was born in Fontenay, was trained as a lawyer, and spent most of his life in public service. He was, however, a reputable mathematician and devoted much of his leisure time to mathematics. His main interests lay in algebra and geometry. He knew how to write multiple angle formulas for sines and was adept at manipulating algebraic forms.

His major work was on the application of algebraic techniques to problems in geometry. His skill in algebra was probably helped by his insistence on using notations that clearly indicated a power, instead of the custom of assigning a different letter for each power. Much of his later work was on roots of equations by factoring, and he devised a closed-form method for computing the roots of cubic equations.

Appendix B

Summary of Pseudocode Structures

The grammar for pseudocodes is explained in Chapter 1. Here we give the major constructions, and we provide additional details on modularity.

B.1 ELEMENTARY CONSTRUCTIONS

Assign: Assign the value of a quantity described on the right-hand side to the memory location that is identified on the left-hand side.

LeftSide ← RightSide

Comment: Use for documentation either on its own line or trailing a statement.

\ Comment

Declare: Indicate the use of arrays by assigning subscripts.

Declare: Array

Initialize: Give an initial value to a quantity, typically by input from an external source.

Initialize: Quantity

367

Write: Output a quantity.

Write: Quantity

B.2 SELECTION

The **If** structure for selection is as follows and is explained by the flowchart of Fig. B–1.

If [Condition$_0$] **then**
    Group$_0$
Else If [Condition$_1$] **then**
    Group$_1$
      .
      .
      .
Else If [Condition$_k$] **then**
    Group$_k$
Else
    Alt–Group
End If

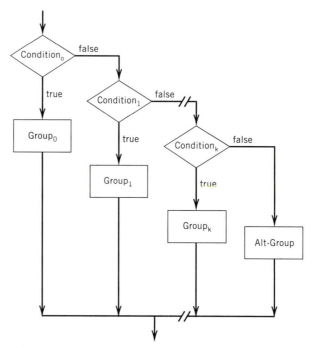

Figure B–1 Flowchart for the **If** structure.

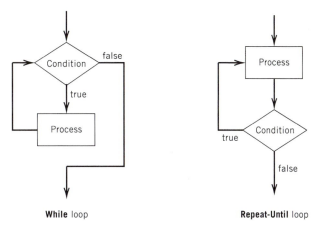

Figure B–2   Flowcharts for **While** and **Repeat–Until** loops.

B.3   LOOPS

The **While** and **Repeat-Until** loop structures are given as follows and are explained in the flowchart of Fig. B–2. These loops are typically used in cases where the number of iterations cannot be determined *a priori*.

While [Condition]	**Repeat**
Process	Process
End While	**Until** [Condition]

The **For** loop structure is a counter-controlled loop. It is given as follows and is explained in the flowchart of Fig. B–3.

For $[k = k_1, k_2, \ldots, k_n]$ \ constant increment $\Delta k = (k_n - k_1)/(n - 1)$
    Process
End For

B.4   MODULES

Modules refer to segments of code that are dedicated to a specified task. The name is generic and is not to be confused with the structure of the same name that is used in the FORTRAN 90 programming language. A module is self-contained except for information that is passed through its substitute list. The general form of the module is as follows.

Module Module Name(Substitute$_1$,Substitute$_2$, $\ldots$, Substitute$_n$)
    ModuleBody
End Module ModuleName

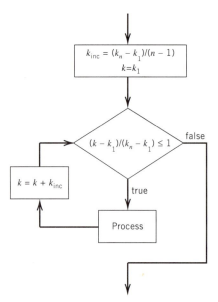

Figure B–3 Flowchart for the **For** loop.

There are two major ways of using a module in a nonrecursive fashion. The first is to treat it in the same way as we would treat a mathematical function; that is, the module returns a value for a set of arguments that is identified by the substitute names. This interpretation corresponds to a function subprogram in FORTRAN, to a function with a return type other than **void** in C, and to a function in Pascal.

The second interpretation is that the module performs its task and returns required values, if any, through its substitute list. This interpretation corresponds to a subroutine subprogram in FORTRAN, to a function with a **void** return type in C, and to a procedure in Pascal.

The exchange of information between the module and the code that invokes it varies with language. To illustrate the differences, we consider the problem of a projectile that is launched with an initial speed v at an angle α to the horizontal. We assume that air resistance is negligible, and that the gravitational acceleration g is constant. We wish to determine the maximum height $h_{\max}$ that the projectile attains, the time of flight t (at which the projectile returns to its initial elevation), and the horizontal range R. These quantities are given as follows.

$$h_{\max} = (v \sin \alpha)^2 / (2g)$$
$$t = (2v \sin \alpha)/g; \quad R = tv \cos \alpha$$

A code for the problem is given in Pseudocode B–1. In this pseudocode, Module Hmax is treated as a function that returns a result for given arguments v and α, while

Pseudocode B–1 Code for the Projectile Problem

Program:

\ Code to compute the maximum height h_{max}, the time of flight t, and the range
\ R of a projectile that is launched with an initial speed v at an angle α to the
\ horizontal. Drag is assumed to be negligible, and the gravitational acceleration
\ g is assumed to be constant at 9.81 m/s^2.

Initialize: v, α \v in m/s and α in radian
RT(v, α, R, t)   \get R and t from **Module RT**

Write: 'Maximum height [m] $= '$, **Hmax**(v, α)
Write: 'Time of flight [s] $= '$, t
Write: 'Range [m] $= '$, R

End Program

\ — — — — — — — — — —

Module Hmax(Speed, Angle):

\ Function to compute the maximum height

    $g \leftarrow 9.81$
    result $\leftarrow$ [Speed $\times$ sin (Angle)]^2/(2g)   \value of module
    return result

End Module Hmax

\ — — — — — — — — — —

Module RT(Speed, Angle, Range, Time):

\ Routine to compute the range and time of flight

    $g \leftarrow 9.81$
    Time $\leftarrow 2 \times$ Speed $\times$ sin(Angle)/g
    Range $\leftarrow$ Time $\times$ Speed $\times$ cos(Angle)

End Module RT

Module RT is treated as a code that returns the required values of t and R through its substitute list.

Real codes corresponding to the pseudocode are given as Code B–1, Code B–2, and Code B–3 in FORTRAN, ANSI C, and Turbo Pascal, respectively. We shall use these codes to discuss the passing of information.

In the FORTRAN and Pascal codes, Module Hmax is identified as a function. The function is assigned a value by using its name on the left-hand side of an assignment statement. All code segments in C are called functions. Module Hmax in the C code is a function with a **double** return type, and the function value is sent to the main function by the **return** statement.

Module RT exchanges information through the substitute list. It is identified as a SUBROUTINE in the FORTRAN code, as a function with a **void** return type in the C code, and as a procedure in the Pascal code.

In the Fortran code, all values are passed **by reference**; that is, the variable names **SPEED**, **ANGLE**, **RANGE**, and **TIME** in the subroutine refer to the same memory locations as **V**, **ALPHA**, **R**, and **T**, respectively, in the program. Thus, any change in the value of a variable that appears in the subroutine header is reflected in the main program.

All values are passed **by value** in C. The quantities **rptr** and **tptr** in the function are **pointers** that correspond to the values of the **addresses** of **r** and **t**, respectively. These addresses are denoted by **&r** and **&t** when the function is invoked from the main function. A value that is assigned to ***rptr** or to ***tptr** is actually assigned to the target of the pointer; namely, to the memory location that is identified by **r** or by **t**. The variables **speed** and **angle** take on the values of **v** and **alpha**, respectively, but do not share memory spaces. Changes in the values of **speed** and **angle** would **not** affect the values of **v** and **alpha**.

The variables in the header of the Pascal code behave similarly to those in the C code. Changes to values of **Range** and **Time** cause changes to **R** and **T**; however, the values of **V** and **Alpha** would be unaffected by changes to **Speed** and **Angle**.

Code B–1   FORTRAN Code for Pseudocode B–1

```
*************************************************************
*    Code to compute the maximum height HMAX, the time  *
*    of flight T, and the range R of a projectile that  *
*    is launched with an initial speed V at an angle    *
*    ALPHA to the horizontal. Drag is assumed to be     *
*    negligible, and the gravitational acceleration G   *
*    is assumed to be constant at 9.81 m/s/s.           *
*************************************************************

      PROGRAM FLIGHT
      REAL ALPHA, V, HMAX, T, R

      WRITE(6,*) 'Please give V [m/s] and ALPHA [rad]'
      READ(5,*) V, ALPHA

      CALL RT(V, ALPHA, R, T)

      WRITE(6,*) 'Maximum height [m] =', HMAX(V, ALPHA)
      WRITE(6,*) 'Time of flight [s] =',  T
      WRITE(6,*) '          Range [m] =',  R

      STOP
      END
```

```
*-----------------------------------------------------------
        REAL FUNCTION HMAX (SPEED, ANGLE)

*   Function to compute the maximum height

        REAL SPEED, ANGLE, G

        G = 9.81
        HMAX = (SPEED*SIN(ANGLE))**2/(2.0* G)

        RETURN
        END


*-----------------------------------------------------------
        SUBROUTINE RT(SPEED, ANGLE, RANGE, TIME)

*   Subroutine to compute the range and time of flight

        REAL SPEED, ANGLE, RANGE, TIME, G

        G = 9.81
        TIME = 2.0*SPEED*SIN(ANGLE)/G
        RANGE = TIME*SPEED*COS(ANGLE)

        RETURN
        END
```

Code B–2   ANSI C Code for Pseudocode B–1

```
#include <stdio.h>
#include <math.h>

#define G 9.81

double hmax(double, double);
void rt(double, double, double *, double *);
/*********************************************************/
/* Code to compute the maximum height hmax, the time */
/* of flight t, and the range r of a projectile that */
```

```c
/*  is launched with an initial speed v at an angle  */
/*  alpha to the horizontal. Drag is assumed to be   */
/*  negligible, and the gravitational acceleration G */
/*  is assumed to be constant at 9.81 m/s/s.         */
/***************************************************/

main()
{
   double alpha, v, t, r;

   printf("Please give v [m/s] and alpha [rad]\n");
   scanf("%lf %lf", &v, &alpha);

   rt(v, alpha, &r, &t);

   printf("Maximum height [m] = %f\n", hmax(v, alpha));
   printf("Time of flight [s] = %f\n", t);
   printf("          Range [m] = %f\n", r);
}

/*----------------------------------------------------------*/

double hmax(double speed, double angle)

/* Function to compute the maximum height */

{
   return pow(speed*sin(angle), 2.0)/(2.0*G);
}

/*----------------------------------------------------------*/

void rt(double speed, double angle, double *rptr,
                                    double *tptr)

/* Function to compute the range and time of flight */

{
   *tptr = 2.0*speed*sin(angle)/G;
   *rptr = *tptr*speed*cos(angle);
}
```

Code B–3   Turbo Pascal Code for Pseudocode B–1

```
(********************************************************)
{   Code to compute the maximum height Hmax, the time   }
{   of flight T, and the range R of a projectile that   }
{   is launched with an initial speed V at an angle     }
{   Alpha to the horizontal.  Drag is assumed to be     }
{   negligible, and the gravitational acceleration g    }
{   is assumed to be constant at 9.81 m/s/s.            }
(********************************************************)

program Flight;

   var Alpha, V, T, R:Real;

{-------------------------------------------------------}
   function Hmax(Speed, Angle:Real):Real;

   {  Function to compute the maximum height }

      var g:Real;

   begin  { Hmax }
      g := 9.81;
      Hmax := Sqr(Speed*Sin(Angle))/(2.0*g);
   end;  { Hmax }

{-------------------------------------------------------}
   procedure RT(Speed, Angle:Real; var Range, Time:Real);

   {  Procedure to compute the range and time of flight }

      var g:Real;

   begin  { RT }
      g := 9.81;
      Time := 2.0*Speed*Sin(Angle)/g;
      Range := Time*Speed*Cos(Angle);
   end;  { RT }

{-------------------------------------------------------}
```

```
begin  { Flight }
   WriteLn('Please give V [m/s] and Alpha [rad]');
   ReadLn(V, Alpha);

   RT(V, Alpha, R, T);     { Procedure call }

   WriteLn('Maximum height [m] =', Hmax(V, Alpha));
   WriteLn('Time of flight [s] =', T);
   WriteLn('         Range [m] =', R);

end.  { Flight }
```

Appendix C

Useful Mathematical Relations

Various mathematical relations are presented here. They are arranged alphabetically by topic.

C.1 MATRIX NORMS AND SPECTRA

Matrix norms and spectral radii were introduced in Chapter 2. In particular, the norms $\|\mathbf{A}\|_1$ and $\|\mathbf{A}\|_\infty$ of an $(n \times n)$ matrix $\mathbf{A}$ were defined as

$$\|\mathbf{A}\|_1 = \max_j \left[\sum_{i=1}^{n} |a_{ij}| \right] = \text{maximum absolute column sum}$$

$$\|\mathbf{A}\|_\infty = \max_i \left[\sum_{j=1}^{n} |a_{ij}| \right] = \text{maximum absolute row sum}$$

For the matrix

$$\mathbf{A} = \begin{bmatrix} 2 & -1 & 0 \\ -1 & 3 & -2 \\ 0 & -4 & 8 \end{bmatrix}$$

we have

$$\|\mathbf{A}\|_1 = 10; \quad \|\mathbf{A}\|_\infty = 12$$

The *spectrum* of a matrix is the set of eigenvalues λ for which the determinant $|\lambda\mathbf{I} - \mathbf{A}|$ is zero. The eigenvalues of our matrix $\mathbf{A}$ are given in Chapter 8 and are

$$\lambda_1 = 9.2984; \quad \lambda_2 = 2.7697; \quad \lambda_3 = 0.9319$$

Thus, we have

$$\text{Spectrum of } \mathbf{A} = \{9.2984, 2.7697, 0.9319\}$$

The *spectral radius* of $\mathbf{A}$ is denoted by $\rho(\mathbf{A})$, and it is the maximum absolute eigenvalue of $\mathbf{A}$. For our case, we have

$$\rho(\mathbf{A}) = 9.2984$$

The *Euclidean norm* of a matrix is

$$\|\mathbf{A}\|_2 = \sqrt{\rho(\mathbf{A}^{\mathbf{T}} \cdot \mathbf{A})}$$

In our example,

$$\mathbf{A}^{\mathbf{T}} \cdot \mathbf{A} = \begin{bmatrix} 5 & -5 & 2 \\ -5 & 26 & -38 \\ 2 & -38 & 68 \end{bmatrix}$$

is a symmetric matrix whose characteristic equation for its eigenvalues μ is

$$\mu^3 - 99\mu^2 + 765\mu - 576 = 0$$

The solutions of the characteristic equation are

$$\mu_1 = 90.6291; \quad \mu_2 = 7.5264; \quad \mu_3 = 0.8444$$

Therefore,

$$\|\mathbf{A}\|_2 = \sqrt{\rho(\mathbf{A}^{\mathbf{T}} \cdot \mathbf{A})} = \sqrt{\mu_1} = 9.5199$$

The Euclidean norm is roughly the same as the other norms. The similarity in the values of the norms is to be expected because a norm is a measure of the magnitude of the matrix.

The Euclidean norm is sometimes called the *spectral norm*. If $\mathbf{A}$ is a symmetric matrix, the eigenvalues λ of $\mathbf{A}$ are related to the eigenvalues μ by

$$\lambda^2 = \mu$$

In such cases, the spectral radius of $\mathbf{A}$ is identical to the spectral norm. The matrix $\mathbf{A}$ in our example is not symmetric; nevertheless, we see that the eigenvalues μ have roughly the same values as λ^2.

C.2 NEWTON–COTES INTEGRATION FORMULAS

Closed Formulas

$$\int_a^{a+h} f(x)\,dx = (h/2)(f_0 + f_1) - (h^3/12)\,d^2f(\xi)/dx^2;$$
$$f_n = f(a + nh), \quad a < \xi < a + h \qquad \textit{[trapezoidal rule]}$$

$$\int_a^{a+2h} f(x)\,dx = (h/3)(f_0 + 4f_1 + f_2) - (h^5/90)\,d^4f(\xi)/dx^4;$$
$$f_n = f(a + nh), \quad a < \xi < a + 2h \qquad \textit{[Simpson's 1/3 rule]}$$

$$\int_a^{a+3h} f(x)\,dx = (3h/8)(f_0 + 3f_1 + 3f_2 + f_3) - (3h^5/80)\,d^4f(\xi)/dx^4;$$
$$f_n = f(a + nh), \quad a < \xi < a + 3h \qquad \textit{[Simpson's 3/8 rule]}$$

$$\int_a^{a+4h} f(x)\,dx = (2h/45)(7f_0 + 32f_1 + 12f_2 + 32f_3 + 7f_4)$$
$$-(8h^7/945)\,d^6f(\xi)/dx^6; \quad f_n = f(a + nh), \quad a < \xi < a + 4h$$

Open Formulas

$$\int_a^{a+2h} f(x)\,dx = 2hf_1 + (h^3/3)\,d^2f(\xi)/dx^2;$$
$$f_n = f(a + nh), \quad a < \xi < a + 2h \qquad \textit{[midpoint rule]}$$

$$\int_a^{a+3h} f(x)\,dx = (3h/2)(f_1 + f_2) + (h^3/4)\,d^2f(\xi)/dx^2;$$
$$f_n = f(a + nh), \quad a < \xi < a + 3h$$

$$\int_a^{a+4h} f(x)\,dx = (4h/3)(2f_1 - f_2 + 2f_3) + (28h^5/90)\,d^4f(\xi)/dx^4;$$
$$f_n = f(a + nh), \quad a < \xi < a + 4h \qquad \textit{[Milne's rule]}$$

C.3 QUADRATIC EQUATIONS

The familiar solutions of the quadratic equation

$$ax^2 + bx + c = 0; \quad a, b, c \text{ real}$$

are given by

$$x = \left[-b \pm \sqrt{b^2 - 4ac} \right] \Big/ (2a)$$

A less familiar form of the solutions is

$$x = 2c \Big/ \left[-b \pm \sqrt{b^2 - 4ac} \right]$$

If $4ac$ is very small compared to b^2, the discriminant (the square root term) is very nearly equal to b, and one of the roots from either set of equations may be computed inaccurately because of subtractive cancellation. The roots x_1 and x_2 may be found from a sequence of computations that avoids subtractive cancellation as follows.

$$q = \left[b + \text{sgn}(b) \sqrt{b^2 - 4ac} \right] \Big/ 2; \quad x_1 = -q/a; \quad x_2 = -c/q$$

The computation of the discriminant is itself subject to cancellation error when $4ac$ differs only slightly from b^2. However, the magnitude of the discriminant is small compared to b in such cases, and the error in the solution is at worst on the order of b times the square root of machine epsilon.

C.4 SERIES EXPANSIONS FOR COMMON FUNCTIONS

Binomial

$$(x + y)^n = x^n + nx^{n-1}y + n(n-1)x^{n-2}y^2/2! + \cdots + y^n; \quad y^2 < x^2$$

Exponential and Logarithmic

$$e^x = 1 + x + x^2/2! + x^3/3! + \cdots$$
$$\ln x = 2(z + z^3/3 + z^5/5 + \cdots); \quad z = (x - 1)/(x + 1), \ x > 0$$
$$\ln(1 + x) = x - x^2/2 + x^3/3 - x^4/4 + \cdots; \quad -1 < x \le 1$$

Hyperbolic

$$\sinh x = (e^x - e^{-x})/2 = x + x^3/3! + x^5/5! + \cdots$$
$$\cosh x = (e^x + e^{-x})/2 = 1 + x^2/2! + x^4/4! + \cdots$$

Trigonometric

$$\sin x = x - x^3/3! + x^5/5! - x^7/7! + \cdots; \quad x \text{ in radians}$$
$$\cos x = 1 - x^2/2! + x^4/4! - x^6/6! + \cdots; \quad x \text{ in radians}$$
$$\tan x = x + x^3/3 + 2x^5/15 + 17x^7/315 + 62x^9/2835 + \cdots; \quad x \text{ in radians}$$

C.5 SPECIAL FUNCTIONS

Error Function

$$\mathrm{erf}(x) = (2/\sqrt{\pi}) \int_0^x e^{-t^2} dt$$

$$\mathrm{erf}(0) = 0; \quad \mathrm{erf}(\infty) = 1; \quad \mathrm{erf}(x) = -\mathrm{erf}(-x)$$
$$\gamma(1/2, x^2)/\Gamma(1/2) = \mathrm{erf}(x); \quad x \geq 0$$

Gamma Function

$$\Gamma(\alpha) = \int_0^\infty e^{-t} t^{(\alpha-1)} dt$$

$$\Gamma(1) = 1; \quad \Gamma(1/2) = \sqrt{\pi}$$
$$\Gamma(\alpha + 1) = \alpha\Gamma(\alpha); \quad \Gamma(\alpha)\Gamma(1-\alpha) = \pi/\sin(\pi\alpha)$$

The **_Weierstrass_** definition of the gamma function is

$$1/\Gamma(\alpha) = \alpha \exp(\alpha C) \prod_{n=1}^{\infty} [(1 + \alpha/n) \exp(-\alpha/n)]$$

where C is the **_Euler-Mascheroni constant_** given by the limiting value of

$$C = \left[\sum_{k=1}^{n} (1/k) \right] - \ln n; \quad n \to \infty$$

The constant C is often called ***Euler's constant*** and is also denoted by the symbol γ; however, γ is sometimes used to indicate $\exp(C)$. The approximate value of C is given by

$$C \simeq 0.577\,215\,664\,901\,532$$

Stirling's asymptotic series for the gamma function with a large positive argument is

$$\Gamma(\alpha + 1) \simeq \sqrt{2\pi\alpha}(\alpha/e)^{\alpha}\{1 + 1/(12\alpha) + 1/(288\alpha^2)$$
$$- 139/(51840\alpha^3) - 571/(2488320\alpha^4)\}; \quad \alpha \gg 1$$

Incomplete Gamma Function

There is more than one function known as the incomplete gamma function. The form that follows is the form described in Chapter 4.

$$\gamma(\alpha, x) = \int_0^x e^{-t} t^{(\alpha - 1)} dt$$

$$\gamma(\alpha, \infty) = \Gamma(\alpha)$$

$$\gamma(\alpha, x)/\Gamma(\alpha) = e^{-x} x^{\alpha} \sum_{n=0}^{\infty} [x^n/\Gamma(\alpha + 1 + n)]$$

Standardized Normal Distribution

$$\Phi(z) = (1/\sqrt{2\pi}) \int_{-\infty}^{z} e^{-(u^2/2)} du$$

$$\Phi(-\infty) = 0; \quad \Phi(\infty) = 1; \quad \Phi(0) = 0.5; \quad \Phi(z) = 1 - \Phi(-z)$$

$$\Phi(z) = [1 + \gamma(1/2, z^2/2)/\Gamma(1/2)]/2; \quad z \geq 0$$

C.6   SUMS OF POWERS OF INTEGERS

$$\sum_{i=1}^{n} i = n(n + 1)/2$$

$$\sum_{i=1}^{n} i^2 = n(n + 1)(2n + 1)/6$$

$$\sum_{i=1}^{n} i^3 = n^2(n + 1)^2/4$$

C.7   TAYLOR SERIES

Single Variable

$$f(x + h) = f(x) + \sum_{n=1}^{\infty} \{(h^n/n!)(d^n f(x)/dx^n)\}$$

$$f(x + h) = f(x) + \sum_{n=1}^{k-1} \{(h^n/n!)(d^n f(x)/dx^2)\}$$
$$+ (h^k/k!)(d^k f(\xi)/dx^k); \quad x < \xi < x + h$$

Several Variables

$$f(x_1 + h_1, x_2 + h_2, \ldots, x_m + h_m) = f(x_1, x_2, \ldots, x_m)$$
$$+ \sum_{n=1}^{\infty} \{\mathcal{D}^n f(x_1, x_2, \ldots, x_m)/n!\}; \quad \mathcal{D} = \sum_{j=1}^{m} (h_j \, \partial/\partial x_j)$$

Appendix D
Physical Models

Physical models that pertain to many of the applied examples and problems in the text are presented here to provide additional background. Topics are arranged alphabetically.

D.1 ARCS AND BODIES OF REVOLUTION

Consider an arc that is described in Cartesian coordinates by

$$y = y(x)$$

The length s of the arc between the x positions x_1 and x_2 is obtained from

$$ds^2 = dx^2 + dy^2 = dx^2(1 + [dy/dx]^2)$$
$$\rightarrow s = \int_{x_1}^{x_2} \sqrt{1 + [dy/dx]^2} \, dx$$

The radius of curvature ρ at a point (x, y) on the arc is given by

$$\rho = (1 + [dy/dx]^2)^{1.5}/(d^2y/dx^2)$$

If the arc is expressed in polar coordinates by $[r = r(\theta)]$, corresponding expressions for s and ρ are given by

385

$$s = \int_{\theta_1}^{\theta_2} \sqrt{r^2 + [dr/d\theta]^2} \ d\theta$$

$$\rho = (r^2 + [dr/d\theta]^2)^{1.5}/(r^2 + 2[dr/d\theta] - r[d^2r/d\theta^2])$$

Let us now consider a body of revolution that is formed by revolving the arc $[y = y(x)]$ about the x axis (with y assumed to be positive or zero). The surface area A and volume V of the body between the planes $(x = x_1)$ and $(x = x_2)$ are given by

$$A = 2\pi \int_{x_1}^{x_2} y \ \sqrt{1 + [dy/dx]^2} \ dx; \quad V = \pi \int_{x_1}^{x_2} y^2 \ dx$$

D.2   PARTICLE DYNAMICS

A particle of mass m that is subjected to a net (vector) force $\mathbf{F}$ has a (vector) acceleration $\mathbf{a}$ that is given by

$$\mathbf{F} = m\mathbf{a} = m \ d\mathbf{v}/dt = m \ d^2\mathbf{r}/dt^2$$

according to Newton's Second Law of Motion. Here, t denotes time, $\mathbf{v}$ is the (vector) velocity, and $\mathbf{r}$ is the (vector) position. If the vectors are expressed in Cartesian coordinates with unit vectors $\mathbf{i}$ and $\mathbf{j}$ in the x and y directions, respectively, we have

$$\mathbf{F} = F_x\mathbf{i} + F_y\mathbf{j}; \quad \mathbf{a} = a_x\mathbf{i} + a_y\mathbf{j}; \quad \mathbf{v} = v_x\mathbf{i} + v_y\mathbf{j}; \quad \mathbf{r} = x\mathbf{i} + y\mathbf{j}$$

and the resulting scalar equations of motion are

$$F_x = ma_x = m \ dv_x/dt = m \ d^2x/dt^2 = mv_x \ dv_x/dx$$
$$F_y = ma_y = m \ dv_y/dt = m \ d^2y/dt^2 = mv_y \ dv_y/dy$$

In an intrinsic coordinate system, the tangential unit vector $\mathbf{t}$ is tangent to the path of the particle and positive in the direction of motion, and the normal unit vector $\mathbf{n}$ is perpendicular to $\mathbf{t}$. These unit vectors change directions relative to an absolute coordinate system as the particle moves along a generally curved path. The vectors $\mathbf{F}$, $\mathbf{a}$, and $\mathbf{v}$ are expressed by

$$\mathbf{F} = F_t\mathbf{t} + F_n\mathbf{n}; \quad \mathbf{a} = a_t\mathbf{t} + a_n\mathbf{n}; \quad \mathbf{v} = v\mathbf{t}$$

and the scalar equations of motion are

$$F_t = ma_t = m \ dv/dt = m \ d^2s/dt^2 = mv \ dv/ds$$
$$F_n = ma_n = mv^2/\rho$$

In the scalar equations, s denotes the arc length and ρ is the radius of curvature given in Section D.1. The radius of curvature is positive if the center of curvature is in the same direction as **n** or negative if the center of curvature is opposite to **n**.

Forces have a variety of sources; among them are gravity, friction, and linear springs. The gravitational force component in the direction of the gravitational acceleration g is the weight mg of the particle. The friction force associated with the dynamic coefficient of friction μ has a component μN opposite to the velocity, where N is the magnitude of the normal reaction on the particle by the surface on which it is sliding. A spring force has a component $k\Delta x$, in which k is the spring stiffness and Δx denotes the amount that the spring is stretched or compressed from its normal length. The direction of the force exerted by the spring is along the spring and opposite to the extension or compression.

D.3   RESISTOR NETWORKS

The basic laws for analyzing resistor networks are Ohms's Law, Kirchhoff's Voltage Law, and Kirchhoff's Current Law. These are stated as follows.

> **Ohms's Law:** The voltage drop across a resistance R is (iR) in the direction of the current i.
>
> **Kirchhoff's Voltage Law:** The net voltage drop in a closed loop is zero.
>
> **Kirchhoff's Current Law:** The net current leaving a node of a circuit is zero.

We illustrate the use of these laws by considering the circuit of Fig. D–1.

The diagram on the left of Fig. D–1 is used to find the loop currents i_1, i_2, and i_3 when the voltage V and the resistances R_1 through R_5 are known. The analysis in this case is known as a *loop* analysis. For each of the three loops, we use Ohm's Law and Kirchhoff's Voltage Law to write the following.

Loop 1 : $\qquad\qquad\qquad i_1R_1 + (i_1 - i_3)R_2 - V = 0$

Loop 2 : $\qquad\qquad\qquad\quad i_2R_3 + (i_2 - i_3)R_4 = 0$

Loop 3 : $\qquad (i_3 - i_1)R_2 + (i_3 - i_2)R_4 + i_3R_5 = 0$

The resulting system of equations is

$$
\begin{bmatrix}
R_1 + R_2 & 0 & -R_2 \\
0 & R_3 + R_4 & -R_4 \\
-R_2 & -R_4 & R_2 + R_4 + R_5
\end{bmatrix}
\cdot
\begin{bmatrix}
i_1 \\ i_2 \\ i_3
\end{bmatrix}
=
\begin{bmatrix}
V \\ 0 \\ 0
\end{bmatrix}
$$

For V equal to 2 and R_j equal to 1, the loop currents are

$$
i_1 = 1.25 \quad ; i_2 = 0.25; \quad i_3 = 0.5
$$

The diagram on the right of Fig. D–1 is used to find the voltages v_1, v_2, and v_3 by a *nodal* analysis. A node is simply a point on the circuit at which a branch in the current occurs. For convenience, we designate the voltage at one of these nodes as a zero reference, and we typically choose a node that connects many branches as the reference. For our circuit, we have chosen v_3 to be zero. We now use Ohm's law and Kirchhoff's Current Law to write the following at the remaining two nodes.

Node 1: $(v_1 + V - v_3)/R_1 + (v_1 - v_3)/R_2 + (v_1 - v_2)/R_5 = 0$

Node 2: $(v_2 - v_3)/R_3 + (v_2 - v_3)/R_4 + (v_2 - v_1)/R_5 = 0$

The system of equations with v_3 set to zero is

$$\begin{bmatrix} 1/R_1 + 1/R_2 + 1/R_5 & -1/R_5 \\ -1/R_5 & 1/R_3 + 1/R_4 + 1/R_5 \end{bmatrix} \cdot \begin{bmatrix} v_1 \\ v_2 \end{bmatrix} = \begin{bmatrix} -V/R_1 \\ 0 \end{bmatrix}$$

For V equal to 2 and R_j equal to 1, the solutions are

$$v_1 = -0.75; \quad v_2 = -0.25$$

The currents from the loop analysis and the voltages from the nodal analysis are, of course, related. For example, the loop current i_3 through R_5 is related to the voltage drop $(v_2 - v_1)$ across R_5 through Ohm's Law according to

$$v_2 - v_1 = i_3 R_5$$

D.4 ROCKET THRUST

The thrust on a rocket is due to the reaction between the rocket and the propellant that is exhausted. Consider a system consisting of a rocket and its propellant acted

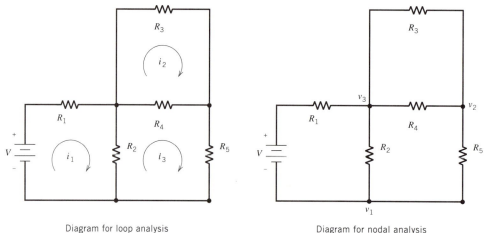

Diagram for loop analysis   Diagram for nodal analysis

Figure D–1 Circuit diagrams for loop and nodal analyses.

on by a net external force $\mathbf{F}$. The mass and (vector) velocity of the rocket at time t are m and $\mathbf{v}$, respectively. At time $(t + \Delta t)$, the mass is reduced by an amount Δm that is equal to the mass of the propellant that is expelled, and the velocity is $(\mathbf{v} + \Delta \mathbf{v})$. The propellant is exhausted at a velocity $\mathbf{v}_e$ ***relative*** to the rocket; therefore, the propellant velocity is $(\mathbf{v} + \Delta \mathbf{v} + \mathbf{v}_e)$. A schematic of the problem is shown in Fig. D–2.

The change in momentum is equal to the impulse; therefore,

$$\mathbf{F}\Delta t = (m - \Delta m)(\mathbf{v} + \Delta \mathbf{v}) + \Delta m(\mathbf{v} + \Delta \mathbf{v} + \mathbf{v}_e) - m\mathbf{v}$$

Let q be the mass rate at which the propellant is exhausted; then Δm is equal to $q\Delta t$, and the impulse–momentum equation is rewritten as

$$\mathbf{F}\Delta t = (m - q\Delta t)(\mathbf{v} + \Delta \mathbf{v}) + q\Delta t(\mathbf{v} + \Delta \mathbf{v} + \mathbf{v}_e) - m\mathbf{v}$$

or

$$m\Delta \mathbf{v} = (\mathbf{F} - q\mathbf{v}_e)\Delta t$$

In the limit, as Δt tends to zero, we have

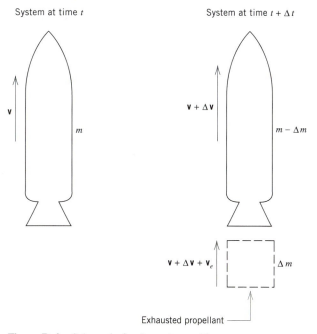

Figure D–2   Schematic for the rocket problem.

$$m(d\mathbf{v}/dt) = m\mathbf{a} = \mathbf{F} - q\mathbf{v}_e$$

in which $\mathbf{a}$ is the (vector) acceleration of the rocket. The term $(-q\mathbf{v}_e)$ is the thrust, and we may model the rocket at time t as a body of mass m acted on by an external force that is equal to $(\mathbf{F} - q\mathbf{v}_e)$.

In the case of forward rectilinear motion of the rocket with the exhaust velocity opposite to the motion, we denote the unit vector in the direction of motion by $\mathbf{i}$ and write

$$\mathbf{F} = F\mathbf{i}; \quad v = v\mathbf{i}; \quad \mathbf{a} = a\mathbf{i}; \quad \mathbf{v}_e = -v_e\mathbf{i}$$

The scalar equation of motion is then given by

$$m(dv/dt) = ma = F + qv_e$$

D.5 STREAM FUNCTIONS AND STREAMLINES

Conservation of mass in an incompressible, two-dimensional flow in terms of the velocity components u and v in the x and y directions, respectively, is expressed by

$$\partial u/\partial x + \partial v/\partial y = 0$$

A function ψ of x, y, and time t may be defined by

$$u = \partial\psi/\partial y = \psi_y; \quad v = -\partial\psi/\partial x = -\psi_x$$

This function satisfies the equation for conservation of mass and is known as the stream function.

A **streamline** is a curve that is tangent to the flow velocity at every point on the curve at a given instant of time. Thus, the equation for a streamline is

$$dy/dx = v/u$$

or

$$u \, dy - v \, dx = 0$$

If we replace u and v by the partial derivatives of ψ, we obtain

$$\psi_y \, dy + \psi_x \, dx = 0$$

The left-hand side of the preceding expression is the change $d\psi$ at a given instant. This change is zero on a streamline; therefore, a streamline is identified by a constant value of the stream function ψ at a given instant.

The definition of a streamline means that there is no flow across the streamline. If we have two streamlines identified by $(\psi = \psi_1)$ and $(\psi = \psi_2)$, the flow rate across any two curves connecting the two streamlines must be equal. Indeed, the volume flow rate Q of the fluid is given by

$$Q = |\psi_2 - \psi_1|$$

If, in addition to being incompressible, the flow is irrotational (for example, because fluid viscosity is neglected), the irrotationality condition is expressed by

$$\partial v / \partial x - \partial u / \partial y = 0$$

When we substitute ψ_y and $(-\psi_x)$ for u and v, respectively, we obtain the Laplace equation in the stream function

$$\nabla^2 \psi = \psi_{xx} + \psi_{yy} = 0$$

as the governing equation for the flow.

D.6 TRUSS EQUILIBRIUM

A simple plane truss consists of m members and j joints such that

$$m = 2j - 3$$

The basic form of the truss is a triangle whose sides are the members and whose vertices are the joints. A general truss is built up from the basic form by adding two members and one joint at a time.

Consider the truss with m equal to 9 and j equal to 6 as shown in Fig. D–3. Each horizontal member is 9 ft long, and each vertical member is 12 ft long.

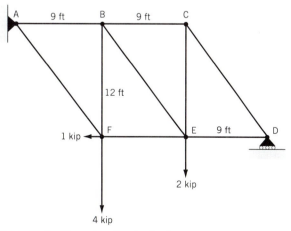

Figure D–3 Geometry of a simple plane truss.

The support at joint A is a pin support, which can carry a two-component force. The support at joint D is a roller support, which can carry only a vertical reaction. The known external loads are the 1-kip, 2-kip, and 4-kip forces that are shown at joints E and F. The problem is to determine the external forces at A and D and the internal forces in the members.

A free-body diagram for the joints of the truss shows both internal and external forces as depicted in Fig. D–4. The internal forces are represented by tensions t_1 through t_9, and the external supporting forces are denoted by t_{10}, t_{11}, and t_{12}.

The forces at each joint act through the same point — the joint. In such cases, the equilibrium equations are

$$\text{Sum of forces in the } x \text{ direction} = 0$$
$$\text{Sum of forces in the } y \text{ direction} = 0$$

We therefore have two equations at each joint, and we can solve for the member forces plus three other forces from the $2j$ or $(m + 3)$ equations.

We denote the unit vectors in the x and y directions by $\mathbf{i}$ and $\mathbf{j}$, respectively, and we note that the unit vector $\mathbf{n}$ in the direction from joint E to joint B is given by

$$\mathbf{n} = (-9\mathbf{i} + 12\mathbf{j})/ \mid -9\mathbf{i} + 12\mathbf{j} \mid$$
$$= (-9\mathbf{i} + 12\mathbf{j})/ \sqrt{9^2 + 12^2} = -0.6\mathbf{i} + 0.8\mathbf{j}$$

The equations at the joints are then written as follows.

Joint A : i component $t_1 + t_{10} + 0.6t_2 = 0$
            j component $-0.8t_2 + t_{11} = 0$
Joint B : i component $-t_1 + t_4 + 0.6t_5 = 0$
            j component $-t_3 - 0.8t_5 = 0$

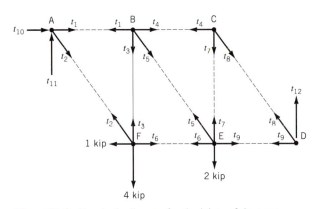

Figure D–4 Free-body diagram for the joints of the truss.

Joint C : **i** component $-t_4 + 0.6t_8 = 0$
 j component $-t_7 + 0.8t_8 = 0$

Joint D : **i** component $-0.6t_8 - t_9 = 0$
 j component $0.8t_8 + t_{12} = 0$

Joint E : **i** component $-0.6t_5 - t_6 + t_9 = 0$
 j component $0.8t_5 + t_7 - 2 \text{ kip} = 0$

Joint F : **i** component $-0.6t_2 + t_6 - 1 \text{ kip} = 0$
 j component $0.8t_2 + t_3 - 4 \text{ kip} = 0$

The twelve equations at joints A through E form a system of linear algebraic equations whose solutions give the required forces.

It is clear that the coefficient matrix for the system of equations is sparse. We now look at the **method of joints**, which can help us to take advantage of the sparseness. The method of joints requires us to determine some or all of the external forces first. Then we use those solutions to write equilibrium equations at the joints in a sequence that contains no more than two unsolved forces at any stage.

The external forces may be found by considering the external forces on the entire truss. We then have a general equilibrium problem for which the conditions are

Sum of forces in the x direction = 0

Sum of forces in the y direction = 0

Sum of moments about any fixed point = 0

Let $(r_x\mathbf{i} + r_y\mathbf{j})$ be the vector from a fixed point to any point on the line of application of a force vector $(F_x\mathbf{i} + F_y\mathbf{j})$. The moment of the force about the fixed point is given by the **cross product** relation

$$\mathbf{M} = M\mathbf{k} = (r_x\mathbf{i} + r_y\mathbf{j}) \times ((F_x\mathbf{i} + F_y\mathbf{j}) = (r_xF_y - r_yF_x)\mathbf{k}$$

A positive value of M indicates that the force has a counterclockwise turning effect about the fixed point; a negative value of M indicates a clockwise turning effect. A judicious choice of the fixed point can simplify the solution process for the general equilibrium problem.

The solution of the external support forces proceeds as follows. We use the diagram of Fig. D–4 (ignoring the internal forces) to write the following.

i component : $t_{10} - \text{kip} = 0$
j component : $t_{11} + t_{12} - 2\text{kip} - 4\text{kip} = 0$
M about A : $(27 \text{ ft})t_{12} - (18 \text{ ft})(2 \text{ kip}) - (9 \text{ ft})(4 \text{ kip}) - (12 \text{ ft})(1 \text{ kip}) = 0$

The three equations are easily solved for t_{10}, t_{11}, and t_{12}; however, only t_{10} and t_{11} are required for us to proceed with the method of joints.

We now apply the equilibrium equations at the joints to solve two of the remaining forces at a time. One sequence that allows us to do so is

Joint A to find t_1 and t_2
Joint F to find t_3 and t_6
Joint B to find t_4 and t_5
Joint C to find t_7 and t_8
Joint D to find t_9 and t_{12}

Note that t_{12} (although available as an external force) is included in our system as an unknown so that we have an even number of forces. If we order the solution vector **t** in the given sequence so that

$$\mathbf{t}^{\mathbf{T}} = [t_1 \; t_2 \; t_3 \; t_6 \; t_4 \; t_5 \; t_7 \; t_9 \; t_{12}]$$

the resulting coefficient matrix for the system of equations will have a lower triangular block form with each block consisting of a (2×2) submatrix. It is considerably more efficient to use a block method to solve the system than it is to use one of the more general matrix solvers.

References

1. Brookshear, J. Glenn, *Computer Science — An Overview,* 3rd ed., Benjamin-Cummings, Redwood City, CA, 1991.
2. McKeown, Patrick G., *Living with Computers,* 3rd ed., Harcourt Brace Jovanovich, San Diego, CA, 1991.
3. Wear, Larry L., Pinkert, James R., Wear, Larry C., and Lane, William G., *Computers — An Introduction to Hardware and Software Design,* McGraw-Hill Book Co., New York, 1991.
4. Isaacson, Eugene, and Keller, Herbert Bishop, *Analysis of Numerical Methods,* John Wiley & Sons, New York, 1966.
5. Hostetter, Gene H., Santina, Mohammed S., and D'Carpio-Montalvo, Paul, *Analytical, Numerical, and Computational Methods for Science and Engineering,* Prentice Hall, Englewood Cliffs, NJ, 1991.
6. Mitchell, A. R., and Griffiths, D. F., *The Finite Difference Method in Partial Differential Equations*, John Wiley & Sons, Chichester, U.K., 1980.
7. Kreyszig, Erwin, *Advanced Engineering Mathematics,* 5th ed., John Wiley & Sons, New York, 1988.
8. Press, William H., Flannery, Brian P., Teukolsky, Saul A., and Vetterling, William T., *Numerical Recipes,* Cambridge University Press, Cambridge, U.K., 1986.
9. Andrews, Larry C., *Special Functions for Engineers and Applied Mathematicians,* Macmillan, New York, 1985.
10. Yakowitz, Sidney, and Szidarovszky, Ferenc, *An Introduction to Numerical Computations,* Macmillan, New York, 1986.
11. Buchanan, James L., and Turner, Peter R., *Numerical Methods and Analysis,* McGraw-Hill Book Co., New York, 1992.

12. Gear, C. William, *Numerical Initial Value Problems in Ordinary Differential Equations,* Prentice Hall, Englewood Cliffs, NJ, 1971.

13. Milne, William Edmund, *Numerical Solution of Differential Equations,* John Wiley & Sons, New York, 1953.

14. Atkinson, L. V., Harley, P. J., and Hudson, J. D., *Numerical Methods with Fortran 77,* Addison-Wesley, Wokingham, U.K., 1989.

15. Faddeeva, V. N., (Trans. Curtis D. Benster), *Computational Methods of Linear Algebra,* Dover, New York, 1959.

16. James, M. L., Smith, G. M., and Wolford, J. C., *Applied Numerical Methods for Digital Computation,* 3rd ed., Harper & Row, New York, 1985.

17. Vemuri, V., and Karplus, Walter J., *Digital Computer Treatment of Partial Differential Equations,* Prentice Hall, Englewood Cliffs, NJ, 1981.

18. Buck, R. Creighton, *Advanced Calculus,* 3rd ed., McGraw-Hill Book Co. New York, 1978.

19. Ball, W. W. Rouse, *A Short Account of the History of Mathematics,* Dover, New York, 1960.

20. Debus, Allen G. (Ed.), *World Who's Who in Science from Antiquity to the Present,* Marquis-Who's Who, Chicago, 1968.

21. Pickover, Clifford A., *Mazes for the Mind: Computers and the Unexpected,* St. Martin's Press, New York, 1992.

Bibliography

Carnahan, Brice, Luther, H. A., and Wilkes, James O., *Applied Numerical Methods,* John Wiley & Sons, New York, 1969.

Cavicchi, Thomas J., *Fundamentals of Electrical Engineering Principles and Applications,* Prentice Hall, Englewood Cliffs, NJ, 1993.

Crandall, Stephen H., Dahl, Norman C., and Lardner, Thomas J., *An Introduction to the Mechanics of Solids,* 2nd ed., McGraw-Hill Book Co., New York, 1992.

Fox, Robert W., and McDonald, Alan T., *Introduction to Fluid Mechanics,* 2nd ed., John Wiley & Sons, New York, 1978.

Greenwood, Donald T., *Principles of Dynamics,* 2nd ed., Prentice Hall, Englewood Cliffs, NJ, 1988.

Hoffman, Joe D., *Numerical Methods for Engineers and Scientists,* McGraw-Hill, New York, 1992.

Noble, Ben, *Applied Linear Algebra,* Prentice Hall, Englewood Cliffs, NJ, 1969.

Pearson, Carl E. (Ed.), *Handbook of Applied Mathematics,* 2nd ed., Van Nostrand Reinhold, New York, 1983.

Richtmyer, Robert D., and Morton, K. W., *Difference Methods for Initial-Value Problems,* 2nd ed., John Wiley & Sons, New York, 1967.

Ritger, Paul D., and Rose, Nicholas J., *Differential Equations with Applications,* McGraw-Hill, New York, 1968.

Shames, Irving H., *Engineering Mechanics — Statics and Dynamics,* 3rd ed., Prentice Hall, Englewood Cliffs, NJ, 1980.

Answers to Selected Problems

Chapter 1

1. (1) $R_{eq} = 3226\ \Omega$;   (3) $R_{eq} = 662\ \Omega$

3. $n = 4$

5. (1) $S_{max} = 59867$ psi, $\alpha_{max} = 5.73°$;   (3) $S_{max} = 32904$ psi, $\alpha_{max} = 3.06°$

6. (a) $\alpha = 0.03818$

9. (a) $B = 11.532\epsilon(2 + \epsilon)$;   (c) $B = (2.345/19.28)\epsilon(3 + 2\epsilon)$

12. $h = 0.4$: particle converges to $x = y = 0.3641$

$h = 0.8$: particle oscillates (at large k) on a square with diagonally opposite corners

$x = y = 0.4752$ and $x = y = 0.7976$

Chapter 2

2. $i_1 = 0.1393$ A; $i_2 = 0.5602$ A; $i_3 = 0.1621$ A; $i_4 = 0.0882$ A

5. $h_1 = h_2 = 160.988$ N; $h_3 = 96.593$ N; $h_4 = 0$;

$h_5 = -154.672$ N; $-h_6 = h_7 = 128.790$ N; $h_8 = 193.185$ N

7. $P^{iii} = 126.414$; $P^{iv} = 324.600$; $P^v = 288.225$

9. The inverse of the matrix $\mathbf{A}$ is

$$\mathbf{A}^{-1} = \begin{bmatrix} 0.106215 & 0.050847 & 0.039548 & 0.067797 \\ 0.050847 & 0.112994 & 0.050847 & 0.039548 \\ 0.039548 & 0.050847 & 0.206215 & 0.067797 \\ 0.067797 & 0.039548 & 0.067797 & 0.163842 \end{bmatrix}$$

13. $a_1 = -1.100$; $a_2 = 1.18\bar{3}$; $a_3 = -0.100$; $a_4 = 0.01\bar{6}$

15. $i_1 = -0.9862$ A; $i_2 = -0.1724$ A; $i_3 = -0.0690$ A;

$i_4 = -0.0345$ A

18. $h_1 = h_2 = 128.790$ N; $\quad h_3 = 77.274$ N; $\quad h_4 = 0$;
$h_5 = -123.738$ N; $\quad -h_6 = h_7 = 103.032$ N; $\quad h_8 = 154.548$ N

20. Partial result: $u_{9,2} = 5.474$; $\quad u_{9,3} = 13.365$; $\quad u_{9,4} = 21.657$

Chapter 3

1. **(a)** $\phi = 2.383$ radian; **(c)** $\phi = 5.426$ radian; **(e)** $\phi = 8.027$ radian

2. **(a)** $f = 0.3739$; **(c)** $f = 0.02588$; **(e)** $f = 0.01886$

3. $r = 0.4007$; $\quad c = 0.1011$

5. $L = 3.630$ H

7. $v = 0.05049$ m^3/kg

9. $V = 138.8$ mph; $\quad G = 0.659$

11. $\beta = -4.1694$ s^{-1}; $\quad \omega = 7.2699$ s^{-1}

13. $r = -1.1139, 0.4007, 0.7132$

15. $\lambda = m\omega^2/k = 0.6667, 0.7427, 1.0000, 3.5907$

Chapter 4

3. **(a)** $d_{\text{mean}} = 148.15$ ft, $s = 11.778$ ft; **(b)** Percentage $= 1.13$

5. Median $= 145$ ft; $\quad$ Mode $= 140$ ft

7. The probability distribution function is

$$\phi(z) = \begin{cases} 0; & z < -3 \\ 0.5 + \text{sgn}(z)(|z|/3 - z^2/18); & -3 \le z \le 3 \\ 1; & z > 3 \end{cases}$$

11. $m = 0.0532 \deg^{-1}$; $\quad C_{L_0} = 0.036$

13. $c = 0.06655$ lb^{-1}; $\quad K = 7111978$

15. **(a)** $\lambda = 0.03246$ V^{-1}, $I_s = 27.225$ A;
(b) $\lambda = 0.03198$ V^{-1}, $I_s = 27.341$ A

Chapter 5

1. $z = 1.6448$

3. **(a)** $a_0 = 0.783$, $a_1 = 0.024$, $a_2 = -0.00128$, $a_3 = -0.000256$
(b) Maximum $C_L = 0.896$ at $\alpha = 21.485°$

5. **(1)** $a/b = 1.25$, $c_1 = 4.535$, $c_2 = 0.172$, $S_{\text{max}} = 53145$ psi, $\alpha_{\text{max}} = 4.88°$
(3) $a/b = 3.75$, $c_1 = 3.514$, $c_2 = 0.279$, $S_{\text{max}} = 131781$ psi, $\alpha_{\text{max}} = 24.03°$

8. Partial results:

i	$(d^2x/dp^2)_i$	$(d^2y/dp^2)_i$
2	0.1999	0.0629
3	-0.1997	-0.2515
4	-0.0010	-0.2570
5	-0.3962	0.0796
6	0.3856	-0.0612
7	0.0536	0.1653

p	x	y
1.8	0.1704	0.7570
2.6	0.3216	0.9326
3.4	0.4528	1.0305
4.2	0.5127	0.9698
5.0	0.5000	0.8000
5.8	0.3342	0.6404
6.6	0.2750	0.4728
7.4	0.3366	0.3494

10. Partial results:

$a_0 = 1.33324$, $a_{10} = 0.00448$, $a_{20} = -0.00096$, $a_{30} = 0.00024$

$a_{-k} = a_k$; $b_k = 0$

12. Partial results:

$a_4 = 0.02172$, $a_5 = -0.03147$, $a_6 = 0.01095$; $a_{-k} = a_k$

$b_4 = 0.02539$, $b_5 = -0.03675$, $b_6 = 0.01280$; $b_{-k} = -b_k$

14. $45\tau =$ 4.6951,   9.4178,   14.1366,

18.8478,   23.5619,   28.2773,

32.9866,   37.7062,   42.4313

Chapter 6

1. (a) $a = 10.046$ ft/s^2;   (c) $a = 10.061$ ft/s^2

2. (b) $a = 10.526$ ft/s^2

4. $dh/dt = 270$ ft/min at $t = 24$ min; $dh/dt = 92$ ft/min at $t = 37$ min

7. $x = 221.9$ ft

10. Trapezoidal: $\Delta x = 0.003100$ m;   Simpson's: $\Delta x = 0.003096$ m

13. $(P/E) \times$ (integral from $x = 0.0$ m to $x = 1.5$ m) $= 0.001221$ m

$(P/E) \times$ (integral from $x = 1.5$ m to $x = 3.0$ m) $= 0.001875$ m

16. $A = 0.322$; $x_c = 0.0610$

19. $S = 1.0461$

22. $I_x = 0.00436$; $I_y = 0.03240$

Chapter 7

1. Partial results for the variation of v [m/s] with t [s]:

t	v
2	14.72
6	50.57
10	92.69
14	135.24
18	170.89

4. Partial results for the variation of v [ft/s] with θ [radian]:

θ	v
0.1	19.280
0.3	17.437
0.5	14.964
0.7	11.599
0.9	6.285

7. Partial results for the variation of y [m] with x [m]:

x	y
25	0.793
50	0.784
75	0.773
100	0.759
125	0.740
150	0.713

10. Partial results for the variation of v [m/s] with t [s]:

t	v
2	14.83
6	50.85
10	92.99
14	135.35
18	170.78

13. See partial results for Problem 7.

18. Step size h [m], x [m], and y [m]:

h	x	y
1	1	0.800
2	3	0.799
4	7	0.798
8	15	0.796
16	31	0.791
32	63	0.778
37	100	0.759

23. Actual results depend on the method that is used. The values of x_1, x_2, and x_3 should sum to 1 at any time t. Typical values are as follows:

t	x_1	x_2	x_3
0.1	0.741	0.246	0.013
0.2	0.549	0.405	0.046
0.3	0.407	0.501	0.092
0.4	0.301	0.554	0.145
0.5	0.223	0.575	0.202

26. $T(0.2) = 81.4$, $T(0.4) = 68.5$, $T(0.6) = 60.0$, $T(0.8) = 55.2$, $T(1.0) = 53.7$

Chapter 8

2. Larger moment of inertia $= 0.3056$ kg·m^2

5. $\lambda_{max} = 4$

8. $\lambda_{min} = 0.268$

10. Characteristic equation: $\lambda^4 + 22\lambda^3 + 168\lambda^2 + 525\lambda + 567 = 0$

13. Characteristic equation: $\lambda^4 + 50\lambda^3 + 2250\lambda^2 + 10000\lambda + 250000 = 0$
$\lambda = (-1.0076 \pm 11.0511i), (-23.9924 \pm 38.1382i)$

Chapter 9

1. Explicit method with $\Delta t = 0.004$:

    $$t = 0.012 : \quad u(0.1, t) = 0.0000, \quad u(0.6, t) = 0.7200$$
    $$t = 0.060 : \quad u(0.1, t) = 0.0850, \quad u(0.6, t) = 0.6150$$
    $$t = 0.120 : \quad u(0.1, t) = 0.0987, \quad u(0.6, t) = 0.6013$$

3. $t = 0.5: u(1.0, t) = 0.649, \quad u(1.4, t) = 0.267, \quad u(1.8, t) = 0.104$
    $$u(2.4, t) = 0.038, \quad u(2.6, t) = 0.011, \quad u(3.0, t) = 0.000$$

5. Let the base extend from $(x, y) = (0, 0)$ to $(x, y) = (0.5, 0)$; let $\Delta x = 1/16$, $\Delta y = \sqrt{3}/16$. The ϕ values are symmetrical about the vertical median of the triangle; node values (starting from the median line) are as follows.

0.00000				
0.00440	0.00000			
0.01172	0.00879	0.00000		
0.01318	0.01172	0.00732	0.00000	
0.00000	0.00000	0.00000	0.00000	0.00000

8. Result for $u_t = -t u_x$:

    $$u_{i,n+1} = 0.5[(pt_n)^2 + pt_n + p\Delta t]u_{i-1,n} + 0.5[(pt_n)^2 - pt_n - p\Delta t]u_{i+1,n}$$
    $$+ [1 - (pt_n)^2]u_{i,n}; \quad p = \Delta t / \Delta x$$

11. Define the vector $\mathbf{v}$ as

    $$\mathbf{v} = \begin{bmatrix} u_t \\ u_x \end{bmatrix}$$
    $$\mathbf{v}_t + \begin{bmatrix} c & 1 \\ -1 & 0 \end{bmatrix} \cdot \mathbf{v}_x = \begin{bmatrix} 1 \\ 0 \end{bmatrix} + \begin{bmatrix} -1 & 0 \\ 0 & 0 \end{bmatrix} \cdot \mathbf{v}; \quad c > 2$$

13. **(b)** ξ component $= 2\xi c/h$, η component $= -2\eta c/h$;
    $h = 2\sqrt{\xi^2 + \eta^2}, (\xi, \eta) \neq (0, 0)$

Index

A

ABC computer   5
Abacus   4
Acceleration parameter   99, 303
Ada   7, 9
Adams–Bashforth method   242
Adams, John   355
Adams method   242
Adams–Moulton method   243
Adaptive methods   249
Aiken, Howard   5
Alternating–direction–implicit   302
Analytical Engine   5
Analytical method   44
And   12
Arc   385
Arithmetic/logic unit   5
Array   10
  element   10
Assignment   11, 367
Associative property   46
Asymptotic expansion   164
Atanasoff, John   5

B

BASIC   7, 8
Babbage, Charles   4, 356

Back substitution   62
Bairstow's method   137
Berry, Clifford   5
Bias   29
Biharmonic operator   292, 304
Binary arithmetic   5
Binary number system   23
Birge–Vieta method   148
Bisection method   118
Bit   6
Body of revolution   385
Boolean quantity   12
Boole, George   357
Bottom–up implementation   323
Boundary-value problem   229, 257
Bracketing method   118
Bubble sort algorithm   153
Byron, Ada   4

C

C   7, 8
Cantilevered beam   259
Capacitor–inductor filter   274
Catenary   134
Central processing unit   5
Characteristic   29, 291
Characteristic polynomial   271

Characteristic value   271
Chauvenet's criterion   178
Chebyshev, Pafnuti   357
Chi-squared distribution   167
Class interval   158
Class midpoint   158
Classical Runge–Kutta method   240
Cofactor   49
Column exchange   52
Comment   11, 367
Commutative property   46
Compiled language   7
Computational error   33
Condition number   79
Confidence level   165
Conformable   45
Conjugate   106
Conjugate gradient method   105
Consistency   231, 296
Consistent ordering   99
Constraint   317, 320, 329
Control system model   240
Convergence   130, 231
Coordinate transformation   310
   curl   312
   divergence   311
   gradient   312
   metric   311
   orthogonal   310
   vector components   311
Correlation coefficient   173
Cotes, Roger   357
Courant–Friedrichs–Lewy condition   306
Cramer, Gabriel   357
Cramer's rule   56
Crank–Nicolson method   297
Criterion   317, 320
Criterion function   323
Critical Mach number   120
Cubic spline   187
   natural   189
Cumulative distribution function   162
Curve fitting   181

D

DOS   6
Data type   10
Degrees of freedom   167

Density function   161
Descartes, René   358
Descartes' rule of signs   276
Design   317
   solar concentrator problem   318
Design process   317
Design projects   334
   airship gas envelope   335
   plane truss   337
   four–bar linkage   339
   rack and pinion steering   342
   rocket configuration   351
   software application   349
   ventilation system   348
   water park slide   345
Determinant   49
Difference formula   200
Dirichlet condition   101
Dirichlet, Peter   358
Dirichlet problem   293
Discrete Fourier transform   190
Distributive property   46
Divided differences   181
Domain of dependence   299, 305
Domain of influence   305
Double precision   31
DuFort–Frankel method   299
Dynamic allocation   11

E

ENIAC   5
Eckert, John   5
Eigenproblem   271
Eigenvalue   81, 271
Eigenvalue problem   290
Eigenvector   271
Elliptic equation   290, 303
Elliptic integral   218
Engineer   3, 4
Equilibration   63
Equilibrium problem   289
Error function   381
Euler constant   382
Euler, Leonhard   358
Euler–Mascheroni constant   381
Euler method   231
Exponent   29
Extended precision   31

F

Faddeev–Leverrier method   272
False-position method   122
Family of solutions   52, 230
Field   10
Finite-difference method   259, 289
Fixed-point   10, 23
Fixed-point iteration   115
Floating-point   10, 27
For loop   15, 369
FORTRAN   7
Fourier, Jean   360
Frequency   156
    cumulative   158
    polygon   158
    reduced cumulative   164

G

Gamma function   167, 381
    incomplete   168, 382
Gauss distribution   161
Gauss elimination   57
Gauss–Jordan method   65
Gauss, Karl   360
Gaussian quadrature   213
    Gauss–Chebyshev   220
    Gauss–Legendre   213
Gauss–Seidel iterations   94
Gear methods   256
Gerschgorin's theorem   275
Givens method   285
Global quantity   20
Goodness of fit   170
Go To   17
Gradient   333
Gragg method   248
Grouped data   157

H

Hamming method   247
Hessenberg matrix   285
Heun method   239
High–level language   7
Hilbert matrix   79
Histogram   158
Hollerith, Herman   5
Horner's method   35

Householder method   285
Hyperbolic equation   290, 305
    first-order system   308
    two space dimensions   310

I

IEEE Standard   29
If structure   12, 368
Ill conditioning   36, 77
Implicit Euler method   244
Implicit method   235
Implicit trapezoidal method   254
Improper integral   216
Incremental-search method   114
Inertia tensor   279
Inf   31
Initial–value problem   229, 290
Initialization   11, 367
Input/output unit   5
Integer   10, 23
Interpreted language   7
Interval halving   118
Interval of absolute stability   233
Inverse power method   280
Irregular boundary   304

J

Jacobian   133, 253
Jacobi, Carl   360
Jacobi iterations   92
Jacobi method   282
Jacquard, Joseph   5
Jump   17

K

Kirchhoff's Current Law   387
Kirchhoff's Voltage Law   84, 387
Kutta method   239

L

Lagrange, Joseph   361
Lagrange interpolation   182
Lagrange multiplier   329
Laplace equation   303
Laplace expansion   50
Laplace, Pierre   361
Laplacian operator   292
Lax–Wendroff method   305

Least-squares approximation 171
   linear combinations of functions 175
   nonlinear models 176
Legendre, Adrian 362
Legendre polynomial 213
Leibniz, Gottfried von 4, 362
Leverrier, Urbain 362
Linear algebraic system 43, 51
   direct method 54
   iterative method 90
Linear dependence 52
Linear independence 52
Line iterative method 105
Linear regression 171
Local quantity 21
Logical operator 12
Loop analysis 387
Low-level language 7
LU decomposition 69
   Cholesky 69, 81
   Crout 69
   Doolittle 69, 70

M

MacCormack's predictor–corrector 316
Machine epsilon 40
Machine language 7
Mantissa 29
Mark I computer 5
Matrix 45
   augmented 58
   bidiagonal 86
   block 88
   coefficient 51
   diagonal 47
   diagonally dominant 63, 82
   dimension 45
   element 45
   identity 47
   inverse 48, 275
   Jacobi 86
   lower triangular 49
   nonsingular 52, 271
   operations 45
   rotation 282
   singular 52
   spectrum 378
   square 47

   symmetric 81, 282
   transpose 81
   triangular 54
   tridiagonal 86
   two cyclic 99
   upper triangular 49
   zero 47
Mauchly, John 5
Mean 153, 159
   arithmetic 153, 159
   geometric 154, 160
   harmonic 154, 159
Median 153
Memory 5
Method of joints 393
Midpoint method 237
Midpoint rule 248, 379
Milne method 246
Milne–Simpson method 247
Milne's rule 379
Mixed problem 293
Mode 159
Modified Euler method 237
Modified midpoint method 249
Modularity 18
Module 17, 18, 369
   C 371, 373
   FORTRAN 371, 372
   header 19
   Pascal 371, 375
Multidimensional integral 220

N

NaN 31
Neumann, John von 5
Neumann problem 293
Neville's algorithm 186
Newton–Cotes formula 210, 379
Newton interpolation 183
Newton, Isaac 363
Newton–Raphson method 124, 134, 332
Newton's Second Law 386
Nodal analysis 388
Nonlinear algebraic equation 113
   systems 132
Norm 77
   Euclidean 78, 378
   infinity 78

matrix   77, 377
p   78
spectral   378
vector   77
Normal distribution   161
   standardized   162, 382
Normalization   29
Not   12
Number base   23
Numerical differentiation   199
Numerical integration   205
Numerical method   3

O

Object-oriented language   9
Offset   29
Ohm's Law   84, 387
Operating system   6
Operational count   55, 62, 69
Optimization   324
   global   332
Or   12
Order   234
Ordinary differential equation   229
   system   251
Output   11, 368
Overflow   24, 27, 31

P

Parabolic equation   290, 292
   explicit method   294
   predictor–corrector method   300
Parallel arrays   10
Partial differential equation   289
   classification   291
Partial pivoting   58
Particle dynamics   386
Pascal, Blaise   4, 364
Pascal language   7, 9
Penalty function   333
Pivot element   58
Pivot row   58
Point iterative method   105
Poisson equation   304
Poisson, Siméon   364
Polynomial evaluation   35
Polynomial interpolation   181
Polynomial roots   137

Population   151
Positive definite   81
Postmultiplication   46
Power method   278
   inverse   280
   with shift   281
Predictor–corrector method   236,
      300
   Adams   244
   iterative correction   245
Premultiplication   46
Procedural language   9
Programming language   7
Propagation problem   290
Property A   99
Pseudocode   11, 367

Q

QR algorithm   285
Quadratic equation   140, 380
Quadratic factor   137
Quadrature   213
   compound   213
   interpolatory   213

R

Radius of curvature   385
Radix   27
Real number   10, 27
Record   10
Recursive module   21
Relational operator   12
Relative error   40
Relative stability   233
Relaxation parameter   99
Repeat–Until loop   13, 369
Residual correction   91
Residual vector   91
Resistor network   83, 387
Richardson extrapolation   203
Rocket   251, 389
Romberg integration   210
Root mean square deviation
      156
Root-solving method   114
Round-off error   33, 203, 230
Row pivoting   57
   order vector   71

Row exchange 52
Runge–Kutta–Fehlberg method 250
Runge–Kutta method 237, 238
   stages 237

S

Sample 152
Sample size 152
Scalar 45
Scaling 36, 63
Search method 129
   constraint boundary 327
   exhaustive 324
Secant method 129
Series expansions 380
Shooting method 257
Sign bit 29
Sign-magnitude notation 26
Signal period 190
Significance level 170
Similarity transformation 282
Simpson's rule 209, 379
Simpson, Thomas 364
Simultaneous iterations 94
Single precision 29
Software 6
Source code 7
Special functions 381
Spectral radius 95, 303, 378
Spring-mass system 141, 273
Spur 272
Stability 232, 296, 306
Standard deviation 154
   population 156
   sample 154, 160
Static allocation 11
Statistical rounding 30
Statistics 151
Steepest descent 106
Steepest gradient method 332
Stiff equations 254
Stirling's asymptotic series 382
Storage error 33
Stored-program computer 5
Stream function 390
Streamline 391
Stress tensor 279
Substitute name 19

Subtractive cancellation 34
Successive approximation 115
Successive overrelaxation 98, 303
Summation formulas 382
Synthetic division 137

T

Task tree 323
Taylor, Brook 365
Taylor series 124, 234, 291, 383
Taylor series method 234
Temperature distribution in a tube
   257
Thomas algorithm 86
Top–down analysis 323
Trace 272
Trapezoidal rule 206, 379
Truncation error 203
   local 231, 296
Truss 71, 391
Truss equilibrium 72, 392
Two's complement notation 26

U

Underflow 27, 31
Unix 6

V

Vandermonde matrix 110
Van der Waal's equation 125
Variable 10
Variance 154
   population 156
   sample 154
Vector 45
   column 45
   component 45
   row 45
Viéte, Francois (Vieta) 365

W

Wave equation 309
Wendroff implicit method 307
While loop 13, 369

X

Xor 12